D0970226

Controversies in Science & Technology

SCIENCE AND TECHNOLOGY IN SOCIETY

Controversies in Science & Technology

Volume 2

From Climate to Chromosomes

Edited by

Daniel Lee Kleinman,
Karen A. Cloud-Hansen,
Christina Matta,
and Jo Handelsman

Mary Ann Liebert, Inc. publishers

www.liebertpub.com

ISBN-13: 978-0-913113-42-4 ISBN-10: 0-913113-42-5

Printed in the United States of America.

Contents

Part 5: Biology and Gender: Scientific Careers and Scientific Theories

Acknowledgments

Any essay collection of this size, depth, and breadth is the work of numerous people without whom *Controversies* would not exist. First, we would like to thank our authors for crafting informative and engaging essays and our reviewers for lending their expertise. Abby J. Kinchy, the staff of the University of Wisconsin Press, and Ben Roberts were invaluable sources of advice. Our publisher, Mary Ann Liebert, editor, Vicki Cohn, and their staff we thank for their hard work and flexibility on our behalf. We wish to thank the Brittingham Foundation, the Howard Hughes Medical Institute, and the University of Wisconsin College of Agricultural and Life Sciences for their financial support of this project. Finally, our thanks goes to those scholars who served as reviewers for portions of this volume and members of our advisory board, who, although we have not drawn on them as much as we should have, have always provided helpful advice.

Controversies in Science & Technology

Introduction

From Climate to Chromosomes

Karen Cloud-Hansen, Daniel Lee Kleinman, Christina Matta, and Jo Handelsman

It seems a cliché to describe the current moment as the digital age, to say we live in a high-tech society, or to note that science affects our lives every day. A cliché isn't untrue, however; it is a term or concept that has been sapped of its meaning by overuse. In this volume, we hope to move well beyond the trite. Our authors provide essays that are rigorous and detailed, though readable and engaging, on some of the most crucial and controversial issues related to science and technology in our world today.

Faced with the possibility of global warming, opportunities for stem cell–based medical therapies, and the prospect of having our very notions of self altered by developments in biology and information technology, science literacy is more important than ever. But science literacy must mean more than knowing simply that the Earth orbits the sun, recognizing the structure of DNA, or understanding that water is composed of oxygen and hydrogen. Surely, we must learn the central theories and findings of science, but this is not sufficient. To participate intelligently in debates about research funding, science policies, and the future direction of research and technological development, citizens must be able to understand the underlying issues in *controversies* in science and technology. We must have access to wide-ranging discussions about the risks and benefits of new technologies and the implications of new scientific findings. We must learn about the relationship between our major social institutions (e.g., religion, government, and education) and scientific and technological developments. In this second volume of *Controversies in Science and Technology*, we provide a collection of twenty-five essays on five crucial topics—stem cell research, information technology, space exploration, global warming, and the nature of gender—that we hope will arm readers with arguments and information about several of the most important issues of our time.

Stem Cell Research: From Science to Religion to Science

Stem cells. Stem cells? Stem cells! STEM CELLS! These are electric words in the news. It is rare to read a dispassionate piece about stem cells because they are the center of swirling controversy fueled alternately by scientific novelty, the need for medical treatments and cures, religion, ethics, and politics. But how many of us truly understand the science and positions associated with this volatile field of biology? How many of us could explain why scientists are so enthusiastic about the potential for stem cell treatments? Or why the Catholic Church is so adamantly opposed to them? Or why the Jewish and Islamic faiths are not? Or what motivates President Bush's policy for regulating the use of federal funds to study stem cell lines or why California Governor Schwartzenegger is spending $45 million on stem cell research? Or the biological and ethical differences between embryonic and amniotic stem cells? In this section, the authors explore the rich and varied arguments that fortify the publicized positions.

Scientists predict that stem cell technology will provide treatments for conditions such as Parkinson's disease, diabetes, and those associated with loss of myelin (the outer structure of nerves). This latter category has attracted enormous public attention because diseases such as multiple sclerosis and ALS ("Lou Gehrig's Disease") and conditions including spinal-cord injury are among those that might be treated by stem cell technology. The common feature of these diseases is that they involve malfunction of a particular cell type. Stem cell research aims to replace missing, damaged, or malfunctioning cells with cells produced in the laboratory.

The challenge in the quest to develop replacement tissues is culturing the appropriate cells in the lab. The power of stem cells is that they have the potential to differentiate into many types of cells and form tissues of diverse function. The barrier for many years was how to maintain stem cells in culture and then coax them to form tissues with different functions. The last decade has led to technology that allows the growth of stem cells in culture and now many types of tissues have been regenerated, and many have been tested in animals, a few in humans.

Stem cell research is predicated on the finding that, before cells acquire their dedicated functions as part of, for example, skin, heart, or liver tissue in humans, or leaf or root tissue in plants, they are "undifferentiated," or cells with no particular personality, but with the potential to become any type of cell in the organism. All cells in a plant or animal except for gametes (egg and sperm) *contain* the same complement of DNA; which genes are *expressed* determines whether a cell produces insulin or becomes a neuron or muscle cell. The process of using stem cells to regenerate tissues with dedicated functions involves manipulating the culture conditions to induce expression of the genes that direct the cell to develop a specific "personality" or function.

The idea of creating "immortal" cultures of cells and regenerating tissues from them is not new. The ability of undifferentiated cells to transform into cells of different functions or tissues is known as pluripotency. This phenomenon was exploited in the 1940s by plant biologists who discovered that many plant cells were "totipotent" (a synonym of pluripotent) and could, therefore, produce an entire plant from a single cell. For many decades, plant regeneration has been a routine part of botanical research as well as a technology critical to many plant-based industries. In fact, each of us comes into contact with this process when we eat mashed potatoes or French fries: most potatoes in the United States are derived from single cells in tissue culture that are used to regenerate complete potato plants, which are then propagated for a few generations. Thus, pluripotency is certainly not a new biological concept and it did not generate furious controversy when it was exploited for crop production.

But the acceptance of regenerating tissue from single cells in plants was not predictive of the public reaction to the same phenomenon used with humans. The fierce divide that has developed around stem cells reflects the source of the cells. The cells that are truly pluripotent and immortal in culture have come from five-day-old human embryos derived from in vitro fertilization. The embryos that are used are frozen and would remain so or be incinerated if not used for stem-cell research. Much of the debate focuses on when human life begins and whether we have a responsibility to protect embryos from "exploitation" for medical research. For some, these are largely biological questions and, for others, they are theological. To elucidate the debate at an intimate level and provide readers with the information and arguments to find their own place in the continuum of views, our authors provide a detailed understanding of stem cell applications, modern embryology, and the basis for the Catholic, Jewish, and Islamic views of stem cell research.

In the introductory essay, Carl Gulbrandsen and Jill Ladwig of the Wisconsin Alumni Research Foundation, which holds the patents on multiple stem cell lines and technologies, present an overview of stem cell research, highlighting the breakthroughs that changed the landscape, opening up possibilities in the treatment of human diseases. They report that some of the results with stem cell research have been promising. One researcher found that rats with injured spines were able to walk within months of treatment with stem cell–derived neural cells. Another was able to restore the ability of the spinal cord to conduct electrical impulses. In Europe, patients with Parkinson's disease, which involves loss of dopamine-producing brain cells, experienced great reduction in symptoms after being injected with fetal neurons. These neurons are in short supply and stem cells might provide an abundant substitute. Advances in generating insulin-producing cells from stem cells in culture suggest that treatment of juvenile-onset diabetes will be another early stem cell success.

All of these exciting medical advances are made possible by a seminal result that changed the field—the method developed by Professor James Thomson in 1998 to culture stem cells taken from five-day-old human embryos. The feat that Thomson accomplished was to keep the undifferentiated cells alive and reproducing in culture. This is the first step toward regenerating tissue. Since then, many types of tissue have been produced by manipulations of the medium in which the cells are cultured.

In the essay by Father Tadeusz Pacholezyk, a neurobiologist and Director of Education at the National Catholic Bioethics Center, we learn that the position of the Roman Catholic Church on stem cells is unbendingly against the use of human embryonic cells. The Church is not opposed to stem cell research with cell lines of other derivations, but so far the only cells that have conformed to the requirements for immortality and tissue regeneration are the embryonic ones, so this is where the controversy has focused.

The Church's objection is based on the belief that the human being exists when fertilization (penetration of an egg by a sperm) occurs. In his essay, Father Tad argues that this is a scientific and logical determination, not a religious one. This position is juxtaposed against the one presented by embryologist Karen Downs, professor of anatomy at the University of Wisconsin–Madison, who explains that embryologists consider that conception occurs 14 days after the Catholic Church maintains it occurs. In fact, embryologists do not consider the mass of cells that results from fertilization to be an embryo until dramatic developmental events have occurred coincident with conception. This is a sufficiently interesting disagreement that it is worth looking at the details of embryonic development.

Downs presents an overview of the current scientific understanding of embryonic development. She highlights two key points that are obvious to embryologists, but are often obscured in the debate and are unknown even to many scientists. First, she distinguishes between the "origin of potential life" and the "origin of the human individual." Second, she illuminates the fairly recent finding that fertilization of an egg with a sperm leads to many cell types and that only some of these are part of the embryo. The product of fertilization is the zygote, which produces the embryo as well as tissues that support the embryo's life. The best-known extraembryonic tissue is the placenta, the origin of which is the fertilized egg and the role of which is to support the life of the embryo. The collection of these tissues is known as the "conceptus," which includes all of the products of fertilization.

In the conceptus are two distinct tissues: the outer portion, the "trophectoderm," and the inner tissue, which is the "inner cell mass." The trophectoderm participates in implantation and cells derived from it will become the placenta. The inner cell mass will become the embryo. At this stage, the inner cell mass is "pluripotent," meaning that the cells can follow any of many developmental paths; the inner cell mass contributes to every tissue in the person who is ultimately born. Therefore, it is the inner cell mass at this stage—prior to implantation and prior to formation of the embryo—that is the source of so-called embryonic stem cells.

The conceptus embeds in the uterine wall where more differentiation occurs. The umbilical cord, placenta, yolk sac, and amniotic fluid form at this time and are absolute requirements for fetal life and development. The umbilical cord provides a vascular bridge to the mother, and the placenta will perform diverse physiological functions (alimentary, pulmonary, renal, hepatic, and endocrine) for the fetus right up until birth, but they are not fetal tissue, despite being derived from the same fertilization event the gives rise to the embryo.

Many developmental biologists consider one event to be the most important in life: formation of the "primitive streak." This structure defines the body axis and establishes the coordinates of front and back, left and right, etc, that are the basis for the symmetry of the human form. Once the streak is fully formed, twinning can no longer occur. At this point, 14 days after fertilization, the patterns for development are begun, it is clear which tissues are embryonic and which are extraembryonic, and the potential for a human individual is now established.

Downs stresses the misconceptions, generated by vocabulary confusion, that have entered both scientific and public discussions. For example, the "embryo" does not exist until after formation of the primitive streak and the establishment of the body plan, but the word is used erroneously to refer to a fertilized egg. By this definition, "embryonic stem cells" do not really come from embryos, but from a post-fertilization, preembryonic structure known as the inner cell mass. Similarly, two processes are needed to produce a zygote—penetration of the egg by the sperm (fertilization) and fusion of the egg and sperm genomes into a common nucleus (syngamy). Because most fertilized eggs die naturally before the embryonic stage and the fertilized egg gives rise to extraembryonic tissue (that most definitely does not and cannot become a human being), as well as an embryo, most developmental biologists agree that the *potential for life* occurs at syngamy, but the *potential for a human being* does not exist until the primitive streak is formed and the conceptus contains a true embryo.

Rabbi Elliot Dorff introduces a Jewish perspective on stem cell research based on theological principles and historical views of the embryo. Judaism holds that our bodies are God's and simply on loan to us, and as such we have a responsibility to keep our bodies healthy and cure disease, when possible. Jewish doctrine tells us that all human life is valued equally and it is therefore our responsibility to protect all human beings. Dorff quotes the Hebrew Bible as saying that we are supposed to "work the world and preserve it" and argues that that balance is our divine duty. Jewish law makes a sharp distinction at 40 days of gestation. Prior to this, the ancient doctrine argues, the embryo does not have a human form, and is therefore not a child. Based on this doctrine, Jewish law is consistent with stem cell research. The embryos created in vitro are not considered to be children yet, and saving them from incineration to use them for stem cell research is not only allowed, but encouraged—considered an "act of loyalty and love" (*hesed*) or perhaps even a "commanded act" (*mitzvah*). Perhaps the greatest concern captured in Dorff's essay is that stem cell technology will be more available to some than to others, most likely based on financial means, and that is inconsistent with the Jewish belief in the equality of all human lives, independent of ability, social circumstances, or financial means.

The Muslim perspective is less clear than either the Catholic or Jewish one. Professor Abdulaziz Sachedina explains the ambiguity in Islamic law. Although traditional Islamic writings do not offer a definition of embryo or fetus, it is clear that the fetus achieves personhood status fairly late during gestation. In one version of the Qur'an, the angel that breathes the spirit into the embryo arrives 45 days after conception (uncannily similar to the Jewish teaching of the 40-day turning point to personhood). Since some versions of the Qur'an omit the sentence that mentions the time of ensoulment, and embryos, abortion, and related topics are not addressed directly anywhere in the writings, modern interpreters have a fair amount of leeway.

One of the intriguing issues, which is also the one that challenges the theologians, politicians, and citizenry, is that we are using ancient writings to guide 21st-century decisions. We must use ideas and language written at a time when the concept of fertilization didn't exist to be part of moral judgments about a technology that couldn't have been imagined even a few decades ago. The stem cell issue has been particularly volatile because of the very nature of the questions: When does life begin? When is a human being a person? What rights belong to a mass of cells or to an embryo? How much should decisions be guided by religious definitions and by scientific definitions?

Information Technology: Inequality, Identity, and Invasions of Privacy

Talk of the digital age and the information society has been around for perhaps a decade or two, but you know we are in the midst of that world when average citizens regularly worry that their identities will be stolen by hackers who garner intimate financial and other personal information from government and corporate databases, when the prospective presidential candidates announce their candidacies live on their campaign Web sites, and when Rupert Murdoch and his News Corporation pay $1.65 billion for MySpace, a Web site made up of personal ads, family photos, home video clips, and friendly gossip (Pareles 2006: 2B).

These and other developments have far-reaching implications. Traveler's checks are now virtually a thing of the past, since vacationers can access their money from automatic teller machines almost anywhere in the world. Arranging holidays late into the night, travelers can locate tour companies that perfectly match their interests and compare prices for flights and hotels. Consumers staying at home can purchase goods from far-flung small-scale shops on the Internet.

Perhaps there is convenience in these developments, and boutique shops can reach consumers they could only have imagined years before. But ultimately what do such developments mean for small versus large-scale businesses? The answer is not yet clear. It is certain that with every key stroke we are monitored by companies and governments. Our private information can be bought and sold. It can be used to micromanage our preferences, invade our intimate lives, and steal our money.

At the same time, this new world of digitized information may be democratizing creativity, giving visual artists, poets, filmmakers, and musicians access to audiences they might have only imagined a few short years ago. Politicians, too, can interact with their constituents unmediated—without national news organizations editing and effectively altering their words.

Again, it is not certain whether this means the fundamental democratization of cultural and civic life. What will the purchase of Myspace and YouTube by media and technology giants mean for aspiring artists? What will the pursuit by the giants of the entertainment industry of higher and higher levels of copyright protection mean for the most famous performers and the yet–to-be-discovered? Will consumers gain or lose?

Barak Obama's initial announcement—live on his Web site—that he would explore running for president was seen by millions, and this new variety of retail politics made national news. But, as the *New York Times* announced, "Campaign Blogs Can Cut 2 Ways." This headline captured the promise and peril of the Internet for national politics. John Edwards's campaign was thrown into a tizzy by the work of two "liberal feminist bloggers" who have histories of penning controversial comments on sex, religion, and politics (Broder 2007: A1).

Beyond the market and politics, there is social life. How should we think about electronic dating services, instant messaging, and shared electronic photo albums? Are we less lonely? Are we safer? Do we keep in touch with loved ones more frequently than we would otherwise? Children are lured by electronically sophisticated pedophiles. Electronic discussion may take tones and forms that real-time interaction would not. Perhaps there will be misunderstanding or much needed frankness.

And finally, there is education. Computers and information technology are radically reshaping this area of social life. Education is accessible at times and in places that were out of reach not long ago. Massive amounts of information are available to students, allowing access to a previously unimaginable range of perspectives and making possible the triangulation of evidence in ways that are vastly more efficient than a trip through a bricks and mortar library. Questions of quality abound here. Indeed, Middlebury College's history department, perhaps a leader, recently placed restrictions on the use of Wikipedia by its students, worried about the quality of the information to which the students have access (Read 2007).

We cannot hope to address all of these issues in the information technology section of this volume, but our authors do address many and do so in ways that are thoughtful and, we hope, compelling. The section opens with an overview essay by Travis Kriplean and Daniel Lee Kleinman. These authors take up several of the most important issues that have accompanied the spread of the Internet: new forms of collaboration in knowledge production, the nature of Internet-aided politics, the changing character of information search, and the so-called digital divide. On the first topic, they explore debates about the quality of information produced by amateurs, discussing, among other things, a study of the caliber of the information on the Internet encyclopedia Wikipedia compared with the traditional hard-copy *Encyclopedia Britannica*. On the matter of politics and the Internet, Kriplean and Kleinman consider how Web-based politicking interacts with grassroots real-time politics. Here, they consider the case of the 2004 Howard Dean campaign and efforts by Moveon.org to organize antiwar demonstrations. In their discussion of information search, Kriplean and Kleinman reveal to readers the machinations behind what appears high on topical search engines such as Google. Finally, given the unquestioned importance of the Internet, who has access and in what manner are crucial issues. President Bill Clinton addressed this matter head-on, with efforts to eliminate the so-called "digital divide." The authors explore efforts to provide broad access to information technology and their implications.

In his chapter, Mark Warshauer, the author of *Technology and Social Inclusion: Rethinking the Digital Divide* and a prominent commentator on the digital divide, takes off where Kriplean and Kleinman end. Warshauer sees the virtues of information technology. At the same time, he isn't on the bandwagon with those who think that simply providing all citizens of the world with access to computers and the Internet will solve the vast problems of poverty and inequality that we face. If citizens are not otherwise literate, if the information to which people have access is not in their native language and not of high quality, and if social relations are not thoroughly reformed, then access itself is not particularly meaningful. Warschauer provides a detailed discussion of strategies for successful use of information technology en route to social inclusion.

In their contribution, Eugene H. Spafford and Annie I. Antón explore the difficult balance of privacy and security in the information age. As these authors note, the same technology that allows citizens easy access to a wide range of political and other perspectives also permits the hatching of terrorist conspiracies with an ease not possible in the days before the Internet. As a society, we value the right to privacy—the capacity to express, relatively anonymously, displeasure with our government, to discuss honestly and openly matters of sex, religion, and other sensitive topics without fear of retribution. Spafford and Antón consider what privacy means in the digital age and how and when it should be protected. While outlining a set of recommendations for protecting citizen privacy, they acknowledge vast differences of opinion on how to balance privacy and security and urge careful deliberation in debates about privacy policy.

James X. Dempsey's essay, which immediately follows Spafford and Antón's, also takes up privacy issues, but, while Spafford and Antón offer the perspective of computer scientists, Dempsey writes from the vantage point of a policy analyst and activist. Dempsey, the Policy Director at the Center for Democracy and Technology in Washington, DC, argues that our privacy laws have not kept pace with technological developments. Pre–information technology law was formulated with matters of physical privacy and, later, telephone eavesdropping in mind. Today, however, with the use of cell phones, personal digital assistants (PDAs) and the Internet, obtaining information on citizens' present and past movements, ascertaining information on a person's network of acquaintances, and gathering private financial records data are well within the capacity of government and other interests. Dempsey details how new technologies can be used to monitor the minutiae of citizens' lives and suggests that people may expect privacy in ways that are not guaranteed by law. He concludes by calling for a broad-based dialogue on how new technologies may be undermining traditional privacy expectations and calls on Congress to develop a system of protections appropriate for our new age.

Our information technology section concludes with an essay by Michael Bugeja, a professor of journalism at Iowa State. In his contribution, Bugeja ponders what new information technologies mean for social life and interpersonal relations, especially in university settings. What he sees is an emerging fusion of social life and market culture and a set of technologies that, as often as not, distract from and disrupt actual human interaction. He suggests that, not only are people missing opportunities for face-to-face interaction, "they also are losing their innate capacity to know when, where, and in what venue they should initiate communication." These are important issues, which we too often overlook as we happily make our Internet purchases and contemplate crucial questions of government surveillance and the educational uses of the Internet.

Space Exploration: Reasons and Risks

Human space exploration occupies an almost legendary place in the public psyche. From *Star Trek* to space tourism, the desire to boldly go where no one has gone before is a significant impetus for exploration, as is national pride. Current space programs build from a rich history of US space flights, both with human crews and without, which is the focus of this section's first essay. The other essays discuss the difficulties associated with experimental technology, organizational cultures, and risk assessment. Following the losses of the *Challenger* and *Columbia*, serious questions are raised when the shuttle launches, as to the importance of a given mission and the safety of the craft and crew. NASA now examines the heat shields of the shuttle in space and has plans in place for emergency docking of the craft at the International Space Station if problems are found (Leary 2006). Shuttle launches and landings have been delayed due to safety concerns (Toner 2005; Johnson 2006), illustrating an increased role for safety inspections and a reprioritization of safety concerns over scheduling pressures. As Nancy Leveson writes in this volume, "Space exploration is inherently risky. There are just too many unknowns and requirements to push the technological envelope to be able to reduce the risk level to that of other aerospace endeavors such as commercial aircraft. At the same time, the known and preventable risks can and should be managed effectively." The four essays in this section discuss the history of space exploration, the causes of the shuttle disasters, and why space exploration is worth a possible loss of human life.

The space exploration section opens with a historical overview of the US space program, with a focus on missions with human crews. Roger Launius, chair of the Division of Space History at the National Air and Space Museum, discusses both the changes in technology and the societal and political forces driving the space race during the Cold War, Moon landings, the Shuttle program, the International Space Station, and the future goals of the space program, such as human exploration of Mars. Launius details the rationale for spaceflight: scientific discovery and understanding, national security and military applications, economic competitiveness and commercial applications, human destiny/survival of the species, and national prestige/geopolitics. He also discusses the scientific advances made possible by the space program, including the deployment of the Hubble telescope and missions to Venus, Jupiter, and Mars. Launius goes further than detailing missions, dates, and discoveries. He weaves these into discussions of US culture and political climate at the time as well. The reader is allowed to appreciate that such a technical enterprise does not occur in a political or societal vacuum and to discover how politics and public expectations have shaped space technology and vice versa.

Nancy Leveson, Professor of Engineering Systems at the Massachusetts Institute of Technology, explains the organizational factors as well as the engineering causes of the shuttle disasters in the context of NASA's management structure in her essay. Leveson traces how funding requirements necessitated that the shuttle perform more tasks at lower cost, thus compromising design. She explores questions such as "Why would intelligent, highly educated, and highly motivated engineers engage in such poor decision-making processes and act in a way that seems irrational in retrospect?" and "why [does] the manned space program [have] a tendency to migrate to states of such high-risk and poor decision-making processes that an accident becomes almost inevitable." Leveson takes her readers beyond O-rings and foam, with an analysis of how flaws in organization and management negated known or potential engineering solutions that would have prevented these tragedies.

The organizational and sociological causes of the shuttle losses are discussed in our third essay. Diane Vaughan, a sociology professor at Columbia University and a member of the *Columbia* Accident Investigation Board, shows that these "accidents" were not due to chance, but rather resulted from a series of decisions that led gradually to disaster. Vaughan chronicles the political and economic environment, its organizational structure and culture, and how each contributed to the *Challenger* disaster and how, despite NASA's best intentions, these conditions were repeated preceding the loss of the *Columbia*.

The section closes with an essay by James Lattis, an historian of science who directs a public outreach center focused on astronomy and space exploration. Lattis discusses the reasons for space exploration, given that "risk is inherent in pushing human activities into realms that are only barely accessible at our current state of knowledge and technology." He places the concepts of risk assessment and cost-benefit analysis into contexts familiar to everyone. Lattis touches on the controversial decision by NASA administrators to stop shuttle missions to repair the Hubble Space Telescope, a policy that was reevaluated in October 2006 (Lawley 2006). He maintains that "wherever we probe into the unknown, there are inevitably, and almost by definition, new knowledge and opportunities to be won" and that space exploration expands available physical environments, natural resources, and economic opportunities, inspires and motivates learning, provides "frontier shock" to help drive societal innovation and development, and enhances the survival of human life. Collectively, these essays allow readers to appreciate how engineering, politics, cultural forces, and technology can combine to yield simultaneously important scientific discoveries, national pride, a mandate for a more technologically-literate population, and a spectacular loss of human life.

Global Warming: Scientific Data, Social Impacts, and Political Debate

Few scientific issues in the popular press are as contentious as global warming. Global warming is thought to have contributed to everything from Hurricane Katrina to Al Gore's political muscle. The collision of colossal scale and complicated modeling of incremental change creates many questions in the public's mind. How fast are temperatures rising? Can we slow the process? What will the results be? Articles in the popular press as well as scientific reports have linked global warming to higher probabilities of devastating hurricanes, more frequent and extreme monsoonal rains, and increased intensity and duration of the wildfire season in the western United States (Roach 2005; Goswami, et al. 2006; Westerling, et al. 2006). The essays in this section raise many questions about global warming. Among them: How do we measure change on a global scale, and what are the policy implications of disagreeing with a particular modeling method or interpretation of the data? How can the debate over whether global warming is occurring be separated from discussion of specific interpretations of data or predictions?

The controversy about whether global warming is occurring seems to have slowed. At the 2006 United Nations Climate Change Conference, the focus was not on greenhouse gas emissions, but rather on the establishment of a fund to provide assistance to developing nations in managing the environmental and public health crises caused by global warming (Stone and Bohannon 2006). However, the debate on the time frame available to reverse this trend and whether the political muscle exists in the United States to do so is still contentious (reviewed in Doniger, et al. 2006). Despite a United States Senate resolution in June 2005 calling for mandatory limits on emissions of greenhouse gases, Congressional proposals provide only initial steps to slow emissions growth and delay decisions on future reductions for decades.

Though it would be impossible to present every nuance of the controversy surrounding global warming in one section of one book, we thought it was important to present varied views on this topic, covering the discourse within the scientific, sociological, and political communities as well as touching on some facets of the topic that are not as widely covered, including issues of social justice and the effects of increasing atmospheric carbon dioxide levels unrelated to temperature. The five essays in this section span a range of topics related to global warming.

Clark Miller's essay opens the section, providing social and scientific context for the others. Miller, an associate professor in the Department of Political Science and the Consortium for Science,

Policy, and Outcomes at Arizona State University, traces the origins of the climate change debate through both scientific discoveries and public reactions. While debate exists in peer-reviewed, scientific journals concerning only questions of timing and magnitude, the idea of global warming is presented as much more contentious in the popular press. Miller proposes that climate models are attacked "as inherently uncertain efforts to project possible future risks, not present-day harms." He also examines how these debates shape and are shaped by society at large. Moreover, Miller provides a plan of action—methods to engage technical and policy experts with each other and the public.

In their essay, Steven Schneider, a national expert on climate-change policy and science, and Patricia Mastrandrea, cofounder of Delphi International and a member of Dr. Schneider's research group, discuss the impacts of global warming on developed and developing nations. They detail several consequences of global warming that are not as readily considered, including the movement of species habitats to higher altitudes and latitudes and how climate change triggers "surprises"—global environmental changes that current models cannot predict. Schneider and Mastrandrea describe five numeraries for judging the significance of climate change impact: market impact, human lives lost, biodiversity loss, distributional impacts, and quality of life. They discuss how multiple numeraire helpfully represent the costs of global warming to developing nations, which can be diminished in a market-only analysis. As Schneider and Mastrandrea stress in their essay, "Climate-change policy decisions don't always benefit the most vulnerable countries or groups within countries."

Daniel Sarewitz, Director of the Consortium for Science, Policy, and Outcomes at Arizona State University, and Roger Pielke, Jr., a professor of environmental studies at the University of Colorado, examine how the position of the United States on climate change has affected American credibility across the globe and shaped world emissions policy. Sarewitz and Pielke propose that the failure of the US government to act on global climate change issues has made American politics, rather than the flaws of current climate change policy, the target of the world's attention. The authors discuss how this diffusion of focus has hindered the development of comprehensive climate change policy. "It means that a third option," according to Sarewitz and Piekle, "—that climate change is a serious problem, but that [the Kyoto Protocol] and the broader [United Nations Framework

Convention on Climate Change] are the wrong way to go about addressing it—has not been able to emerge into the marketplace of ideas." It is this option that Sarewitz and Piekle explore. They question why, if technology innovations are touted as a means of stopping global warming, the public investment in energy research and development by industrialized nations has fallen by as much as 65% since the late 1970s.

Interpretation of climate change data can be a contentious topic, as David Legates describes in his essay. He explains how different interpretations of climate change data can influence public policy decisions and scientific discourse. Legates also discusses the desire of many media sources to focus on extreme opinions, while downplaying the majority consensus. He questions how scientist-advocates can simultaneously represent data accurately and bring action to important issues.

Separate from global warming, the ongoing increase in atmospheric carbon dioxide continues to affect the biosphere. The final essay in the section, by Lewis Ziska, an ecologist who studies the impacts of rising carbon dioxide levels and global climate change on weed biology, examines how rising carbon dioxide levels have affected plant diversity and agricultural production. Recent increases of carbon dioxide levels have stimulated growth of most plant species. Ziska examines both the potential benefits and costs of this change to agricultural systems and native plant communities. He also discusses how changes in plant community compositions impact human health by varying allergen production and how rising levels of carbon dioxide may substantially alter phytochemical concentrations in plants. As Ziska explains, many of these compounds serve as the basis of pharmaceuticals used to treat a variety of health concerns. Through their analysis of some of the facets of global warming, all the authors in this section consider the collision of politics and technology in this debate, with serious consequences to the planet.

Biology and Gender: Scientific Careers and Scientific Theories

In 2005, Lawrence Summers, then president of Harvard University, commented in a speech at a conference devoted to diversifying science and engineering that women's slow advancement in these fields was due not to discrimination, but rather to differences in men's and women's aptitude in such subjects (Summers 2005). This assertion drew immediate outrage from many scientists—male and female alike—and sparked a media frenzy. One "expert" after another stepped forward to comment on the validity of Summers's remarks or to offer their own favored hypothesis about women's intellect and resulting ability to "do" math and science. Though Summers later revised his remarks and expressed his regret at having made them, his apparent claim that women are inherently not as good at science as men dogged him throughout the remainder of his career at Harvard. But why were his comments so controversial? In her introduction to this section, Patricia Adair Gowaty explores this question further, considering matters of bias in scientific research, divergent opportunities for young male and female scientists and theories of sex, gender, and evolution.

To be sure, there *are* differences between men and women. On the surface, matters of anatomy, social role, physiology, or even IQ test scores could pose as reliable markers of who is male, who is female, and what their abilities and personalities might be. But the boundaries between male and female are not as clearly delineated as one might assume. Are humans dimorphic—that is to say, do their bodies and attributes fall into only two categories? How do we classify people as one sex or the other—is anatomy the only way, or are there other criteria we use? What happens when someone does not match what we commonly believe to be representative of "male" or "female"? How do our assumptions about "male" and "female" characteristics shape the practice and purpose of scientific research? What social consequences do our answers to these questions have? From Patricia Adair Gowaty's introduction, to Ben A. Barres's commentary on discrimination in science, the six chapters in the biology and gender section challenge how we conceptualize "men" and "women" in biological, behavioral, and social terms.

For centuries, the predominant explanation for sexual differentiation represented "male" as the result of active forces and processes, and "female" as a sort of default—an absence of maleness, rather than its own distinct quality. Although the exact mechanisms in question and the terminologies in which they were couched have changed since Aristotle famously described being female as a "natural deficiency," the social models have not. In her contribution to this volume, Caitilyn Allen, a plant pathologist and professor of women's studies, demonstrates that, as a result, explanations for female development stubbornly held true to the language of male dominance through centuries of commentary by natural philosophers and scientists, and has only very recently begun to fuel research on the basis of female development. Or, as Allen notes, though "extensive research has focused on the developmental pathway leading to a male baby, surprisingly little effort has gone into understanding how a human embryo becomes female."

Human development, therefore, offers a telling case study of the dangers of implicit bias in scientific research: first, because of the assumptions about male and female roles that informed (or misinformed!) hypotheses of sex development and differentiation, and second, because of the lack of interest male researchers showed in understanding female development as the result of its own unique processes. (Though, as Patricia Adiar Gowaty notes, female scientists can be blinded by these expectations as well.) This dual influence was so strong that even now, as Allen describes, introductory-level biology textbooks still tend to explain sexual differentiation in language that prioritizes the Y chromosome as the determining factor, showing how strongly our assumptions about male and female roles influence what we see as true or how we interpret data—even data that blatantly contradict established paradigms.

If social, cultural, and biological assumptions about men and women are inexorably intertwined in scientific research on human sexual categories, they are even more so in our creation of our own sex and gender identities. In her chapter, Alice Dreger, an historian of science and medicine whose pursuit of the historical hermaphrodite led her into medical activism on behalf of individuals with uncommon anatomies, explores this interrelationship by questioning how we construct ourselves as male, female, or another category oft overlooked in a society steeped in dimorphism. She asks, "How do you know if you're a man or a woman? Or put a bit differently, what makes someone a man or a woman? Does being male make you a man? Does being a woman make you female? For that matter, what makes a person male or female?" And, most important, "Why does all this matter?"

Part of the problem inherent in knowing whether to describe someone as male or female lies in the selection of criteria to use. Chromosomes, anatomies, wardrobe, and behavior are all possible criteria for identifying men or women, but Dreger notes that social roles, identity, and biology are not always the same thing. By selecting criteria within these categories, it's possible to arrive at a reasonable means of deciding who is male or who is female, but, as Dreger observes, there will always be at least one individual who does not match. Even anatomy, which would seem a fail-safe mechanism for telling the men from the boys and both from the women, can challenge even the most carefully constructed systems; as Dreger points out, individuals with disorders of sex development (or intersex) have anatomies that are atypical compared to our expectations of male and female, but their genders—that is, how they think of themselves or define their role in society—seem perfectly clear to the outside observer. However we choose to define men and women, Dreger concludes, we should recognize that our fixation on two-ness is a *social* enterprise—it is not a reflection of the nature of human diversity.

As Allen's and Dreger's chapters show, efforts to classify individuals according to dimorphic categories are not a new phenomenon. Nor are medical efforts to "correct" aberrant bodies, for that matter; even the arguments put forth by John Money, who advocated surgery as the healthiest possible option for children whose bodies did not conform to arbitrarily quantified criteria for male or female, had historical precedent. In "Ambiguous Bodies and Deviant Sexualities," Christina Matta, a historian of science, examines the frequency of, and justification for, surgeries meant to define uncommon or ambiguous sexual anatomies in the nineteenth and early twentieth centuries. She argues that the rise in the number of surgeries performed on individuals with unusual reproductive anatomies was *not* the result of improved surgical techniques, but rather a direct product of physicians' interest in their patients' sexual behavior. She draws parallels between the medicalization of homosexuality and that of hermaphroditism to suggest that patients whose bodies were not immediately recognizable as male or female posed a significant challenge to social beliefs in a strict boundary between the two sexes.

These unusual anatomies forced surgeons to reevaluate the relationship a patient's body had to his or her behavior. While all social behavior became fodder for discussion and comparison with an individual's anatomy, it was sexual behavior—and the fear of homosexuality in particular—that prompted physicians to attempt to define male and female through surgeries. But what was "normal" sexual behavior when it was unclear who was male or female? How physicians answered this question, when placed against the history of surgery more generally, demonstrates how fears of "abnormal" sexual behavior led to increased efforts to surgically establish norms for male and female bodies. The precedents set in the late nineteenth and early twentieth centuries therefore established a model for "treatment" that lasted into the late twentieth century.

Stereotypical roles for men and women have also colored traditional understanding of the evolution of sexual characteristics, as Joan Roughgarden, an evolutionary biologist, details in her chapter. Roughgarden describes how biases about gender and sexuality affect both conclusions about sexual selection and the language used to discuss it. She advocates taking a broader view and exploring alternate hypotheses in animal and human evolution. These discussions—of the mechanisms of sex differentiation, mate choice, or defining gender and sex identity—are not purely theoretical, though they may seem to be. The relationship between culture and scientific knowledge is complex, and the two are mutually reinforcing in ways that are often unexpected. Just as stereotypes of men as aggressive or forceful prompted many scientists to interpret showy male traits and dominant behaviors as desirable, the claims these scientists made then reinforced the very stereotypes and bias that had supported their hypotheses in the first place.

When put into social practice, however, this circle becomes even more vicious. The perils of creating or maintaining a sex or gender identity according to its premises have been summed up neatly in a poster meant to help children understand and celebrate the wide range of behaviors and emotions that comprise lived experience, regardless of their usual categorization as "masculine" or "feminine." "For every girl who is tired of acting weak when she is strong," reads the poster, "there is a boy tired of appearing strong when he feels vulnerable. . . . For every boy who is burdened with the constant expectation of knowing everything, there is a girl tired of people not trusting her intelligence" (CrimethInc. 2004).

Ben A. Barres knows both sides of sex and gender bias all too well. His chapter disassembles arguments that women's lack of professional advancement in science and related disciplines is the result of their natural inability compared to men. Barres explains that it is not in fact innate inability that has hampered women's progress, but rather *assumptions* of innate inability that have influenced decisions made by fellowship committees, hiring committees, and other gatekeepers that control access to professional science. Furthermore, unconscious bias often plays a large factor in such decisions, thus placing women, minorities, or individuals who otherwise do not reflect the identity of committee members at an additional disadvantage. People of color face similar challenges, even though popular media have focused predominantly on sex bias in the wake of Summers's remarks.

It is important to note, as Barres does, that many who do hold to ideas of differential abilities are "fair-minded people" who also value merit-based advancement. But he also notes that such beliefs remain prevalent even in the face of evidence to the contrary. His analysis demonstrates how biological arguments—in this case, claims about qualities that are both innate and unique to men—become institutionalized in social practice.

Conclusions

This volume of *Controversies in Science and Technology* grapples with some of the knottiest issues of our time. The knottiness derives from their reach into our hearts and souls. These issues challenge our beliefs about each other, our planet, societal fairness, the concept of "knowing," and the definition of human life itself. Society *should* be confronting these issues, but that means that every citizen should be contributing to the debate. There are few simple answers that are "right" or "wrong"; instead there are judgments that both wise and unwise people will make. To ensure that the best decisions are made for all of society, every one of us needs to participate in the dialogue—indeed, shape the dialogue. We can take our rightful place in society only when we have sufficient knowledge to make judgments based on facts and rational arguments; knowledge is power. We hope that our readers will be empowered by the knowledge they gain from examining these tough issues from many angles.

References

Blackless, M. et al. 2000. How sexually dimorphic are we? Review and synthesis. *American Journal of Human Biology* 12:151–166.

Broder, John M. 2007. Edwards learns campaign blogs can cut 2 ways. *New York Times,* February 9.

CrimethInc. 2004. Every Girl Every Boy. Gender Subversion Kit #69-B. http://www.crimethinc.com/a/gender/poster.pdf 2004 (accessed March 10, 2007).

Doniger, David D. et al. 2006. An ambitious, centrist approach to global warming legislation. *Science* 314:764–765.

Goswami, B. N. et al. 2006. Increasing trend of extreme rain events over India in a warming environment. *Science* 314:1442–1443.

Gould, S. J. 1996. *The mismeasure of man*. New York: W. W. Norton & Co.

Johnson, John, Jr. 2006. Unidentified debris keeping shuttle in space: NASA delays today's landing so astronauts can inspect *Atlantis*' exterior after an object is seen drifting away from the craft. *Los Angeles Times,* Sept. 20.

Lawley, Andrew. 2006. Hubble gets a green light, with other missions on hold. *Science* 314:736.

Leary, Warren E. 2006. Repair drill in spacewalk takes 7 hours. *New York Times,* July 12.

Pareles, Jon. 2006. 2006, brought to you by you. *New York Times,* Arts and Leisure section, 1.

Read, Brock. 2007. Middlebury College history department limits students' use of Wikipedia. *Chronicle of Higher Education*, February 16, A39.

Roach, John. 2005. Hurricanes are getting stronger, study says. *National Geographic News*, Sept. 15.

Stone, Richard, and John Bohannon. 2006. U.N. conference puts spotlight on reducing impact of climate change. *Science* 314:1224–1225.

Summers, Lawrence. 2005. Remarks at NBER Conference on Diversifying the Science & Engineering Workforce. http://www.president.harvard.edu/speeches/2005/nber.html (accessed March 10, 2007).

Toner, Mike. 2005. Faulty sensor forces shuttle delay. *Atlanta Journal-Constitution,* July 14.

Westerling, A. L. et al. 2006. Warming and earlier spring increase western U.S. forest wildfire activity. *Science* 313:940–943.

PART 1

Stem Cell Research: Science, Religion, and Public Policy

1

Human Embryonic Stem Cells and the Future of Medicine

Carl E. Gulbrandsen and Jill O. Ladwig

The derivation and manipulation of human embryonic stem cells (ES cells) have created an emotional public debate, but enthusiasm remains regarding the cells' impacts on biology and the biomedical sciences. Scientists Fiona Watt of the Imperial Cancer Research Fund in London and Brigid Hogan of Vanderbilt University Medical Center, writing about stem cells in the journal *Science*, summed it up this way: "For some critics, acquiring this knowledge and deriving human pluripotential stem cells is tantamount to society leaving the original Garden of Eden. For others, such studies, carried out under appropriate guidelines, hold great promise for not only unexpected insights into biology, but ultimately for the alleviation of human suffering" (Watt and Hogan 2000).

Human ES cells are the body's most primal cells—the tabula rasa on which the more than 200 specialized tissues in the human body are drawn and formed. Early medical texts mention the existence of "wandering cells"—regenerative cells that gravitated near wounds to speed healing (Hall 2003). Such cells have long interested developmental biologists—including Kaspar Wolff, who hypothesized in 1759 that the specialized tissues of the body were formed out of undifferentiated tissue and began to illuminate the process of human development (reviewed in Gillespie 1971). The modern area of stem cell research dates back to the 1960s, with the work of Canadian biologists Ernest McCulloch and James Till, who reported the presence of self-renewing cells in the bone marrow of mice. Their discovery led to the use of bone marrow transplant as an effective treatment for cancer and other diseases. The field was further advanced by scientists who derived mouse embryonic stem cells for the first time in 1981 (Evans and Kaufman 1981; Martin 1981).

In his seminal 1998 *Science* paper, Dr. James Thomson, a researcher at the University of Wisconsin–Madison, described the first successful derivation of human ES cells and speculated as to the far-reaching impact of his work. He wrote, "These cells should be useful in human developmental biology, drug discovery, and transplantation medicine" (Thomson, et al. 1998:1145–1147).

Science magazine called the successful derivation of human ES cells by Thomson the Scientific Breakthrough of the Year in 1998, and later named it one of the most significant innovations in the history of science. Dr. Harold Varmus, a Nobel Laureate who was then the director of the National Institutes of Health, said of Thomson's work: "It is not too unrealistic to say that his research has the potential to revolutionize the practice of medicine and improve the quality and length of life" (Varmus 1998).

What Is So Exciting About Human Embryonic Stem Cells?

All stem cells have the peculiar ability either to make exact copies of themselves or to develop into more specialized cells. Stem cells are broadly classified as adult and embryonic. Adult stem cells, when they differentiate, give rise to the type of cell that makes up the tissue in which they are resident. For example, we have stem cells in the bottom layers of our skin. When we slough off skin cells, the stem cells in the skin give rise to more skin cells. In the stem cell state, the cell is referred to as undifferentiated, and when it makes an exact copy of itself, it is said to replicate. When the stem cell makes a more specialized cell—that is, when the skin stem cell makes a skin cell—it is said to differentiate. In addition to skin stem cells, we have stem cells in many tissues throughout our bodies. There are blood stem cells, neural stem cells, and so forth.

Stem cells are present in all stages of human development: the embryo, the fetus (including its umbilical cord), and the fully-formed human (these are the so-called "adult" stem cells). Human ES cells come from a human embryo, as the name implies. In addition to their origin, human ES cells have two important characteristics that distinguish them from adult stem cells: First, human ES cells are pluripotent, meaning they have the ability to differentiate into many specialized cell types. In the case of the human ES cell, it has the ability, as far as we know, to differentiate into any type of cell in the human body. Adult stem cells are not pluripotent, at least not to the extent of the embryonic stem cell.

The other important characteristic of human embryonic stem cells is that they are said to be immortal; when cultured under appropriate conditions in the laboratory they can replicate or "double" themselves seemingly forever without losing their pluripotency. Adult stem cells, when cultured in the lab, go through only ten or twenty doublings and then stop and eventually die. These characteristics, pluripotency and immortality, are what makes human ES cells so potentially important to biologists.

Capturing a Moment in Time

The human ES cells developed by Thomson were derived from the inner cell mass of embryos obtained through in vitro fertilization procedures that resulted in excess embryos. These embryos were donated for research with informed consent. The cells were derived from the preimplantation, or blastocyst, stage of embryonic development, about five days after fertilization and just before implantation in the uterus.

The blastocyst consists of a trophoblast, which later forms the fetal component of the placenta, and the inner cell mass, which later forms the embryo (Figure 1.1).

Once the embryo is implanted in the uterus, the cells that make up the inner cell mass develop into the various tissues and organs of the body. By figuring out how to capture these cells at precisely the right moment and keep them in their undifferentiated state, Thomson essentially froze in place a precious period of time—after fertilization and before implantation—when the cells' developmental potential is greatest. As he explained, "if the ICM [inner cell mass] is taken out of its normal embryonic environment and cultured under appropriate conditions, ICM-derived cells can proliferate and replace themselves indefinitely, yet maintain the developmental potential to form any cell type" (Thomson, et al. 1998).

In culture, ES cells exhibit and maintain a stable, full, normal complement of chromosomes and can develop into all the fetal tis-

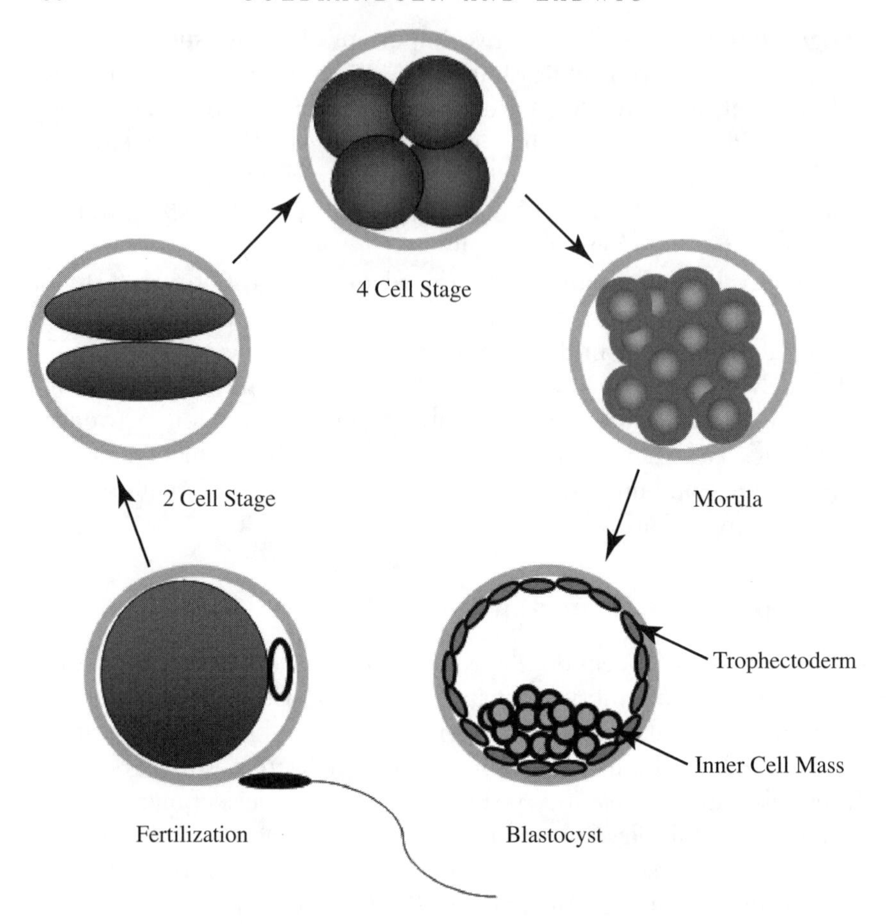

Figure 1.1 Human preimplantation development. After fertilization, the one-cell embryo undergoes a series of cleavage divisions and forms a blastocyst at about six days of development. The blastocyst is composed of an inner cell mass and a trophectoderm. Embryonic stem cells are derived from the inner cell mass. Figure is from: James A. Thomson, "Human embryonic stem cells," in *The human embryonic stem cell debate: Science, ethics, and public policy*, ed. S. Holland et al. (Cambridge, MA: The MIT Press, 2001), 16. Reprinted with permission.

sues. Perhaps the most important characteristic in the lab, however, is their ability to proliferate indefinitely while maintaining their embryonic shape and structure. These cells are capable of colonizing the three germ layers of the human body: the mesoderm, which gives rise to blood, bones, and muscle; the endoderm, which forms the gut, liver, and pancreas; and the ectoderm, which forms neurons, skin, eyes, and ears (National Institutes of Health 2001; and Table 1.1).

Table 1.1 Embryonic Layers from Which Tissues Develop

Embryonic Germ Layer	Differentiated Tissue
Endoderm	Thymus
	Thyroid
	Larynx, trachea, lung
	Urinary bladder, vagina, urethra
	Liver, pancreas
	Lining of the gastrointestinal tract
	Lining of the respiratory tract
Mesoderm	Bone marrow, blood cells
	Adrenal cortex
	Lymphatic tissue
	Muscle
	Connective tissues including bone and cartilage
	Urogenital system
	Heart and blood vessels
Ectoderm	Skin
	Neurons
	Adrenal medulla
	Pituitary gland
	Connective tissue of the head and face
	Eyes
	Ears

Today, leading ES cell scientists are trying to understand the mechanisms underlying pluripotency. What factors control cell differentiation? Why do some cells retain plasticity while others do not? Scientists' efforts to direct differentiation have met with varying degrees of success. Some researchers have figured out ways to nudge the cells down one pathway or another to form specific tissues by manipulating the culture media. The answers to these research questions will inform our understanding of developmental biology.

The Promise of Regenerative Medicine

Many researchers believe the study of human embryonic stem cells will bring about a new era of medical possibilities and ultimately, change the course of human health through regenerative medicine. Loosely defined, regenerative medicine is the replacement of diseased or damaged cells, tissues, and organs to restore lost function. Research in the application of stem cells to treat diseased and damaged organs is moving very quickly; the experiments discussed herein may tomorrow be disproved or may have moved along to the next phase of testing.

The ideal targets for treatment with human ES cells are those diseases and injuries that affect the brain and spinal cord (reviewed in National Academies Press 2002). As far as we know, these organs do not have the natural capacity to renew themselves. The prospect of treating these illnesses and injuries by supplying these organs with healthy cells creates new possibilities for treatments and cures. Because people afflicted with these diseases and injuries currently have few or no treatment options, many hope for a cure. Deriving heart muscle cells, called cardiomyocytes, from ES cells is the first step in rebuilding tissue that's been critically damaged by heart disease. Many forms of the disease, which is still the leading cause of death in the Western world, are caused by the loss of functional cardiomyocytes, and, therefore, heart disease is another good candidate for cell replacement therapies. Stem cell therapies could potentially treat 5 of the 15 leading causes of death in the United States (National Academies Press 2002; Hoyert, et al. 2006 and Table 1.2).

Hans Keirstead, associate professor of anatomy and neural biology and codirector of the Stem Cell Research Center at the University of California–Irvine, is exploring the possibility of using ES cells to restore mobility after spinal-cord injury. Keirstead has succeeded in driving the development of high-purity populations of precursor neural cells, called oligodendrocytes, from ES cells. Keirstead is injecting these cells into rats whose spines have been mechanically damaged and has seen "robust results."[1] The laboratory rats, paralyzed by an injured spine, are injected with the ES-derived neural cells and regain their ability to walk within two months.

Keirstead and others are using these precursor neural cells to treat myelin loss, which is a hallmark of spinal-cord injury, multiple sclerosis, brain injury, and stroke. Myelin is the membrane sheath that coats the neurons in the central nervous system. This sheath serves as an insulator for nerve transmission and efficient signaling. Keirstead and others believe patients in all categories may benefit from "remyelination" therapy.

[1]Unless otherwise indicated, quotations in this chapter were obtained from interviews or e-mail correspondence with the source.

Table 1.2 Potential Health Impact of Stem Cell Therapies

Disease	Number of US Deaths (2003)	Rank among Leading Causes of Death
Cardiovascular disease	685,089	1
Cancer	556,902	2
Diabetes	74,219	6
Alzheimer's disease	63,457	8
Parkinson's disease	17,997	14

Like Keirstead in California, Ian Duncan and colleagues in the School of Veterinary Medicine at the University of Wisconsin–Madison have begun to successfully direct the derivation of the mylenating cells from embryonic stem cells (reviewed in Nistor 2005). Duncan is using oligodendrocytes to remyelinate portions of the spine affected by multiple sclerosis, Amyotrophic Lateral Sclerosis (ALS) and other degenerative neurological disorders, including genetic defects in young children. Applying the cells to form myelin on the spinal lesions created by these maladies appears to restore the ability of the spinal cord to conduct electrical impulses and enable mobility.

Parkinson's disease, a degenerative disease caused by the death of cells in the brain that produce dopamine, could also ostensibly be treated with stem cell therapy. It has been reported that some two dozen patients in Europe have made dramatic progress after being treated with fetal neurons to regenerate the growth of the dopamine-producing cells (Barinaga 2000). Work with fetal cells, while promising, has been thwarted by a lack of supply; it takes six fetuses to treat one patient. Deriving dopamine-producing cells from human ES cells could solve that problem.

Juvenile-onset diabetes is another candidate for a cure with ES cells. Today, the only way for diabetic patients to acquire new insulin-producing cells is via organ donation. Researchers, including University of Wisconsin–Madison scientist Jon Odorico, are having some success in deriving islet cells, the pancreatic cells that produce insulin, from ES cells. Their work could lead to an unlimited supply of islet cells, thus making organ transplantation as a treatment for diabetes a thing of the past.

ES Cell International, a biotech company based in Singapore, is working with ES cells to create cardiomyocytes to treat heart failure, as well as islet cells for diabetes patients. They plan for clinical trials to begin in 2008 (A. Colman, personal communication). Biotechnology companies Geron, Advanced Cell Technology (ACT), and ES Cell International are developing new cells for therapeutic use on a commercial scale. Geron CEO Thomas Okarma says that human ES cells offer the ability to provide replacement cells for any tissue in the body, and he says Geron researchers have figured out how to develop eight different cell types with "high purity" at industry-quality standards.

A "Black Box" to Investigate Degenerative Disease

While the idea of regenerative medicine is intriguing, perhaps the most immediate use of human embryonic stem cells is as a tool for research. The cells provide a powerful mechanism for understanding human development and the mechanics of disease and a new way of testing the safety of drugs and other compounds.

Dr. Kevin Eggan, an assistant professor of molecular and cell biology at Harvard University and assistant investigator with the Stowers Institute for Medical Research, is using the cells to develop treatments for neurodegenerative disease. Degenerative diseases are difficult to study because, as Eggan says, when the patient comes in to the clinic, the destruction has already begun. "It's like investigating a plane crash—coming to the crash site and trying to figure out what happened by putting the fuselage back together," he says. "What you really need is the black box." By allowing researchers to recreate the beginning stages of disease, human ES cells essentially provide the black box.

Eggan and his colleagues use ES cells that are derived from embryos eliminated after preimplantation genetic diagnosis (PGD). PGD is a procedure doctors use to test the health of embryos early in pregnancy in patients who have a genetic predisposition to a disease. In PGD, a doctor takes a few cells from the embryo and tests them for the disease. The embryos with the genetic mutation are eliminated, providing scientists with a way to look at a disease before it actually develops in a person. The Eggan team is also using a technique called nuclear transfer to develop cell lines that carry the genes responsible for the disease. The resulting in vitro model is useful in screening new drugs and in gaining insight into the cellular mechanisms of disease.

Similarly, Dr. Gabriela Cezar, assistant professor of animal science at the University of Wisconsin–Madison, is using human ES cells as a microscale laboratory in which to test the safety of drugs and other chemical compounds. She is specifically interested in identifying compounds that induce fetal abnormalities. "You can throw chemicals at these cells and map the response by looking at the molecules, which are secreted in blood," Cezar explains. Those secretions serve as a "biomarker," a guage that measures a biological process. Detecting the biomarkers for a disease helps researchers identify, diagnose, and eventually develop better treatments for people who have the disease. They may even provide a way to prevent onset of the disease in those who are most at risk.

The ES cells serve as a microcosm of a human body. Cezar is building the next generation of microarrays, using stem cells to test human response directly, instead of testing new compounds on animals to surmise their safety in humans. Animal testing is time-consuming and expensive, and, according to Cezar, only about 50% accurate in predicting the human response to new drugs. "This is a humanized, quick assay that we can do to predict if a new drug will cause birth defects," Cezar said. Like Eggan, Cezar is interested in building an in vitro model of disease. "There is a major need for human models of disease for toxicology testing. This is the best way." She also believes the processes she is developing could one day be used by the US Food and Drug Administration to identify compounds that disrupt human development and also as a means of assessing the risk of environmental chemicals.

Plowing a Stony Field

Major obstacles exist and need to be overcome before this technology can be developed to its full potential, including development of animal product–free, defined cell culture media; generation of functional, differentiated cells; and figuring out how to combat the potential complications of tumor formation and immune rejection. Human ES cells are notoriously finicky and need constant attention. To keep living cells alive and "happy" outside the human body, they must be kept in a dish, awash in a rich bath of nutrients, hormones, and serum. For a number of years, the most successful culture mixture relied on mouse embryo cells and/or other animal products. Although this was effective, scientists were concerned about the possibility of an animal virus or other agent being passed on to the human cells. The "mouse layer" on which the human embryonic cells fed was also very fragile and difficult to maintain.

Researchers began to develop their own special culture media, but these recipes were either haphazard or proprietary. The development of a chemically defined, animal product–free growth medium by Tenneille Ludwig and researchers at the University of Wisconsin–Madison in 2006 (Ludwig, et al. 2006) was an important step toward the widespread study and use of human ES cells. Ludwig's efforts are aimed at optimizing growth media to ensure the pluripotency of the cells and create a pure population of the cells for other researchers. The new medium is easier for researchers to use, resulting in more consistent data. And it can be used to culture new lines of cells or to eliminate animal proteins from cells that were grown on mouse feeders (T. Ludwig, personal communication).

Several companies, including Geron, ACT, and ES Cell International, claim to have a chemically defined, animal product–free culture media. They also say they have already learned to direct the cells to form specific specialized cell types. These companies say the next "big thing" is to figure out how to grow large-scale quantities of the cells for clinical and commercial use. If the cells are only used for customized, individualized cell therapies, they will be cost-prohibitive. So, the leading companies in this field are striving to develop large quantities of specialized cells with industry-standard manufacturing protocols. "The greatest challenge that remains is the derivation of high-purity populations of any cell type. Unless you have high purity, you don't have a good research tool, you don't have a realistic clinical application, and you cannot have a commercial product," says Keirstead.

Since 1995, Geron has supported the work of Keirstead and other researchers, and invested more than $100 million in human embryonic stem cell technology, "always with an eye toward development, specific disease targets, and GMP manufacturing procedures," said the company's chief executive Okarma. He says the company has developed high-purity populations of eight different cell types (T. Okarma, personal communication). Keirstead reports that he has successfully developed populations with a purity of more than 90%, making them clinically and commercially viable (H. Keirstead, personal communication).

There are a number of other issues that need to be addressed if we are ever to see the cells used widely in clinical settings. One is the ability to control the development and proliferation of cells after they're injected into the patient. "We can manipulate these cells in culture, in the lab, but we don't know about the long-term prospects once they're injected into the human body," Alan Colman, chief executive of ES Cell International, points out. "We can try to stimulate the development of new tissue, but there are still gaps in our knowledge about what happens in that process."

Researchers like Thomson, Eggan, and Keirstead are cautious. They believe that if biotech companies move the cells too quickly into the clinic, they may have problems with the process, causing the public to be distrustful and possibly reject this technology. But Colman believes that having audacious goals is essential to keep the science moving forward. "We have to have confidence. We have to set ambitious goals or we will never achieve anything," he says. One checkpoint in the transition from laboratory to clinic will be procedures to test for rare, but serious, communicable diseases that could be transmitted through stem cell therapies (reviewed in Lo, et al. 2005). "First we have to demonstrate safety, then efficacy, and we don't know how long that will take," says Geron's Okarma. "We started in this area by plowing a stony field."

Aside from the pressing scientific questions, the biggest obstacles to the development of stem cell technology in the United States are the lack of federal funding and the limited number of cell lines available for research. According to the NIH Stem Cell Registry, many of the ES cell lines eligible for federal funding were derived in laboratories outside of the United States—specifically, in Israel, Korea, Singapore, and Sweden (reviewed in Yim 2005; see also stemcells.nih.gov). In order to qualify for federal research funding under the president's guidelines, these lines had to meet the same requirements set forth for cell lines derived in the United States—that is, they had to be derived before August 9, 2001; from an embryo created for reproductive purposes that was no longer needed; and with the informed consent of donors.

While some scientists believe there are already enough cells for basic research purposes, there is widespread agreement that more cells should be made available for research in order to capture the genetic variety of the human population. The lack of federal support in the United States has implications worldwide. Germany's stem cell law mirrors US policy, permitting importation of stem cell lines derived only from embryos destroyed before January 1, 2002. World opposition to the use of cloned embryos as a means to generate new stem cell lines was vocalized in passage of the United Nations Declaration on Human Cloning, which urges a ban on all forms of human cloning (reviewed in Cameron 2006). The restrictions on use of ES cells may also be responsible for increased study of pluripotent stem cells derived from other sources including bone marrow, peripheral blood, the inner ear, umbilical cord blood, nasal mucosa, amniotic fluid, and the placental amniotic membrane for their use in clinical applications (reviewed in Prentice 2006).

The Political and Social Landscape

The derivation of the cells from human embryos—even those donated with informed consent or destined for destruction—is a thorny issue. The debate centers on the moral status of the embryo, and includes some of the same language that has emerged from discussions about the morality of in vitro fertilization and preimplantation diagnosis. Dr. Alta Charo, an ethicist at the University of Wisconsin–Madison who helped establish the protocols for research on embryos, says, "For Christians, grace is attained through the acceptance of God's will. So any manipulation of that is a bad thing. In Judaism and in Islam, there is a partnership between God and humans, and manipulation of life to suit our purposes is a positive thing. The issue is whether or not the result is positive."

In the 2000 presidential campaign, human ES cell research was a contentious issue. Candidate Al Gore supported it, while George W. Bush promised he would not support the research. In the first several months of Bush's administration, there was a firestorm of media interest; national and international media ran stories each day about the science, the ethics, the politics, the business impact, and the far-reaching potential of the research. On August 9, 2001, in his first televised address to the nation, President Bush laid out what many at that time considered a brilliant political compromise. The policy he established provided some federal funding for research on ES cells, but only those cell lines created before that day. This solution was immediately attacked by both sides—religious conservatives and patient groups. Researchers hoped the compromise would move the science forward and allow the research community time to prove the value of their work.

Originally, the Bush administration claimed there were more than 70 cell lines available for research. In reality, only 21 of those lines could be distributed to researchers. The first research grants were made available in spring 2002, with money focused primarily on expanding and distributing the government-sanctioned, National Institutes of Health registry cell lines. Five years later, in the summer of 2006, Bush struck down legislation that would have enabled more cell lines derived from embryos to be eligible for federal research funding. While the NIH spends approximately $640 million on stem cell research, only $40 million is spent on studying human embryonic stem cells. Researchers now believe more lines should be made available (Fossett 2007).

Bernard Siegel of the Genetics Policy Institute, a patient advocacy group based in Florida, says the lack of federal support for the research has galvanized the public—especially patients—and "created a pro-cures movement." He believes the attack on this research from religious conservatives is "a proxy battle" for the reversal of the Roe v. Wade Supreme Court decision legalizing abortion. "This is going to help them fight abortion. If people liken these embryonic stem cells to human beings, then they oppose abortion and will oppose stem cell research," Siegel says.

Geron's Okarma agrees. "The association of this medical platform with the abortion debate has stymied the development of this technology into real therapies," he said in a Web interview recorded by the Regenerative Medicine Foundation (http://www.regenerativemedicinesociety.org). However, polling data indicates that many abortion opponents are in favor of embryonic stem cell research (reviewed in Yim 2005).

Furthering Science in an Uncertain World

The dearth of federal funding for research on human ES cells has resulted in a patchwork of laws across the United States and has made the country a battleground over such research. This climate has prompted companies, patient advocacy groups, and states to jump into the void to fund the research.

The basic patents on human ES cells are held by the Wisconsin Alumni Research Foundation (WARF), the patenting and licensing organization that supports the University of Wisconsin–Madison, where Thomson was employed when he made his discovery. Patent applications were filed beginning in 1995, when Thomson succeeded in deriving primate stem cells from rhesus monkey embryos. That research was supported through a grant from the National Institutes of Health and took place at the National Primate Research Center on the University of Wisconsin–Madison campus.

Federally funded research is subject to the requirements of the Bayh-Dole Act, which governs how federally funded inventions must be handled (35 USC 200-212 and 37 CFR, Part 400). Under that law, grant recipients are required to patent federally funded inventions and are entitled to take ownership of those patents. After successfully culturing ES cells from rhesus monkeys, Thomson succeeded in culturing the cells from marmosets. These two types of monkeys are genetically further apart than the rhesus monkey is from humans, so Thomson's success with those species led him to believe that human ES cells could be isolated and cultured as well.

While federal funds were not available for this type of experiment, private money was. Thomson had already been approached by Geron Corporation with an offer to fund studies to produce human ES cells. Geron and the University of Wisconsin entered into a sponsored research agreement, allowing the biotech company to fund the work. In 1997, Thomson disclosed that he had succeeded in being the first scientist to derive and culture human embryonic stem cells. As a consequence of its funding agreement, Geron was granted the first license to the basic human embryonic stem cell discoveries. Today, there is a lot of controversy surrounding those patents and the license to Geron.

The California-based Foundation for Taxpayer and Consumer Rights, and some scientists, have insisted that the patents and WARF's licensing policy have obstructed research (Foundation for Taxpayer and Consumer Rights press release, Sept. 28, 2006). Others, including high-level administrators at the University of Wisconsin–Madison and the State of Wisconsin, disagree. Proponents of WARF's policies point to formation of WiCell Research Institute, which distributes cells to researchers and trains them to culture the cells in their own labs. Elizabeth Donley, chief executive of Stemina BioMarker Discovery, a stem cell company, and former executive of WiCell and WARF, says the patents do not impede the research. She notes that WARF provides licenses for the cells free of charge to researchers and has a flexible program for licensing to private companies based on their size and ability to pay.

According to many analysts, the hodgepodge of regulations in the research arena thwarts progress in the field. Because the federal government will not support the derivation of new cell lines, some states have passed laws to support the conduct of the research. California led the way when it passed its resolution to fund human ES cell research to the tune of more than $3 billion. Connecticut, Illinois, and Massachusetts are drafting resolutions that would allow funds to support such research, but other states, including Michigan, have passed severe restrictions on its conduct.

"I'm afraid state initiatives will complicate the issue," says Okarma of Geron. "Restrictions on the science, based on theology, will have a huge impact on the medical advances made in this country and throughout the world." "This is a huge problem," says Dr. Norm Fost, a pediatrician and ethicist at the University of Wisconsin–Madison. "Research in this field is interdisciplinary. Researchers want to share cells and share information that would facilitate collaboration and cooperation." The beauty of having a federal system to fund this research is that it would allow researchers to collaborate across state lines.

Because of the federal government's lack of support in this area, there are no real federal guidelines governing conduct of this research (A. Charo, personal communication). The National Research Council and Institute of Medicine proposed voluntary guidelines for ES cell research, which include these basic tenets: the ES cells must come from excess embryos resulting from in vitro fertilization procedures; the embryos must be donated with informed consent; and there can be no financial inducements for the patient to donate the embryos (National Research Council and Institute of Medicine Committee for Guidelines on Human Embryonic Stem Cell Research 2005). "In the best of all possible worlds, the federal government would operate with the goal of human health and safety, not with moral judgment," says ethicist Dr. Alta Charo. "There are no regulations, really. The government has not made an effort to regulate new cell lines being developed. The NAS guidelines are being used as regulations."

Dr. James Battey, chair of the National Institutes of Health Stem Cell Task Force, says governing this area of research is difficult. "The challenge is, within the federal regulations set forward, to do as much as we can to promote this research." NIH provided a contract to establish the first National Stem Cell Bank at the WiCell Research Institute in 2005 to serve as a repository for NIH-sanctioned human ES cells. The agency offers short-term training for researchers interested in learning how to culture the cells. Battey adds that, while science is in fact advancing, he believes more functional cell lines should be made available for federally funded research.

Turning the R&D Paradigm on Its Head

Patient advocacy groups and private companies are playing critical roles in ensuring this research moves forward. Some say this dynamic has caused the private sector to go beyond its traditional role in the commercialization of discoveries made in academic labs. They say private companies are in the vanguard for this research, perhaps for the first time. This new paradigm frightens some people. "I think there is a basic distrust of what biotech companies are doing," says Okarma. "People assume we are financially motivated and not socially motivated."

Academic researchers worry that the race to clinical applications is getting in the way of scientific progress. "We cannot overpromise," says John Hearn, provost and professor of physiology at the University of Sydney in Australia and former director of the Wisconsin National Primate Research Center at the University of Wisconsin–Madison. "There is enormous potential for this technology to revolutionize medicine. But we cannot overpromise. The rush for applications is damaging scientific progress."

Hearn and others worry that scientific scandals and backbiting in academia and private industry may eventually weaken society's support for stem cell research. Thomson says the most damaging impact of the political and ethical debate is its chilling effect on research. "Young scientists are shying away from the field," he says. Siegel of the Genetics Policy Institute says the lack of federal support has given an increased role to patient groups who have been able to create urgency and legitimacy around the research by humanizing it. "When they enacted these regulations, they didn't count on the fact that patient advocates, medical scientists, and even some conservatives would ban together to very vocally support this research," Siegel says.

At least one company, ACT, has found a way to create human ES cells that might be blessed by the federal government. ACT has developed two lines of cells that were derived without destroying embryos (Klimanskaya, et al. 2006), prompted by the needs of large pharmaceutical companies who want to initiate research in this area, but who don't want to turn off potential investors—or cross the federal government—in the process (M. West, personal communication).

One of those alternative means is to remove one of the blastomeres of an eight-cell embryo, as is done in preimplantation genetic diagnosis, and then return it to the freezer.

"It's really kind of silly," admits ACT's Michael West. He said his company has invested considerable time and effort into developing these "embryo-free protocols" even though many excess embryos are available through in vitro fertilization clinics. But to fully and finally get around the ethical objections to embryonic stem cell research, the best option is to "reprogram" adult stem cells to their embryonic state and derive new types of cells from that starting point. "That's the new Holy Grail," said Thomson.

Out of Eden? The Future of Medicine

Human ES cells give scientists the potential to create treatments and cures for some of mankind's most intractable and dreaded diseases. Visionaries like Michael West, John Hearn, and others fully believe it is possible to create a world of customized medicine—a world in which a tiny scrape of a patient's skin would capture some DNA, which could then be banked and used to develop genetically matched stem cells, which could be cultured and manipulated to create whatever cells that person might need to solve personal health problems throughout his or her lifetime.

Others, like Thomas Okarma, see large-scale commercialization, and not personalized medicine, as the end goal. Okarma calls himself a realist. "We face an enormous medical dilemma that pits access to medical care against innovation," he said, noting that given the current limitations of our healthcare system, the most innovative procedures and techniques will be very expensive, at least at first. "The system, as it is, wants us to stay the same, keep doing things the same way. It doesn't want to pay for new cures and treatments."

Whatever the future holds, there is no doubt that the continued study of these cells will answer the most fundamental questions in human developmental biology, including, how a single cell—a fertilized egg—can become a complex, multicellular organism. And it will shed light on areas of human health that have been hitherto inaccessible, such as pregnancy loss, and infertility. So what is the future of human health? Have we really cast ourselves out of Eden? "The future of medicine will be guided by a better understanding of the human body, not necessarily transplant therapy and regenerative medicine," says Thomson. The founder of this controversial field believes the highest value of his discovery is simply the ability to study further the earliest processes of human development.

For Further Reading

Holland, Suzanne et al., eds. *The human embryonic stem cell debate: Science, ethics, and public policy*. Cambridge, MA: MIT Press, 2002.

Scott, Christopher Thomas. *Stem cell now: From the experiment that shook the world to the new politics of life*. New York: Pi Press, 2006.

References

Barinaga, Marcia. 2000. Fetal neuron grafts pave way for stem cell therapies. *Science* 287:1421–1422.

Cameron, Nigel M. de S. 2006. Research ethics, science policy, and four contexts for the stem cell debate. *Journal of Investigative Medicine* 54:38–42.

Coalition for the Advancement of Medical Research. Nearly three-quarters of America supports embryonic stem cell research. Press release. May 16, 2006.

Evans, M. J., and M. H. Kaufman. 1981. Establishment in culture of pluripotential cells from mouse embryos. *Nature* 292:154–156.

Gillespie, C. C., ed. 1981. *Dictionary of scientific biography*. New York: Scribner.

Hall, Stephen S. 2003. *Merchants of immortality: Chasing the dream of human life extension*. New York: Houghton-Mifflin.

Hoyert, D. L. et al. 2006. Deaths: Final data for 2003. In *National vital statistics reports*. 54(13). Hyattsville, MD: National Center for Health Statistics.

Keirstead, Hans S. 2005b. Stem cells for the treatment of myelin loss. *Trends in Neurosciences* 28:12.

Klimanskaya, I. et al. 2006. Human embryonic stem cell lines derived from single blastomeres. *Nature* DOI:10.1038.

Lo, Bernard et al. 2005. A new era in the ethics of human embryonic stem cell research. *Stem Cells* 23:1454–1459.

Ludwig, T. E. et al. 2006. Derivation of human embryonic stem cells in defined conditions. *Nature Biotechnology* 24(2): 185–187.

Martin, G. R. 1981. Isolation of a pluripotent cell line from early mouse embryos cultured in medium conditioned by teratocarcinoma stem cells. *Proceeding of the National Academy of Sciences USA* 78:7634–7638.

Milestones of Science. 2005. *Science* vol. 309.

National Research Council and Institute of Medicine, Committee on the Biological and Biomedical Applications of Stem Cell Research. 2002. *Stem cells and the future of regenerative medicine*. Washington, DC: National Academies Press.

National Institutes of Health, Department of Health and Human Services. 2001. *Stem cells: Scientific progress and future research directions*.

National Research Council and Institute of Medicine (U.S.) Committee for Guidelines on Human Embryonic Stem Cell Research. 2005. *Guidelines for human embryonic stem cell research.* Washington, DC: National Academies Press.

Nistor, Gabriel I. et al. 2005. Human embryonic stem cells differentiate into oligodendrocytes in high purity and myelinate after spinal cord transplantation. *Glia* 49:385–396.

Panno, Joseph. 2005. *Stem cell research: Medical applications and ethical controversy*. New York: The New Biology Series, Facts on File.

Prentice, David A. 2006. Current science of regenerative medicine with stem cells. *Journal of Investigative Medicine* 54:33–37.

Thomson, James A. et al. 1995. Isolation of a primate embryonic stem cell line. *Proceedings of the National Academy of Science* 92:7844–7848.

Thomson, James A. et al. 1998. Embryonic stem cell lines derived from human blastocysts. *Science* 282:1145–1147.

Varmus, Harold. 1998. Statement before the Senate Appropriations Subcommittee on Labor, Health and Human Services, Education and Related Agencies, December 2.

Watt, Fiona, and Brigid L.M. Hogan. 2000. Out of Eden: Stem cells and their niches. *Science* 287:1427–1430.

Yim, Robyn. 2005. Administrative and research policies required to bring cellular therapies from the research laboratory to the patient's bedside. *Transfusion* 45:144S–158S.

2

Embryological Origins of the Human Individual

Karen M. Downs

Overview

The biology of the fertilized egg is the subject of this brief essay. While many different arguments have been made as to when life begins, developmental biologists base their views on careful and systematic experimental analyses of the fertilized egg and the derivative products of conception. The single most important biological fact about human reproduction, generally ignored by scientists and the public at large, is that the fertilized egg produces many more cells and tissues than just the "embryo." Lack of such general knowledge severely limits wider public debate about reproduction, the consequences of which affect each one of us. Sadly, little of the biology of human development is taught as part of a standard curriculum in schools in America. As a result, poorly informed leaders are making decisions on our behalf. Recently, the debate concerning human embryonic stem cells, in particular, has been severely hampered by ignorance of basic concepts of developmental biology. It is difficult to understand the differences in derivation techniques, the range in developmental potential of stem cells from various sources, or the nuances of the ethical debates surrounding this issue with little or no understanding of what the fertilized human egg produces. Moreover, ethical issues are further confused by imprecise terminology. Thus, it is toward a collective wisdom concerning human reproduction that this chapter was written. Two central themes, based on developmental biology, will be put forth. The first is the origin of potential life. The second is the origin of the human individual.

Potential Life

Nearly everything that is currently known about mammalian development was acquired by observation and experimentation in a variety of animals. The results provide a consistent framework across species. Like most mammals, developing humans exhibit an extreme form of viviparity, which means that, as development progresses, the fertilized egg becomes entirely dependent upon an internal uterine environment and external blood supply for development and survival to birth. Dependency is mediated by a battery of extraembryonic tissues formed by the conceptus, and not by the mother.

Fertilization is a process, and not a single moment. During fertilization, a sperm and egg come together, paternal and maternal genetic complements now confined to a single entity, the fertilized egg. The egg and sperm each contain one half of the amount of genetic material required for human life. Reduction in genetic content began prior to fertilization in a process called meiosis, which in females takes place predominantly within her ovaries, and in the male, within his testes (Figure 2.1).

After coitus, a spermatozoan travels through the female's reproductive tract to reach the mature egg that is moved to the oviduct from the ovary (Figure 2.1A). Subsequently, the sperm head penetrates the egg's tough outer coating to fuse with it, and releases its parcel of genetic material into the egg's cytoplasm (Figure 2.1B, 2.1C). Male and female genetic materials ultimately intermingle to create a single combined packet, the nucleus, in a process called syngamy (Figure 2.1D). The full amount of genetic material will be maintained in all cells until specialized processes once again halve it in the specialized sperm or eggs in the new organism. On the basis of these observations, the consensus among developmental biologists is that potential life begins at syngamy.

We talk about "potential" life because there is no other time in an individual's development, except perhaps old age, that is so hazardous and prone to death as prenatal development. Many fertilized eggs fail to implant, or are reabsorbed, that is, break down and degrade because they are not healthy enough to progress. It is thought

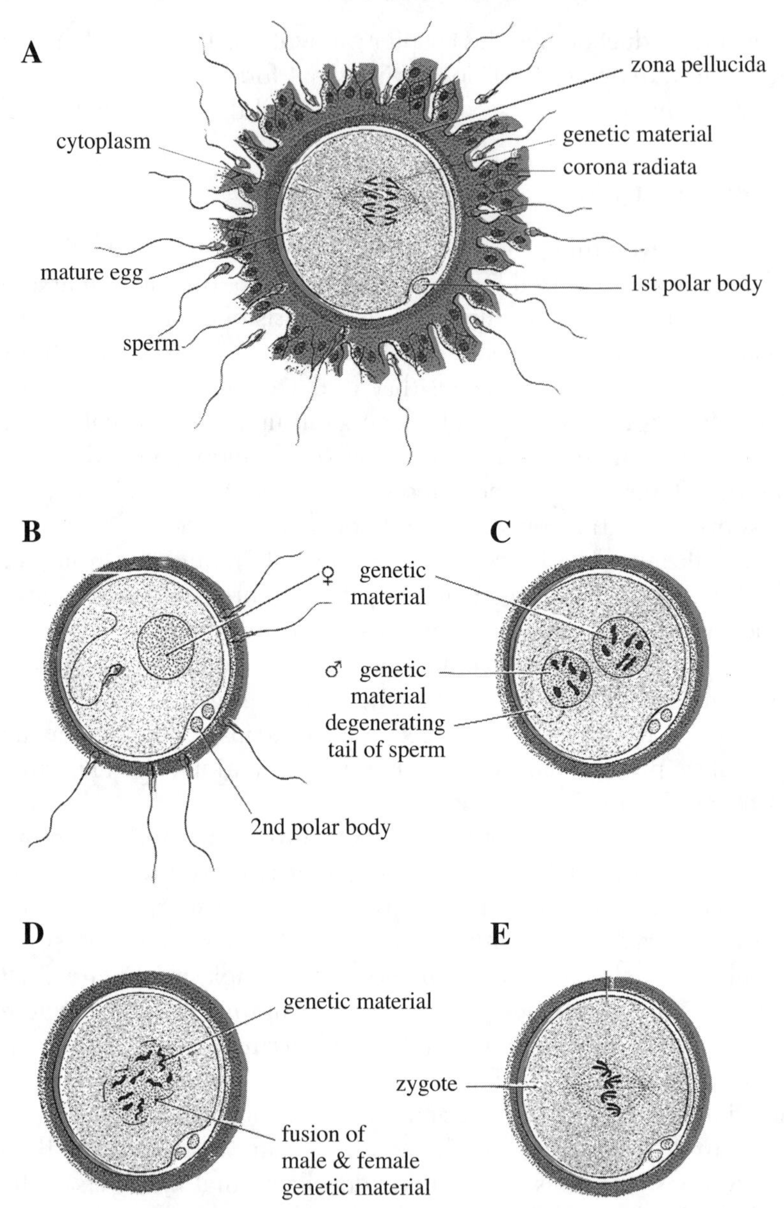

Figure 2.1 Diagram of the process of fertilization. (A) A large number of sperm are swarming the mature egg in the oviduct. Under normal conditions, only one of these will penetrate the tough zona pellucida and fuse with the egg's surface. (B) The remnants of a single sperm within the cytoplasm of the egg. (C) The casing of the sperm has degenerated within the egg's cytoplasm, leaving only its intact genetic information. (D) Syngamy is occurring, in which the genetic material from both the male and female intermingles to form a single packet. (E) The fertilized egg, called the conceptus, or zygote, prior to first cleavage stage. Modified from Moore (1982), Figure 2-14. Used with permission.

that, prior to implantation, mortality is as high as 70%. If a fertilized egg successfully implants, it faces perils thereafter, its demise appearing as a miscarriage or stillbirth. These account for about 20% of established pregnancies. Thus, more than half of all conceptions fail to produce surviving newborns. It is on the basis of such precarious development and morbidity that syngamy sets into motion a potential life.

The Human Individual

A basic problem in mammalian reproduction is the terminology applied to the entire developing structure after fertilization. It is generally agreed that the fertilized egg is called the zygote, from the Greek *zygotos*, meaning the yolking together of two similar things, here, the sperm and the egg. Thereafter, imprecise terminology has, to the detriment of clarity, become unwittingly widespread. Upon division of the zygote, the resulting two-cell entity is misleadingly called an "embryo." While such terminology would be accurate in a frog, for example, which does not depend upon its mother for subsequent survival, it is not appropriate in the mammal. To survive and develop within its mother, a large portion of the zygote's descendent cells must form extraembryonic supporting tissues and organs. They are thus called because they do not become appendages of the adult but will be shed at birth. The most commonly recognizable supporting organ is the placenta (Figure 2.2).

The partitioning of the zygote into embryonic and extraembryonic tissues is a biological fact that was not fully recognized until relatively recently. The nature of the fertilized egg's products was determined by experimental investigation in the latter half of the twentieth century in a variety of mammals. Thus, the term "embryo," used by philosophers and scientists alike, is an outdated terminology. In mammals, an embryo is truly the embryonic body. Descendent cells of the zygote are more appropriately referred to as the "conceptus," meaning all the products of conception, encompassing embryonic and extraembryonic tissues.

In biological terms, the human is an individual when all of its parts belong to it alone. In other words, the onset of human individuation occurs when embryonic cells become entirely separate from extraembryonic ones. As described below, this criterion poses some difficulties, as the end point of compartmentalization is not clear. The current well-established criterion, formation of the primitive streak, will be discussed later in this essay.

Figure 2.2 The human infant at birth, still connected to its placenta. The umbilical component of the placenta forms a vascular bridge between the chorionic disk, into which mother's blood infiltrates, and the fetus. The remnant of our connection to our mother during gestation is manifest in our navel, or "belly button." Originally from the anatomical plates of Julius Casserius, which were later included in a book by author Adrianus Spigelius in 1626; reproduced in Corner (1942; renewed 1970), Plate XX. Used with permission.

Development of the Conceptus: Overview

In this section, an overview of the three major phases of human development will be presented, after which key biological processes within each phase that contribute to our understanding of the developmental origin of the individual will be highlighted.

During the first phase, called preimplantation, the zygote becomes parsed into a number of cells that, after several days, form an intermediate entity called the blastocyst. The blastocyst contains an outer layer of cells, the trophectoderm, and a small cluster of inner cells, the inner cell mass, or ICM.

During the second phase of development, called implantation, trophectoderm excavates a space in the mother's uterus that enables the conceptus to embed inside of it. Implantation involves developmental synchrony between the conceptus and its mother.

During the third and most prolonged phase, postimplantation, trophectoderm's descendents establish one of two major placental components, the chorionic disk. The ICM's derivatives, primitive endoderm and epiblast, elaborate further extraembryonic tissues, including the umbilical component of the placenta and the yolk sac, and the embryo/fetus.

Preimplantation Development

Figure 2.3 depicts the series of events that take place during preimplantation development, which culminates in the formation of the blastocyst. Shortly after syngamy, the fertilized egg, or conceptus, typically called a zygote (or, unfortunately, the embryo), duplicates its genetic material, which is then parsed into two descendent cells, called "daughters." The result is a two-cell conceptus (Figure 2.3.1, 2.3.2A), each cell of which is referred to as a blastomere. The blastomeres, loosely attached to each other, are held together within the egg's tough outer coating, the zona pellucida. If the zona is removed, the blastomeres easily separate.

Each of the two blastomeres divides and produces two cells, now making a total of four (Figure 2.3.1, 2.3.2B). Each of the four blastomeres divides, producing an 8-cell conceptus (Figure 2.3.1, 2.3.2C) and so on, until the multicelled conceptus arrives in the lumen, i.e., the space inside the uterus (Figure 2.3.1). The conceptus is now a rounded mass that, because of its resemblance to a mulberry, is referred to as a morula (Figure 2.3.1, 2.3.2D). Outer cells of the morula flatten, and adhere tightly to each other, creating trophectoderm (Figure 2.3.1, 2.3.2E). The remaining inside cells become the inner cell mass, pushed to one side by a fluid-filled cavity; the blastocyst is created. The blastocyst enlarges, "hatching" or shedding its zona pellucida and,

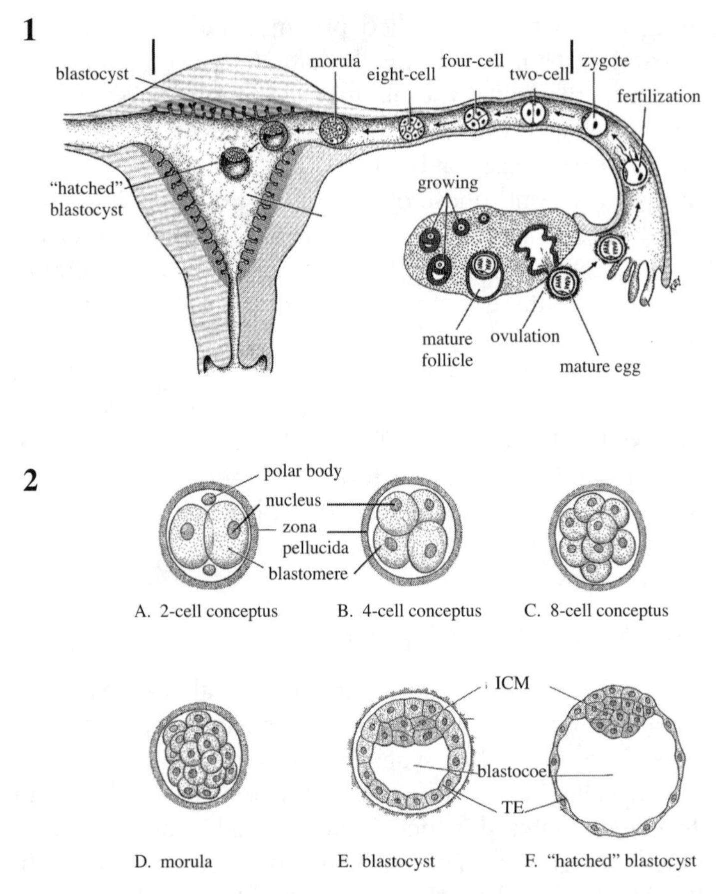

Figure 2.3 Summary of preimplantation development. (1) Schematic drawing of the female reproductive tract. The ovary contains growing eggs, within follicles. As they mature, the eggs begin to halve their genetic information within the ovary. At the appropriate time, a mature egg ruptures its follicle and moves into the oviduct where it may be fertilized by sperm that have traveled through the cervix and uterus. Fertilization results in syngamy, 2.1E, after which the fertilized egg, or conceptus, divides and ultimately forms a blastocyst, which arrives in the uterus. There, it hatches out of its zona pellucida and implants. The stages of preimplantation development shown in (2) are located between the two vertical black lines above the reproductive tract. Modified from Moore (1982), Figure 2-18. Used with permission. (2) Detail of preimplantation development. The fertilized egg, which was depicted in Figure 2.1, has now undergone first cleavage, and formed the two-cell conceptus contained within its zona pellucida. Each cell is called a "blastomere," and each blastomere's nucleus contains the full amount of genetic material. The polar bodies are the remnants of reduction in genetic material in the female and will not participate in further development of the conceptus. (B) Four-cell conceptus. (C) Eight-cell conceptus. (D) Morula. The outer cells of the morula will form the trophectoderm, and the inner ones will form the inner cell mass. (E) Blastocyst. Trophectoderm (TE) cells encircle the blastocoel cavity and inner cell mass (ICM). (F) "Hatched blastocyst." The blastocyst has shed its zona pellucida and is ready to implant in the uterus. Modified from Moore (1982), Figure 2-15. Used with permission.

via the action of trophectoderm, begins to implant in the uterus (Figure 2.3.1, 2.3.2F).

Preimplantation Stages and the Origin of the Individual

The majority of fertilized eggs that ultimately survive the treacheries of development produce a single human individual. However, about 1% of these zygotes may, for reasons that are not clear, produce identical twins; thus, two individuals.

Human twins are generally monozygotic or dizygotic, i.e., identical or nonidentical, the latter often referred to as fraternal twins. Dizygotic twins are the result of independent release and fertilization of two different eggs by different sperm. However, as their name implies, monozygotic twins are derived from a single fertilized egg. It is well established that monozygotic twinning ceases during early postimplantation. Until then, more than one individual may emerge from the conceptus. The process of twinning will be summarized in Figure 2.9.

In addition to identical twinning, another criterion for individuality has emerged from experimental studies, particularly in the mouse. Results have demonstrated that, when experimentally aggregated, that is, when placed together, blastomeres, morulae, and the inner cell mass of the same stage, can mix with each other. The component cells assimilate to form a single chimera (Figure 2.4). In Greek mythology, Chimaera was a fearsome creature made from three animals, the lion, goat, and serpent. In zoology, a chimera contains cells from genetically different conceptuses.

In experimentally produced chimeras, aggregated cells reorganize so that duplicity or multiplicity disappears (Figure 2.4.2, 2.4.3). When returned to the uterus of a surrogate mother (Figure 2.4.4), the aggregated conceptus develops as a single unit. At birth, the chimeric individual contains a mixture of cells throughout its body from both entities (Figure 2.4.5). This extraordinary flexibility is taken as further embryological proof that the conceptus has not yet individuated at the time of aggregation.

Chimeric human individuals are found in nature. They appear to be the result of fusion between two nonidentical twin sisters, brothers, or sister-brother combinations, because, as mentioned above, more than one egg may be fertilized at coitus. These rare cases typically come to light under circumstances where the chimeric individual undergoes blood tests later in life.

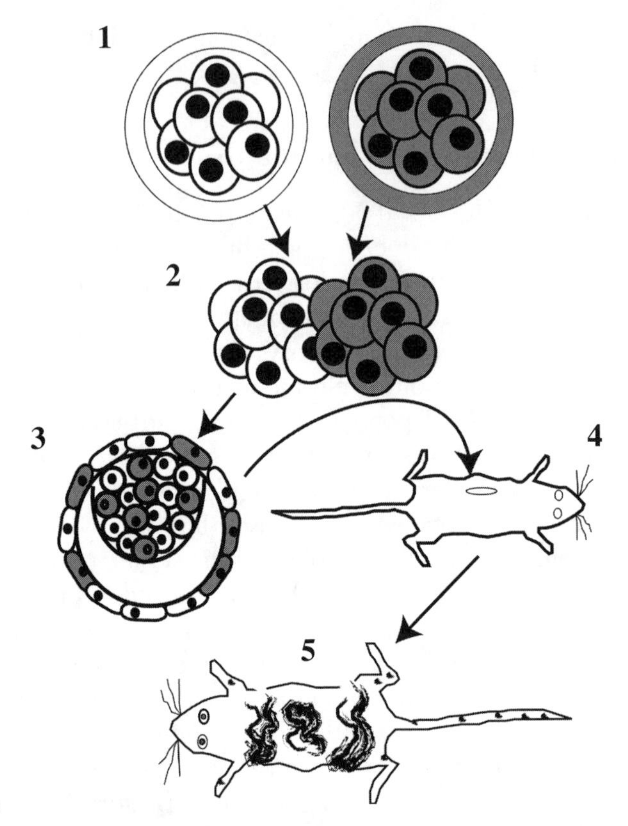

Figure 2.4 Morula aggregation. Schematic diagram showing two genetically distinct (grey and white) morulae inside their zona pellucida (1). In (2), the zona pellucida was removed from the morulae, after which they were placed together in a dish and aggregated. (3) The aggregated morulae have now developed into a single blastocyst, their cells mixed throughout both the trophectoderm and ICM. In (4), the chimeric blastocyst is surgically inserted into the uterus of a foster mother and carried to term. At birth (5), the chimeric pup contains cells throughout its body that are derived from both morulae, represented by the striping on its fur.

The Blastocyst

We will now give further attention to the blastocyst, and its two distinct cell types. The blastocyst stage is a turning point on the way to individuation. Experimental manipulations have revealed that, once the blastocyst has formed, its trophectoderm and inner cell mass are partitioned, or segregated, from each other. In other words, when trophectoderm is experimentally placed into the ICM of a host blastocyst, it will not assimilate into and contribute to the inner cell mass. Similarly, ICM cells will not assimilate into trophectoderm.

Thus, trophectoderm and inner cell mass are distinct from one another. However, only the trophectoderm seems to be developmentally fixed. Its derivatives will form a related set of tissues involved in implantation and placentation. The ICM is still developmentally labile, its cells often referred to as "pluripotent."

Such remarkable plasticity was demonstrated experimentally by "blastocyst injection" in the mouse (Figure 2.5). An inner cell mass from a donor blastocyst was transplanted into the blastocoel cavity of a host (Figure 2.5.1, 2.5.2b). There, donor cells assimilated into the host's ICM (Figure 2.5.3, 2.5.4). Operated blastocysts were then transferred to a surrogate mother whose uterus was primed to receive them. After birth, it was observed that the donor ICM had contributed to all cell types in the newborn.

Derivation of Embryonic Stem (ES) Cells from the Blastocyst

A fundamental question in developmental biology is: How does a single cell, the fertilized egg, give rise to its dazzling array of diverse cell types? In other words, how does the zygote produce both embryonic and extraembryonic tissues, and when do these developmental decisions occur?

Given its extraordinary ability to contribute to so many different cell types, the ICM was recognized as a potentially valuable entity in addressing this question. It could be successfully propagated in culture, thereby yielding enough biological material to address this major question. Further, large numbers of potentially identical cells would permit uniformity and reproducibility between experiments, a gold standard in all scientific work. Finally, cell lines would have the added benefit of reducing the numbers of animals required for this research.

Embryonic stem (ES) cells, as they were named, were thus grown out of the mouse ICM (Figure 2.5.2a). ES cells have even recently been derived from preblastocyst stage conceptuses. Under appropriate conditions, ES cells can be maintained indefinitely in culture dishes. They can be manipulated by gene replacement, aggregated with a morula, or injected into a blastocyst (Figure 2.5.3). When ag-

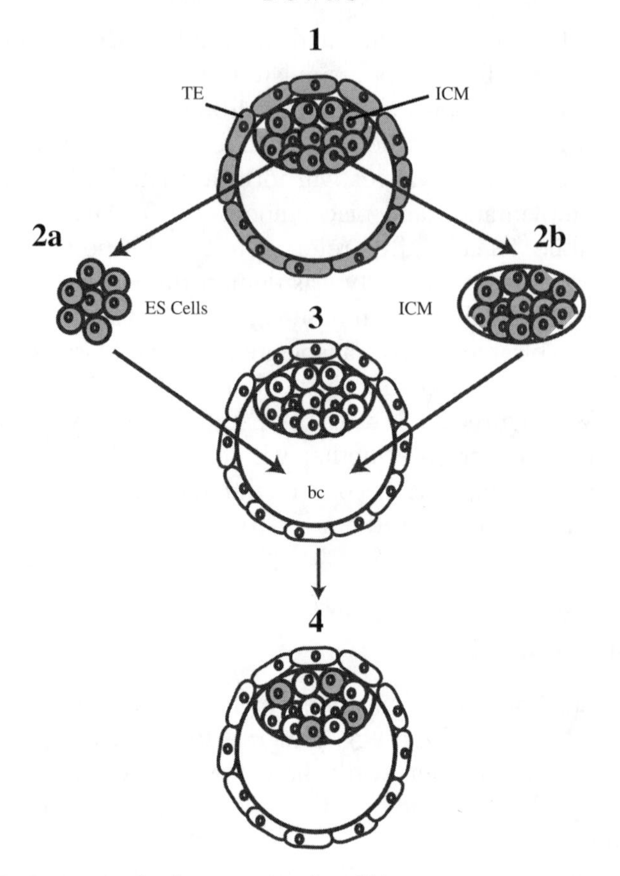

Figure 2.5 Derivation of embryonic stem cells and blastocyst injection. Two genetically distinct blastocysts, indicated by the grey and white colors, are represented in this schematic diagram. (1) The inner cell mass (ICM) of the grey blastocyst can either be propagated to form embryonic stem (ES) cells (2a, left), or removed from the surrounding trophectoderm (TE) as an isolated ICM (2b, right). (3) Either the ES cells, which can be further genetically manipulated, or the ICM, can be placed into (3) the blastocoel cavity (bc) of the white blastocyst, where they will assimilate into one chimeric blastocyst, represented by the mixture of grey and white cells (4). Notice that the donor ES and ICM cells contribute only to the host's ICM and not to its trophectoderm. As described in the text, this is because the TE and ICM are now distinct, noninterchangeable tissue types.

gregated or injected, ES cells contribute to a chimeric animal. When ES cells are altered, and then aggregated with a blastocyst, the alteration can be passed on to the next generation. Large numbers of mouse models of human disease have been created in this way, permitting assessment of the role of specific gene products in the organism. Now, several decades after the first isolation of mouse ES cells, human ES cells have been derived from supernumerary blastocysts resulting from Assisted Reproductive Technology (ART).

In light of their ability to contribute to almost any cell type in any environment in which they are placed, ES cells represent an invaluable resource for investigation into fundamental questions in reproductive and developmental biology. With further understanding through basic research, they also represent a potential adjunct avenue for treating a number of debilitating conditions recognized prior to, at, or after birth.

Implantation

Until now, the zygote and its cellular descendents have been free-floating in the oviduct and uterus. No direct contact has been made with the mother's tissues. Stored nutrients in the egg as well as maternal substances secreted into the uterine space have provided the preimplantation conceptus with its nutrition. Gases were obtained via diffusion within the mother's reproductive system.

But the blastocyst must enlarge. Enlargement comes under a mathematical law which states that the surface available for diffusion increases as the square of the diameter, while the bulk of the tissues that must be nourished grows as the cube. Thus, a larger and more efficient surface of exchange must be provided. This is accomplished first through trophectoderm-mediated implantation into the uterine wall.

A series of well-coordinated hormonal steps that began when the egg was expelled from its home in the ovary now culminate in the transformation of the uterus, the "womb," into an environment hospitable to receive the conceptus.

The process of implantation is summarized in Figure 2.6. With the mother's uterus prepared, the trophectoderm sets to work on hollowing out the uterine wall (Figure 2.6A). Although the final juxtaposition of conceptus and uterus varies among mammals, mouse and human trophectoderm cells are highly invasive, thereby forging an intimate relationship with the mother (Figure 2.6B). As trophectoderm, now called "trophoblast" carves out a space and breaches the walls of the uterine blood vessels, maternal blood flows over the blastocyst, supplying it with nutrients and oxygen (Figure 2.6C). Thus,

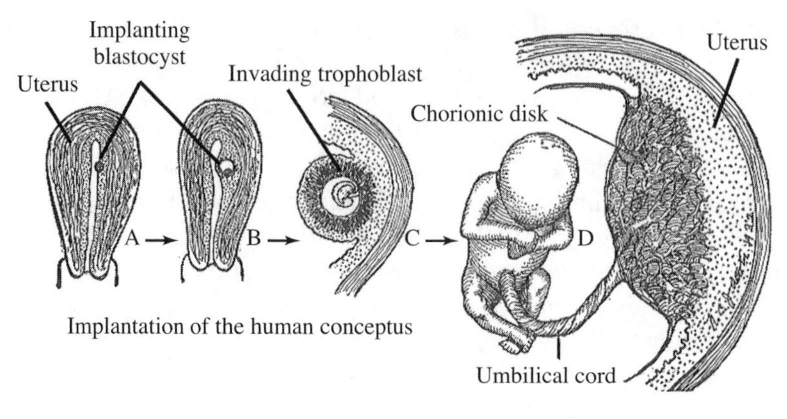

Figure 2.6 Overview of implantation and postimplantation developmental phases. This schematic diagram depicts implantation (A-C), and the later stages of postimplantation (D). (A) The hatched blastocyst has made contact with the surface of the uterus. (B) As a result of trophectoderm excavation and a receptive uterus, the blastocyst is now completely embedded in the uterus. (C) The trophectoderm, now called trophoblast, penetrates farther into the uterine wall, breaching maternal blood vessels, whose blood bathes the implanted conceptus. (D) During postimplantation, the embryo/fetus has formed and is connected to its mother by the chorio-allantoic placenta, composed of the chorionic disk, in which mother's blood is pooled, and the umbilical cord, which, as a result of its direct connection to the fetus, shuttles fetal blood to and from the disk for exchange with the mother. Adapted from Corner (1952), Figure 14. Used with permission.

through trophoblast activity, the blastocyst nestles down within its mother, and prepares for rapid and large-scale growth.

While trophectoderm is juxtaposing mother and conceptus, the inner cell mass continues to develop, its cells segregating into two distinct types, the primitive endoderm and the epiblast. Primitive endoderm forms a major component of the extraembryonic yolk sac, a temporary placenta within which blood vessels and blood form, thereby delivering oxygen to the developing embryo until the definitive and more robust chorio-allantoic placenta forms. Epiblast will give rise to both extraembryonic and embryonic cells.

Postimplantation

The Placenta

For development to progress, the blastocyst must now establish an enduring means by which the embryo/fetus can acquire its nutrient and gas supply from maternal blood throughout the remainder of gestation. Toward this end, the trophoblast and epiblast conspire to form a remarkable organ that serves as the interface between mother and fetus. This life support system is called the chorio-allantoic placenta.

In ancient times, the placenta maintained a spiritual connection with its owner. The Egyptians, particularly in dynastic families, preserved it in receptacles throughout its owner's life. In the Hmong culture, the placenta is regarded as the individual's first garment, and is buried. At death, the placenta is "reunited" with its owner for safe journey into the afterworld. In Western cultures, the placenta is accorded little more than a cursory look at birth, and then incinerated. Thus, our shared origins with the placenta and other extraembryonic tissues formed from the fertilized egg are largely forgotten or ignored. Perhaps it is for that reason, and our natural tendencies toward egocentrism, that use of the term "embryo" rather than "conceptus" is so widespread.

The chorio-allantoic placenta is a compound organ, composed of both the chorionic disk and the "allantois"-derived umbilical cord (Figure 2.2 and Figure 2.6D). Each component develops independently of the other.

Shortly after implantation, trophoblast further diversifies into well-organized layers of cell types that contribute to most of the chorionic disk. These cells are ultimately responsible for permitting diffusion and transport of maternal nutrients and oxygen to the fetus. Thus, trophoblast cells of the chorionic disk communicate with maternal blood.

The umbilical cord is derived from the epiblast. Through anatomical changes too complex to elaborate on here, the future chorionic disk and epiblast are separated by a cavity. The umbilical precursor, called the allantois, or sometimes the body stalk, enlarges through this cavity toward the chorion and forms blood vessels that are ultimately secured onto the chorionic disk. Thus, the chorionic disk and allantois, initially physically separated, fuse to become a single unit. The umbilical cord is now continuous with the fetus's circulatory system at one end, and with the chorionic disk where maternal blood is pooled, at the other. Thus, the umbilical cord forms a perfectly aligned vascular bridge between the embryo and the mother's blood supplies. The placenta is now in place; it matures and thereafter provides the developing embryo with a uniquely comprehensive range of services, including alimentary, pulmonary, renal, hepatic, and endocrine functions until birth.

Epiblast

The epiblast establishes both embryonic and extraembryonic cells. During early postimplantation development, the epiblast puts forth a very special structure called the primitive streak. It has been said by the developmental biologist, Lewis Wolpert, that, more than birth, marriage, and death, formation of the primitive streak is the single most important event in an individual's life.

Appearance of the streak identifies the fetus's future posterior end. In humans, it forms by 14 days. From its posterior site, the streak elongates. Thus, the streak establishes the embryonic body axis. Anterior and posterior coordinates in place, the others, left and right, dorsal and ventral, naturally follow, resulting in correct placement of all organ systems in the fetus, and possibly, the conceptus as a whole.

For reasons that are not understood, the streak may split shortly after it appears, thereby creating two individuals within a single amniotic sac, each of which possesses its own umbilical cord while sharing a single chorionic disk. Sometimes, erroneously, only portions of the streak split, resulting in individuals conjoined at various levels of the body axis (Figure 2.8).

As the streak is laid down, the conceptus is more or less compartmentalized into distinct embryonic and extraembryonic regions. The epiblast makes three definitive germ layers, called ectoderm, mesoderm, and endoderm. All three contribute to embryonic and some extraembryonic structures. Within the embryonic compartment of the conceptus, the germ layers interact to form all of the organ systems of the fetus, placing them in the correct place by cues from the coordinates established by the streak. In the extraembryonic compartment, they are referred to as "extraembryonic," and complete the formation of the yolk sac and placenta, and form the entire amnion, which surrounds the embryo and provides it with a fluid in which it floats and is cushioned. Thus, the embryo becomes a fetus, the transition to the latter term sometimes made when the reproductive organs are definitively established. At this point, the fetus is supported by an array of extraembryonic tissues. Twinning has ceased; the individual has been established.

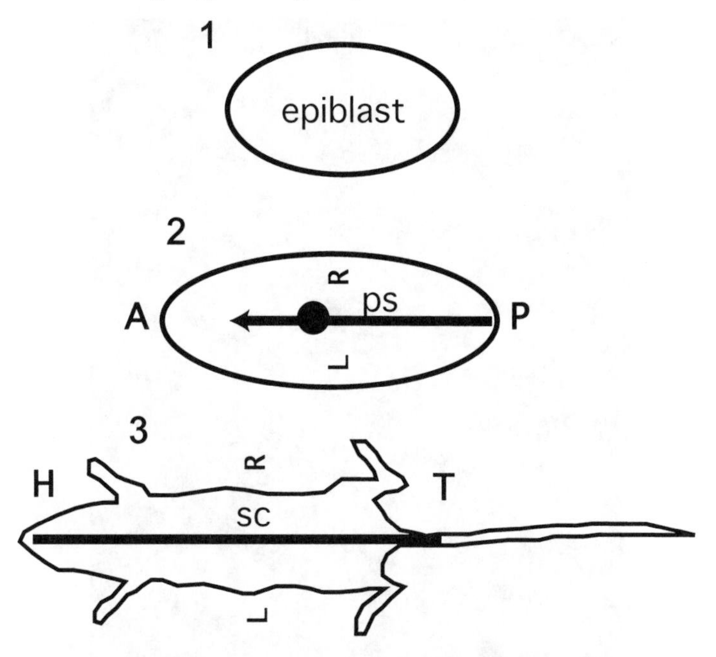

Figure 2.7 The primitive streak. In this highly stylized schematic drawing, the epiblast establishes the primitive streak and, thus, all polar coordinates in the embryo. (1) The epiblast is a blank slate. Until the primitive streak appears in (2), it is not known where any organ system will form. (2) The appearance of the primitive streak (ps) defines the future posterior end of the embryo. The streak then elongates in an anterior direction (shown by the arrow affixed to it), thereby establishing the left and right coordinates and ultimately, dorsal-ventral. (3) The primitive streak is ultimately replaced by the spinal cord (sc); the result is a perfectly patterned adult mammal, whose head/tail, left/right, and dorsal/ventral coordinates formed the basis for establishing its entire body plan. If the streak splits along its length prior to completion of formation, conjoined twins result, an example of which is shown in Figure 2.8.

Summary

The human fertilized egg is not an embryo, but, more accurately, a conceptus, which contains all the information required to establish both the embryo and extraembryonic supporting tissues. The fertilized eggs of placental mammals, including humans, are entirely dependent upon a female's uterine environment for development to birth. At least half, possibly more, fertilized eggs, or potential lives, do not survive to birth, the greatest loss thought to occur during preimplantation and implantation. Which conceptuses will be lost and which will progress to birth cannot be predicted. The preimplanta-

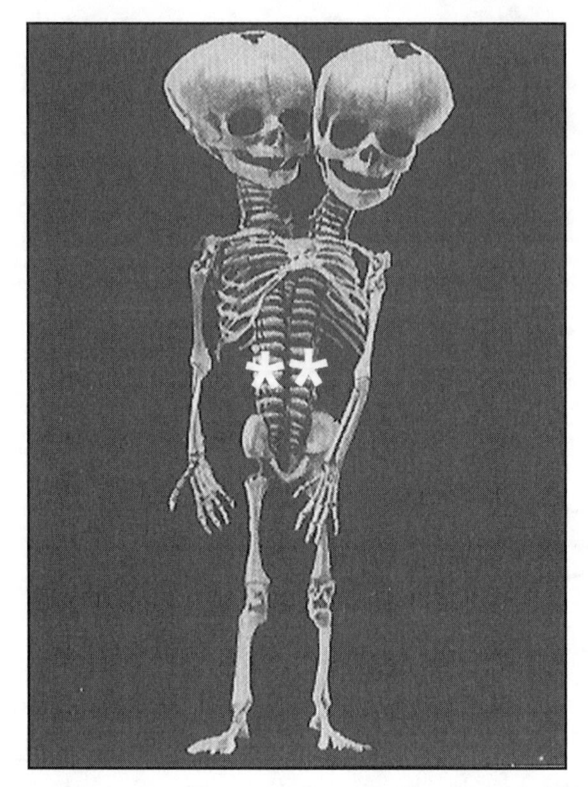

Figure 2.8 Conjoined twins. Photomicrograph revealing the skeleton of incomplete, conjoined twins fused at the pelvis, with duplication of the vertebral column (asterisks), which was laid down by the primitive streak. From Leroi (2003). Used with permission.

tion conceptus exhibits extreme developmental lability. Importantly, twinning, summarized in Figure 2.9, can occur throughout the preimplantation and implantation phases, and thus, a single human individual has not emerged from the conceptus during this time period. Once the primitive streak is complete during early postimplantation development, identical twinning no longer occurs and the individual emerges.

Thus, as summarized in Figure 2.10, emergence of the human individual is a process. No single event thus far known is more important than any other prior to formation of the primitive streak. Formation of the streak is a defining moment in the origin of the individual. All further organization of the fetus occurs around this midline. Thus, by 14 days in the human, the body plan for life is established.

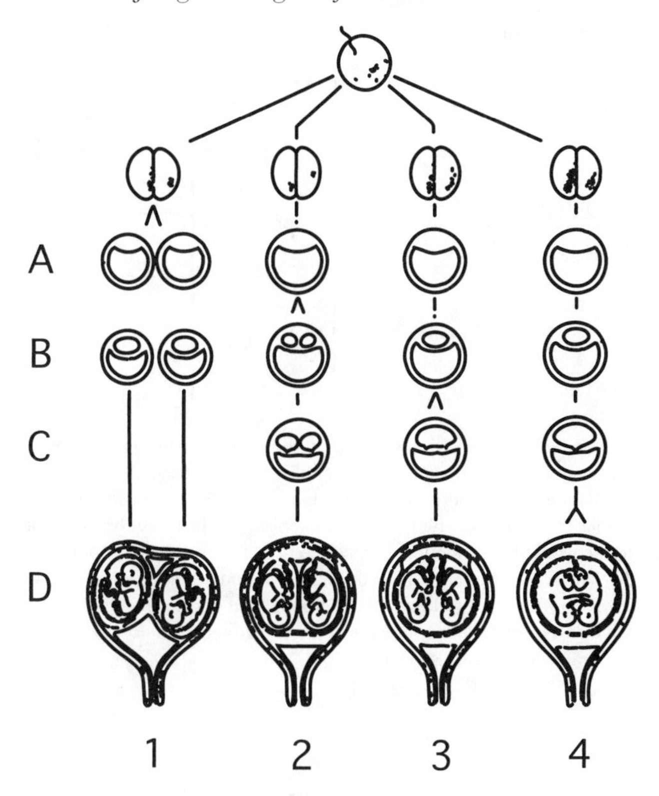

Figure 2.9 Summary of twinning through blastocyst stage. The fertilized egg cleaves into two cells, called the two-cell conceptus. As described in the text, identical twinning to form two complete but identical individuals may take place until the formation of the primitive streak is complete, during early postimplantation development. In Column 1, row A, the two-cell conceptus has formed two identical blastocysts. In 1D, the fetuses are supported by their own, but identical, placentae, and are thus called "dichorionic twins." In 2B, the single ICM of the blastocyst splits, to form two identical twins that share a chorionic placental component, but which have distinct amniotic encasings and umbilical cords (2D). These are called monochorionic diamniotic twins. In 3C, the ICM/epiblast splits after the amnion has formed, and thus, the identical twins in 3D are called [monochorionic] monoamniotic twins. In 4C, the epiblast, derived from the ICM of the blastocyst, does not split until the primitive streak begins to form, resulting in conjoined twins. Modified from Ford (1989), Figure 5.1. Used with permission.

Acknowledgments

My time was supported by the National Institutes of Health (RO1 HD042706) and the University of Wisconsin School of Medicine and Public Health. I thank the anonymous reviewers for their most valuable and constructive comments on the manuscript.

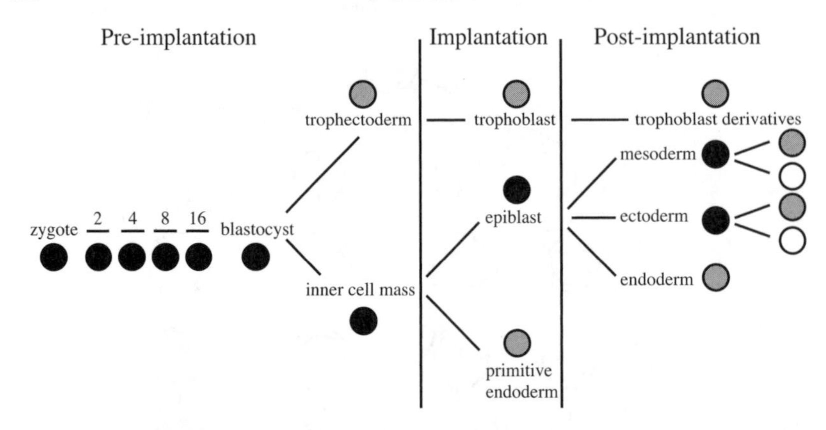

Figure 2.10 Summary of tissue diversification in the conceptus through the primitive streak stage. The three major developmental phases subdivide this schematic diagram. The circles represent major tissues discussed in this article, and their sequence of formation. Black indicates ability of that tissue to contribute to both embryonic and extraembryonic cells. Grey indicates contribution only to extraembryonic cells, and white represents contribution only to embryonic cells. The white ectoderm, mesoderm, and endoderm circles will join the primitive streak (not shown) in establishing the embryonic body plan and future fetus. Primitive endoderm will further differentiate into components of the extraembryonic yolk sac, which would be grey, but for simplicity, are not shown.

Glossary of Terms

Not all of the following terms were used in this essay; those that were not are often used in other more technical publications, and are thus included here for reference.

Blastocyst—a conceptus that contains an outer layer of cells, the trophectoderm, and an internal population of cells, the inner cell mass

Blastomere—a cell found in a conceptus at the two-cell through early morula stages, i.e., prior to the blastocyst stage

Chimera (British, Chimaera)—a single organism that is derived from two or more genetically distinct conceptuses

Chorionic disk—one of two major components of the chorio-allantoic placenta; derived from trophectoderm, its trophoblast cells permit diffusion and transport of oxygen and nutrients from mother to fetus

Conceptus—a fertilized egg or the developmental product of a fertilized egg

Diploid—having two sets of chromosomes as in a normal body cell

Dizygotic—fraternal twins; formed from two zygotes (a result of two ovulatory events within the same menstrual cycle)

Ectoderm—one of three primary germ layers formed from the epiblast

Embryo—strictly speaking, should refer only to the embryonic body plan when it is established. However, it is often used synonymously with "fertilized egg," zygote, two- and multicelled preimplantation entities, morula, blastocyst, postimplantation conceptus, etc.

Endoderm—one of three primary germ layers formed from the epiblast

Epiblast—one of two derivatives of the inner cell mass. Its descendents will contribute to both the embryo as well as extraembryonic tissues, including the umbilical cord, yolk sac mesoderm, chorionic mesoderm, and the amnion.

Gamete—egg or sperm

Haploid—having one set of chromosomes or half the number of chromosomes in a typical body cell, as in gametes

Inner cell mass (ICM)—one of two cell populations forming the blastocyst

Meiosis—the process by which chromosome number is reduced from diploid to haploid. This biological process only occurs in cells as they develop into gametes and insures that each gamete gets a complete chromosomal set.

Mesoderm—one of three primary germ layers formed by the epiblast; a major derivative of mesoderm is the circulatory system.

Monozygotic—from one zygote, identical twins

Morula—the conceptus just prior to becoming a blastocyst

Multipotent—capable of giving rise to several cell types from a single germ layer

Placenta (also, "chorio-allantoic" placenta)—a structure that enables exchange of nutrients, metabolic wastes, and gases between the fetus and the mother

Pluripotent—capable of giving rise to many cell types from more than one germ layer

Primitive endoderm (Hypoblast)—derivative of the inner cell mass that forms part of the yolk sac, an extraembryonic structure

Primitive streak—embryonic body axis that provides spatial cues for the arrangement of organs in the conceptus

Pronucleus—the packet in which half the amount of male or female genetic material is found in the egg's cytoplasm prior to syngamy

Syngamy—intermingling of the genetic material coming from an egg and a sperm into a single packet, the nucleus within the fertilized egg or zygote

Totipotent—capable of giving rise to all cell types or which can develop into a complete, functional organism

Trophectoderm (TE)—the outer population of cells of a blastocyst. The cells of the trophectoderm participate in implantation; cells descending from the trophectoderm form the chorionic component of the placenta.

Viviparity—dependence of the fertilized egg on its mother's internal environment and blood supply for nutrients during development; giving birth to live young

Zona pellucida—a thin, transparent membrane that surrounds the preimplantation conceptus of a mammal, which includes humans

Zygote—a fertilized egg

Recommended Reading

The following resources provided the basis for this essay. Asterisked contributions are technical in nature; while the vocabulary may seem daunting, the reader now has a working command of developmental terms and should feel confident to grasp major concepts.

Beaconsfield, P. et al. 1980. The placenta. *Scientific American* 243:94–102.

Bulmer, M. G. 1970. *The biology of twinning in man*. Oxford: Clarendon Press.

Corner, G. W. 1942. *The hormones in human reproduction*. Princeton: Princeton University Press.

Corner, G. W. 1944. *Ourselves unborn: An embryologist's essay on man*. New Haven: Yale University Press.

Corner, G. W. 1952. *Attaining manhood*. 2nd ed. New York: Harper and Row.

Fadiman, A. 1997. *The spirit catches you and you fall down. A Hmong child, her American doctors, and the collision of two cultures.* New York: Farrar, Straus, and Giroux.

*Evans, M. J., and M. H. Kaufman. 1981. Establishment in culture of pluripotential cells from mouse embryos. *Nature* 292:154–156.

Ford, N. M. 1989. *When did I begin? Conception of the human individual in history, philosophy and science.* Cambridge: Cambridge University Press.

Gardner, R. L. 1998. Contributions of blastocyst micromanipulation to the study of mammalian development. *Bioessays* 20:168–180.

Green, R. M. 2001. *The human embryo research debates: Bioethics in the vortex of controversy*. Oxford: Oxford University Press.

*Lawson, K. A. et al. 1991. Clonal analysis of epiblast fate during germ layer formation in the mouse embryo. *Development* 113:891–911.

Leroi, A. M. 2003. *Mutants. On genetic variety and the human body.* New York: Viking.

*Martin, G. R. 1981. Isolation of a pluripotent cell line from early mouse embryos cultured in medium conditioned by teratocarcinoma stem cells. *Proceedings of the National Academy of Sciences USA* 78:7634–8.

Moore, K. L. 1982. *The developing human. Clinically oriented embryology.* 3rd ed. Philadelphia: W. B. Saunders Company.

Needham, J. 1959. *A history of embryology*. New York: Abelard-Schuman.

Steven, D. H. 1975. *Comparative placentation: Essays in structure and function*. London: Academic Press.

*Tarkowski, A. K. 1961. Mouse chimaeras developed from fused eggs. *Nature* 190:857–860.

*Tesar, P. 2005. Derivation of germ-line-competent embryonic stem cell lines from preblastocyst mouse embryos. *Proceedings of the National Academy of Sciences USA* 102:8239–8244.

*Thomson, J. A. et al. 1998. Embryonic stem cell lines derived from human blastocysts. *Science* 282:1145–1147.

Wolpert, L., Beddington, R., Brockes, J., Jessell, T., Lawrence, P.A., and Meyerowitz, E. 1989. Principles of development. Oxford: Oxford University Press.

3

The Catholic Church and Stem Cell Research

Tadeusz Pacholczyk

Stem cell research is an area of intense modern controversy, and the Roman Catholic Church has provided a significant presence and voice in this debate. It can be instructive to examine a few of the most common misconceptions regarding the Church's position on this issue. One of the most widespread misconceptions is the idea that the Catholic Church stands opposed to stem cell research. Several distinct varieties of stem cell research can be identified (see Box 3.1).

Of the various forms listed (Figure 3.1), only research involving pluripotent stem cells derived from human embryos would be unreservedly opposed by the Catholic Church. As Pope John Paul II once summarized the matter:

> This moral condemnation also regards procedures that exploit living human embryos and fetuses—sometimes specifically 'produced' for this purpose by in vitro fertilization—either to be used as 'biological material' or as *providers of organs or tissue for transplants* in the treatment of certain diseases. The killing of innocent human creatures, even if carried out to help others, constitutes an absolutely unacceptable act" (1995).

Pluripotent germ cells derived from human fetuses would be considered morally problematic only if the cell extraction occurred from an induced abortion. Naturally miscarried fetuses could provide em-

Box 3.1. TYPES OF STEM CELL TECHNOLOGIES

1. Pluripotent stem cells derived from human embryos (Thomson et al., 1998)

   The embryos could be cloned or created during in vitro fertilization procedures.

2. Pluripotent germ cells derived from human fetuses (Shamblott et al., 1998)

3. Human adult stem cells (reviewed in Prentice, 2006)
   a. Human umbilical cord stem cells (reviewed in Mayani and Lansdorp, 1998)
      i. Unrestricted somatic stem cells (Kogler et al., 2004; Kogler et al., 2005)
      ii. Cord-blood-derived embryonic-like stem cells (McGuckin et al., 2005)
   b. Pluripotent stem cells derived from adult germ cells (Guan et al., 2006)
      i. Multipotent adult germ line stem cells (Guan et al., 2006)
      ii. Germ line stem cells (PrimeGen Biotech)
   c. Pluripotent stem cells derived from blood, bone marrow, or other adult sources
      i. Marrow isolated adult multilineage inducible cells (D'Ippolito et al., 2004)
      ii. Tissue committed stem cells (Kucia et al., 2006a)
      iii. Very small embryonic like cells (Kucia et al., 2006)
      iv. Multipotent adult progenitor cells (Jiang et al., 2002; Tolar et al., 2006)
      v. Blastomere-like stem cells (Moraga Biotechnology)
      vi. Other forms (Yoon et al., 2005)

4. Stem cells derived from nonhuman animals (Evans and Kaufman, 1981; Martin, 1981; Matsui, 1992)

5. Dedifferentiation or reprogramming strategies to yield pluripotent stem cells (Taranger et al., 2005; Takahashi and Yamanaka, 2006)

6. Altered nuclear transfer – oocyte assisted reprogramming (Arkes et al., 2005; Pacholczyk, 2005)

bryonic germ cells without raising any moral red flags, assuming the parents gave informed consent, much like their providing consent for an organ donation from their recently deceased child. Hence, under the appropriate circumstances, the Catholic Church would support almost all forms of stem cell research.

A second common misconception is that the Catholic Church opposes the destruction of human embryos because such embryos are persons. While it is true that the Church teaches that the intentional and direct destruction of human embryos is always immoral, it is false to suggest that the Church teaches that zygotes or other early-stage embryos are persons (or that they have immortal, rational souls). The magisterium of the Church has never definitively stated when the ensoulment of the human embryo takes place. It remains an open question. The *Declaration on Procured Abortion* from the Congregation for the Doctrine of the Faith in 1974 phrases the matter with considerable precision:

> This declaration expressly leaves aside the question of the moment when the spiritual soul is infused. There is not a unanimous tradition on this point and authors are as yet in disagreement. For some it dates from the first instant; for others it could not at least precede nidation [implantation in the uterus]. It is not within the competence of science to decide between these views, because the existence of an immortal soul is not a question in its field. It is a philosophical problem from which our moral affirmation remains independent.

And the moral affirmation of the Church is simply this: that the human embryo must be treated *as if* it were already ensouled, even if it might not yet be so. It must be treated *as if* it were a person from the moment of conception, even if there exists the possibility that it might not yet be so. Why this rather subtle, nuanced position, instead of simply declaring outright that zygotes are ensouled, and therefore are persons? First, because there has never been a unanimous tradition on this point; and second, because the precise *timing* of ensoulment/personhood of the human embryo is irrelevant to the question of whether we may ever disaggregate such embryos for research purposes.

Interestingly, the question of the timing of ensoulment has been discussed for centuries, and so-called *delayed ensoulment* was probably the norm for most of Christian history, with *immediate ensoul-*

ment gaining some serious momentum of its own only in the 1600s. Thomas Aquinas, for example, held that ensoulment occurred not right at the first instant, but at a timepoint removed from the beginning, in order to allow the matter of the embryo to undergo development and become "apt" for the reception of an immortal soul from God. Augustine seemed to shift his opinion back and forth during his lifetime between immediate and delayed ensoulment. Even today in various quarters, the discussions continue, with new embryological details like twinning and chimaerization impinging upon the debate, and new conceptual questions arising from the intricate biology surrounding totipotency and pluripotency.

We must recognize that it is God's business as to precisely *when* He ensouls the human embryo. We may never categorically resolve the matter from our limited vantage point. It is also important to understand that we do not need an answer to this fascinating and speculative question in order to grasp the Church's moral teaching that human embryos are absolutely inviolable and deserving of unconditional respect at every stage and moment of their existence. The Church's ethical stance cannot, and does not, stand or fall on a point of theological speculation, like trying to count the number of angels that can fit on the head of a pin. Rather, the Church's teachings follow on the heels of the scientific and embryological data regarding early human development. And this is precisely the point in the argument where it becomes clear that the Catholic Church's defense of embryonic humans is in fact not fundamentally a matter of religious dogma, but rather a matter of reason.

All of us began as embryonic human beings, and such human beings should not be instrumentalized for stem cell extraction or other destructive ends. The Church stresses that we need not tarry about the fine details of the timing of personhood or ensoulment in a misguided attempt to identify a basis for the moral question. We need to recognize instead that once you are constituted a human being at fertilization (or at an event that mimics fertilization like cloning), you are an embryonic member of the human race who is deserving of unconditional respect and protection. The human zygote, thus, is already a being that is human, and such beings are sacrosanct entities, because that's what each of us directly springs from at the root level. What the human embryo actually is, even at its earliest and most undeveloped stage, makes it the only kind of entity capable of receiving the gift of an immortal soul from God; no other animal or plant embryo can receive this gift. Hence, the early human embryo is never merely biological tissue; at a minimum, it is the privileged sanctuary of someone meant to develop as a human person, and to be treated and respected as such. Once you are a *human* being, you are a bearer of *human* rights, even if your personhood/ensoulment might end up coming further along in the sequence of things. This teaching may well be one of the strongest declarations of the Church's belief in the absolute primacy of the value of personhood over all other considerations. The human person, even in his or her most incipient and precursorial instantiation in the embryonic human being, is to be safeguarded in an absolute and unconditional way.

Embryonic stem cell researchers typically marshal several arguments to encourage public approval and funding for their research, which requires the direct destruction of five- to seven-day-old human embryos. One of the most common arguments runs as follows: "Well, that's your feeling about embryos, your narrow religious viewpoint. Your sentiments about embryos are different than mine, and we're all entitled to our own sentiments and opinions." One time when I testified in Virginia at legislative hearings about embryonic stem cell research, one of the senators adverted to precisely this line of analysis when she asked, "By arguing against embryonic stem cell research, don't you see how you are trying to impose your Catholic beliefs on others, and shouldn't we as elected lawmakers avoid imposing a narrow religious viewpoint on the rest of society?" The senator's question was an example of the fuzzy thinking that has become standard fare in recent years within many state legislatures and among many lawmakers.

Two major errors were incorporated into the senator's question. First, the senator failed to recognize the fact that law is fundamentally about imposing somebody's views on somebody else. Imposition is the name of the game. It is the very nature of law to impose particular views on people who don't want to have those views imposed on them. Drug dealers don't want laws imposed on them which make it illegal to sell drugs. Yet our lawmakers are elected precisely to craft and impose such laws all the time. So the question is not whether we will impose something on somebody. The question is instead whether whatever is going to be imposed by the force of law is reasonable, just, and good for society and its members.

The second logical mistake the senator made was to suppose that because religion happens to hold a particular viewpoint, that implies that such a viewpoint should never be considered by lawmakers or enacted into law. Roman Catholicism teaches very clearly that stealing is immoral. Would it follow that if I support laws against robbery, I am imposing a narrow religious viewpoint on society? Clearly not. Rather, the subject of stealing is so important to the order of society that religion also feels compelled to speak about it. Religion teaches many truths that can be understood by people who aren't religious at all. Atheists can understand as well as Catholics how stealing is wrong, and most atheists are just as angry as their Catholic neighbors when their house is broken into and robbed. What is important is not whether a proposed law happens to be taught by religion, but whether that proposal is just, right, and good for society and its members.

To be more coherent, of course, the senator could have addressed the substance of the argument offered during my stem cell testimony which, interestingly, did not depend on religious dogma at all. It depended rather on an important *scientific* dogma, namely, that adult humans come from embryonic humans. Given the fact that we were all once embryonic humans, it becomes immediately clear why destructive embryonic research is an immoral kind of activity. Exploiting the weak and not-yet-born in the interests of the powerful and the well-heeled should not be permitted in a civilized society, and this argument can be clearly grasped by atheists, not just Roman Catholics.

An example from the animal kingdom reminds us how this issue is properly contextualized by science and reason. In the United States we have a stringent federal law, passed in 1940, which protects not only the national bird, the bald eagle, but also that bird's eggs. If you were to chance upon some of those eggs in a nest out in the wilderness, it would be illegal for you to destroy them. If you were to do so, you would suffer the same penalties and sanctions as if you had shot the adult bird out of the air. By the force of law, we acknowledge the scientific truth that the eagle's egg (that is to say, the *embryonic eagle* inside that egg) is the same creature as the beautiful bird that we witness flying overhead. Therefore we pass laws to safeguard not only the adult but also the very youngest member of that species. Even atheists can see how a bald eagle's eggs ought to be protected; it's not a religious question at all. If bald eagles are valuable (in this case, for pragmatic reasons of conservation), then it is right and fitting to protect them at all stages of their existence. The same logic holds for humans who are valuable not for pragmatic, but for intrinsic reasons. What is troubling, then, is how we are able to understand the importance of protecting the earliest stages of various forms of animal life, but when it comes to our own human life, we convince ourselves to carry out some sophisticated mental gymnastics so that a kind of mental disconnect can occur regarding our own origins. Our thinking and moral judgment soon become murky and obtuse when we desire to do certain things that are not good, like destroying young humans for their stem cells.

When lawmakers suggest that an argument in defense of sound morals is nothing but imposing a religious viewpoint, we need to look more deeply at what may really be taking place. Such elected officials may not be so concerned about avoiding the imposition of a particular view on others—more likely, they are jockeying simply to be able to impose *their* view, a view that is ultimately less tenable and defensible in terms of sound moral thinking. Hence, they seek to short-circuit the discussion by stressing religious zealotry and imposition without confronting the substantive ethical or bioethical argument itself. Once the religious imposition card is played, and enough of the lawmakers are rendered mute regarding the defense of human life and sound morals, advocates of embryo-destructive research then feel free to do the imposing themselves, without having expended too much effort confronting the essence of the moral debate itself.

This misguided approach to debating and legislating has embedded itself in the modern American system to a remarkable degree and has been used quite effectively to justify embryonic sacrifice by stem cell researchers themselves. At its core, they take a scientific question and simply declare it a religious one. Once it falls into the category of religious mystagogy, it can be dismissed out of hand as irrelevant to public policy and discourse. Embryonic stem cell researcher Dr. Douglas Melton at Harvard took exactly this tack when he spoke with the *New York Times* in 2006: "This is all about differing religious beliefs. I don't believe I have the right to tell others when life begins. Science doesn't have the answer to that question; it's metaphysical" (quoted in Dreifus, 2006). With that sleight of hand, he sought to transform embryology into theology. The fact is, of course, that the statement, "a human embryo is a human kind of being" does not depend on religion any more than does the statement "a cow embryo is a cow kind of being." Science, quite apart from any narrow, dogmatic religion, affirms dogmatically that human embryos are human beings, rather than cat, caterpillar, or cow beings. Science, quite apart from religious dogma, affirms dogmatically that every person walking around in the world was once an embryo. This scientific dogma is absolute and admits of no exceptions. In sum, then, we know exactly what the embryo is, namely, a *human being*, a *being* that is clearly and unmistakably *human*. It is not a zebra type of being, a plant type of being, or some other kind of being. This scientific affirmation does not ultimately depend on religion, value systems, or imposing anything on anyone. It is a simple matter of empirical observation quite apart from any ecclesiastical or papal declaration.

Embryos, of course, are remarkably unfamiliar to us. They lack hands and feet. They don't have faces or eyes for us to look into. Even their brains are lacking. They look nothing like what we expect when we imagine a human being. But they are precisely as human as each of us. When we look at a scanning electron micrograph of a human embryo, a small cluster of cells, sitting on the point of a sewing pin, we need to ask ourselves a very simple question: "Isn't that exactly what a young human is supposed to look like?" The correct answer to that question doesn't depend on religion, revelation, or theology, but only on embryology. Embryos seem unfamiliar to us at first glance, and we have to make an explicit mental effort to avoid the critical error of disconnecting from what we once were in our own humble embryonic origins.

So, while science makes it clear that human embryos are *human* beings, religion steps in *after* that fact to speak to the question of whether it is correct that all human beings should all be treated the same way, or whether it is allowable to discriminate against some human beings in the interests of others. Even though it is a fundamental embryological truth that each of us was once an embryo, the advocates of embryonic stem cell research are eager to portray human embryos as different from the rest of us, unable to make the grade, and hence fair game for destruction at the hands of those lucky enough to have already passed through those early and vulnerable stages. Will fundamental injustices and ethical transgressions like these become systemic and promoted as the societal norm? Will advocates be permitted to get away with confusing embryology and theology in the public square? Will the powerful be permitted to violate and instrumentalize the weak? These are questions with enormous implications for the future of our society, and they are questions that the Catholic Church continues to insist deserve the most serious consideration by all men of good will.

Mr. Rogers, the famous children's TV personality, once gave a talk where he mentioned his favorite story from the Seattle Special Olympics. Here's how he described it:

> Well, for the 100-yard dash there were nine contestants, all of them so-called physically or mentally disabled. All nine of them assembled at the starting line and at the sound of the gun, they took off. But not long afterward one little boy stumbled and fell and hurt his knee and began to cry. The other eight children heard him crying; they slowed down, turned around and ran back to him. Every one of them ran back to him. One little girl with Down Syndrome bent down and kissed the boy and said, "This'll make it better." And the little boy got up and he and the rest of the runners linked their arms together and joyfully walked to the finish line. They all finished the race at the same time. And when they did, everyone in that stadium stood up and clapped and whistled and cheered for a long, long, time. People who were there are still telling the story with great delight. And you know why. Because deep down, we know that what matters in this life is more than winning for ourselves. What really matters is helping others win too. (Rogers, 2002)

This beautiful story of everyone turning around and looking after the interests of the weakest and the most vulnerable reminds us of exactly the kind of society we should aspire to build, one where every life, even the weakest embryonic life, is embraced as a gift and treasure of infinite and irreplaceable value. The Catholic Church seeks to remind us that with God's help and our determined efforts, we can and ought to construct just such a society, where every human being is unconditionally protected and esteemed.

References

Arkes, H. et al. 2005. Production of pluripotent stem cells by oocyte-assisted reprogramming: Joint statement with signatories. *National Catholic Bioethics Quarterly* 5:579–583.

Congregation for the Doctrine of the Faith. Declaration on procured abortion (November 18, 1974), Nos. 12–13. *Acta Apostolicae Sedis* 66:738.

D'Ippolito, G. et al. 2004. Marrow-isolated adult multilineage inducible (MIAMI) cells, a unique population of postnatal young and old human cells with extensive expansion and differentiation potential. *Journal of Cell Science* 117:2971–2981.

Dreifus, Claudia. 2006. A conversation with Douglas Melton: At Harvard's Stem Cell Center, the barriers run deep and wide. *New York Times*, January 24.

Evans, M. J., and M. H. Kaufman. 1981. Establishment in culture of pluripotential cells from mouse embryos. *Nature* 292:154–156.

Guan, Kaomei et al. 2006. Pluripotency of spermatogonial stem cells from adult mouse testis. *Nature* 440:1199–1203.

Jiang, Y. et al. 2002. Pluripotency of mesenchymal stem cells derived from adult marrow. *Nature* 418:41–49.

Kogler, G. et al. 2004. A new human somatic stem cell from placental cord blood with intrinsic pluripotent differentiation potential. *Journal of Experimental Medicine* 200:123–135.

Kogler, G. et al. 2005. Cytokine production and hematopoiesis supporting activity of cord blood-derived unrestricted somatic stem cells. *Experimental Hematology* 33:573–583.

Kucia, M. et al. 2006a. A population of very small embryonic-like (VSEL) CXCR4(+) SSEA-1(+) Oct-4(+) stem cells identified in adult bone marrow. *Leukemia* 20:857–69.

Kucia, M. et al. 2006b. Cells enriched in markers of neural tissue-committed stem cells reside in the bone marrow and are mobilized into the peripheral blood following stroke. *Leukemia* 20:18–28.

Martin, G. R. 1981. Isolation of a pluripotent cell line from early mouse embryos cultured in medium conditioned by teratocarcinoma stem cells. *Proceedings of the National Academy of Sciences USA* 78:7634–7638.

Matsui, Y. et al. 1992. Derivation of pluripotential embryonic stem cells from murine primordial germ cells in culture. *Cell* 70:841–847.

Mayani, Hector, and Peter M. Lansdorp. 1998. Biology of the human umbilical cord blood-derived hematopoietic stem/progenitor cells. *Stem Cells* 16:153–165.

McGuckin, C. P. et al. 2005. Production of stem cells with embryonic characteristics from human umbilical cord blood. *Cell Proliferation* 38:245–255.

Moraga Biotechnology Corp. discovers a primitive embryonic-like stem cell in adult tissues. http://www.moragabiotech.com.

Pacholczyk, T. 2005. Ethical considerations in oocyte assisted reprogramming. *National Catholic Bioethics Quarterly* 5:446–447.

Pope John Paul II. 1995. *On the Gospel of Life*. Encyclical Letter of Pope John Paul II, March 25.

Prentice, D. A. 2006. Current science of regenerative medicine with stem cells. *Journal of Investigative Medicine* 54:33–37.

PrimeGen Biotech presents proprietary method for deriving human adult-derived pluripotent stem cells. http://www.primegenbiotech.com.

Rogers, Fred McFeely. 2002. Commencement Address at Dartmouth College, June 9. http://www.dartmouth.edu/~news/releases/2002/june/060902c.html.

Shamblott, M. J. et al. 1998. Derivation of pluripotent stem cells from cultured human primordial germ cells. In *Proceedings of the National Academy of Sciences USA* 95:13726–13731. Erratum: 1999. In *Proceedings of the National Academy of Sciences USA* 96:1162.

Takahashi, K. and S. Yamanaka. 2006. Induction of pivripotent stem cells from mouse embryonic and adult fibroblast cultures by defined factors. *Cell* 126:1–14.

Taranger, Christel K. et al. 2005. Induction of dedifferentiation, genomewide transcriptional programming, and epigenetic reprogramming by extracts of carcinoma and embryonic stem cells. *Molecular Biology Cell* 16: 5719–5735.

Thomson, J. A. et al. 1998. Embryonic stem cell lines derived from human blastocysts. *Science* 282:1145–1147.

Tolar, J. et al. 2006. Host factors that impact the biodistribution and persistence of multipotent adult progenitor cells. *Blood* 107:4182–8.

Yoon, Y. S. et al. 2005. Clonally expanded novel multipotent stem cells from human bone marrow regenerate myocardium after myocardial infarction. *Journal of Clinical Investigation* 115:326–338.

4

Judaism and Stem Cells

Elliot N. Dorff

Our analysis of human embryonic stem cell research is guided by four fundamental theological principles (reviewed in Dorff 1998). First, our bodies belong to God; we have them on loan during our lease on life. God, as owner of our bodies, can and does impose conditions on our use of our bodies. Among those is the requirement that we seek to preserve human life and health (*pikkuah nefesh*). As a corollary to this, we have a duty to seek to develop new cures for human diseases. Second, the Jewish tradition accepts both natural and artificial means to overcome illness. Physicians are the agents and partners of God in the ongoing act of healing. Thus the mere fact that human beings created a specific therapy rather than finding it in nature does not impugn its legitimacy. On the contrary, we have a duty to God to develop and use any therapies that can aid us in taking care of our bodies, which ultimately belong to God. Third, at the same time, all human beings, regardless of their levels of ability and disability, are created in the image of God and are to be valued as such. Finally, we are not God. We are not omniscient, as God is, and so we must take whatever precautions we can to ensure that our actions do not harm ourselves or our world in the very effort to improve them. A certain epistemological humility, in other words, must pervade whatever we do, especially when we are pushing the scientific envelope, as we are in stem cell research. We are, as Genesis 2:15 says, supposed to work the world *and* preserve it; it is that *balance* that is our divine duty.

Jewish Views of the Sources of Human Embryonic Stem Cells

Our analysis of the morality of human embryonic stem cell research is also influenced by Jewish views of genetic materials.

Frozen Embryos

Early-stage embryos develop into every cell in the human body. The hope, then, is that harvesting the "pluripotent" (that is, very plastic) cells of embryos will enable scientists to form the cells necessary to repair cells damaged by such diseases as Alzheimer's, Parkinson's, and strokes. Further, embryonic cells turn off, so that we get one head, for example, and not more. If we can determine how that happens, we would have the ultimate cure for cancer, for cancer is cells that have gone haywire. The Jewish mandate to heal would then urge us to embark on such research, but only if we are not harming human beings in the process. This, then, raises the question of the status of early embryos in Jewish law.

According to the Talmud, during the first 40 days of gestation, the embryo is "simply water" (*Yevamot* 69b). That is because, as the Mishnah, the authoritative second-century statement of the Oral Tradition, asserts, "a woman who miscarries [up to or] on the 40th day need not worry that she has delivered a child [for which she has to observe the special period of impurity after the birth of a child, for] . . . the Sages say, the creation of both the male and the female takes place on the 41st day" (*Niddah* 3:7 [30a]). Furthermore, according to the Mishnah, "Anything that does not have the form of a child is not a child," (*Niddah* 3:2 [21a]), and thus an embryo before the 41st day, which is without a form, is not a child. Maimonides (1140–1204) calls upon his medical experience in saying that even on the 41st day the figure of a human being is "very thin," and that within 40 days "its shape is not yet finished" (*Laws of Forbidden Intercourse* 10:2, 17). Similarly, the *Shulhan Arukh*, an authoritative code (1565), specifies that it is a vain prayer (*tefillat shav*) if a man prays after the 40th day that his pregnant wife be carrying a boy, for by the 41st day the gender of the child has already been determined (*Orah Hayyim* 230:1). The Mishnah, on which this is based, however, does not mention the 40-day limitation (*Berakhot* 9:2 [54a]).

As it happens, modern science provides good evidence to support the Talmudic Rabbis' understanding of fetal development. As

Rabbi Immanuel Jakobovits noted long ago, the Rabbis' "40 days" actually is approximately 56 days of gestation, for the Rabbis counted from the woman's first missed menstrual flow (Jakobovits 1959, p. 275). By 56 days of gestation, the basic organs have already appeared in the fetus. Moreover, we now know that it is exactly at eight weeks of gestation that the fetus begins to develop bone structure and therefore looks like something other than liquid. The 40-day marker comes originally from Aristotle, and it was adopted by none other than Augustine and Aquinas. In fact, the Catholic Church itself did not hold that a fertilized egg immediately became a person until 1869, when, at the first Vatican Council, they wanted to strongly affirm the virgin birth of Jesus, and so they needed to see him as a person immediately upon conception by the Holy Spirit. That change did not occur in Canon Law until 1917. The Rabbis probably came to their conclusion about the stages of development of the fetus because early miscarriages indeed looked like "merely water," while those from 56 days on looked like a thigh with flesh, bones, and even hair. For that matter, even the Rabbis who proclaimed the embryo in the first 40 days to be "simply water" clearly were announcing an analogy and not an equivalence, for they certainly knew that from that water a child might develop, unlike any glass of drinking water!

Thus, while we should have respect for gametes and embryos in a petri dish as potential building blocks of life, they may be discarded if they are not going to be used for some good purpose. If an embryo during the first 40 days of gestation is "simply water," an embryo situated outside a woman's womb, where it cannot with current technology ever become a human being, surely has no greater standing; it is *at most* "simply water." Therefore, when a couple agrees to donate such embryos for purposes of medical research, our respect of such gametes and embryos outside the womb should certainly be superseded by our duty to seek to cure diseases. Finally, because the embryo in the first 40 days is "simply water" and "not a child," and all the more so in the first 14 days, when stem cells would be removed for research, our duty to seek to cure diseases provides ample warrant, in my opinion, for removing the inner cellular mass in the first place so that stem cell research can go forward. In doing so, we are not killing a human being, as we would be if we were to remove a person's heart before death; we are rather taking a part of an object that has not yet achieved the status of a formed fetus, let alone a human being.

What would happen, though, if we could gestate a human being entirely outside a woman's womb in some sort of machine? Infertile couples who can produce sperm and eggs but cannot carry a baby to term might indeed be highly interested in such a possibility. Would that change our perception of the fetus during its first days of gestation?

The problem is more theoretical than real, for we do not currently have such a machine. Moreover, given the options of surrogate mothers and adoption, and given the inevitably great cost of developing and then using such a gestation machine, it is not likely that such machines will be available for quite some time.

Second, it is important to note that the wisdom and authority of moral and legal decisions depend critically on their context. *At this time*, at least, an embryo outside a woman's womb is different in relevant ways from an embryo within a woman's womb. If and when we develop the ability to gestate a person outside a woman's womb, then the physical location of an embryo in a petri dish may cease to have as much import as it does now, but that would be a *different* context requiring a new weighing of the evidence.

The most important thing to note, though, is that the argument for seeing the embryo as less than a person is not based solely on the grounds of where the embryo in a petri dish happens to be. Rather, characteristics of the early embryo itself argue for assigning it the status of "mere water." Specifically, to procure embryonic stem cells scientists can use only embryos during the first 14 days of gestation, for then, with the appearance of the neural streak, which later develops into the spine, the stem cells begin to differentiate and lose some of their ability to become cells of all sorts, which is why scientists are interested in them in the first place. During that early period before 14 days, the embryo in a petri dish can be distinguished from a human being not only according to its location outside of a womb and its resulting inability to develop into a human being (that is, its lack of human potential), but also by *its low level of cell organization, the short period of time that it will remain in this state, and its incapacity to live without further development.*

Thus, if very good scientific reasons support the Talmudic precedent to classify an embryo of up to 40 days as "mere water," an embryo of 14 days of gestation or less is even more justifiably classified that way, even if it were within a woman's uterus; the more so outside one. At no point during those 14 days, then, do stem cells become a human, and so stem cell research represents an enormous potential good at no human price. (In contrast, we regularly use full human beings in medical research because animal research alone cannot guarantee the safety and efficacy of medications for human beings. While we take safeguards to protect human subjects, people in recent cases at the University of Pennsylvania (Wade 1999) and Johns Hopkins University (Altman 2001) have died as part of such research. Stem cell research poses no such risks to human beings.) Thus, even if it were possible to gestate a human being mechanically, we would still have good reason to classify an embryo during the first 40 days as "simply water" and to use it for stem cell research.

In sum, then, couples should be encouraged to donate their extra embryos to stem cell research. Doing that is minimally an act of *hesed,* of loyalty and love, and possibly, given its goal of cure, even a *mitzvah,* a commanded act.

Creating Embryos, Harvesting Eggs, or Parthenogenesis Specifically Intended for Medical Research and Ultimately Therapy

Couples, though, are often reticent to donate their extra frozen embryos for research (Kolata 2001). Further, some scientists worry that embryos produced by couples with infertility problems may have flaws that would harm their research. This has led scientists to investigate other possibilities of obtaining embryonic stem cells.

One is creating embryos from donated sperm and eggs specifically for the purpose of doing medical research. This lacks the justification of using materials that would just be discarded anyway, but it would nevertheless be permissible under one condition. Unlike the Catholic view, the problem for the Jewish tradition in doing this is *not* that it would amount to murder to create and then destroy an embryo outside the uterus. In that state, an embryo has no greater claim to protection than an embryo in its first 40 days in utero, much less that of a person. The issue is not even classical Jewish law's ban on "wasting seed" (*hashatat zera*). The Mishnah and Talmud forbid a man from touching his penis lest he induce it to become hard and ejaculate (Babylonian Talmud *Niddah* 13a-13b), and later Jewish sources use the phrase *hashatat zera* (Tosafot on Babylonian Talmud *Yevamot* 12b, 32b and *Ketubbot* 39a; *Shulhan Arukh Even Ha-ezer* 23). Masturbating to procure the sperm for "farmed" embryos would not constitute "wasting seed," for here the purpose would be specifically to use the man's sperm for the consecrated purpose of finding ways to heal illnesses.

Procuring eggs from a woman for this purpose, however, does pose a problem. It is not so much that this requires subjecting her to the risks of surgery, for now eggs can be procured without surgery and with minimal risk or pain through laparoscopy. To produce the eggs, though, the woman must be exposed to the drugs that produce hyperovulation (release of multiple eggs at one time), and there is some evidence that repeated use of such drugs increases a woman's risk of ovarian cancer and other maladies (Spirtas et al. 1993). After the 1992 Stanford study, which suggested that fertility drugs might raise the risk of ovarian cancer, "later research cast doubt on that finding—but only after thousands of women were terrified" (Lemonick 2002, 69). However, a 1988 Congressional report stated that a number of other possible complications are caused by commonly used drugs to stimulate the ovaries, including early pregnancy loss, multiple gestations, ectopic pregnancies, hair loss, pleuropulmonary fibrosis, increased blood viscosity and hypertension, stroke, and myocardial infarction (Office of Technology Assessment, 1988, 128-129). Such risks may be undertaken to overcome a woman's own infertility or even, I have held (Dorff 1996, 48-50; Mackler 2000, 81-84; Dorff 1998, 106-107), to donate eggs once or twice to infertile couples; assuming such risks for medical research is less warranted, especially because embryos can also be obtained from frozen stores that couples plan on discarding. Still, the demonstrated risks for her to do this once or twice is minimal, especially if she is prescreened and deemed safe to undergo that procedure, and so a woman may donate eggs for this purpose with those limitations, along with monitoring to intervene should health complications arise.

The same concerns about the risks in procuring human eggs would apply to using eggs to obtain stem cells from cloning procedures or from parthenogenesis, if either proves to be possible. Thus, while it is best to obtain embryonic stem cells from frozen embryos that would otherwise be discarded, embryos may also be specifically created, and eggs may be cloned or tricked into producing stem cells through parthenogenesis for purposes of medical research on the condition that the woman providing the eggs for such efforts is prescreened to insure her safety and, even then, that she does this only once or twice.

Some have raised two other objections to creating embryos intentionally for research. Some object that the embryo in a petri dish is, after all, potential life in that it could be implanted in a woman's uterus and someday we may even be able to grow it in a machine. Some also worry that allowing the use of embryos specifically created for research creates a slippery slope in that human genetic materials will then be diminished in our estimation as just a means to a practical end, that human creation will lose its mystery and holiness. In Kantian terms, this smacks of violating the second version of the categorical imperative—that is, never treat a person merely as a means. Or, in the terms of more modern theorists, even though I am assuming that the donors are not paid, this seems awfully close to commodifying people in that we are looking at both men and women as (merely?) sources for genetic products. Without articulating it precisely this way, it is this concern that often underlies how the average person responds to the prospect of doing research on embryos.

Stem cell research does, of course, entail the destruction of potential life, but one must remember that the embryo in a petri dish remains potential life only through considerable scientific interventions to provide an environment where the zygote will remain alive in that state. Its hold on life is, at best, tenuous; indeed, since 80% of conceptions miscarry, usually in the first month, the Rabbis were right in classifying such early embryos as "merely water"—and that is in utero. Thus, it seems to me that we need to realize how weak the potentiality of that life is. In contrast, the potentiality of stem cell research rests on a solid foundation of successful attempts to use adult stem cells in humans and embryonic stem cells in animals. Thus when we speak of an embryo in a petri dish, we must remember that we are, at most, balancing *potential* life against what we have good reason to hope will be actual treatments for serious diseases; that is much easier to justify than balancing *actual* lives against that hope, as we do whenever we use human subjects in medical research. If we do the latter—and we must, albeit under stringent controls, if we are ever going to have medications that are safe and effective—then we should do the former with yet greater warrant.

As for the second objection, human creation must surely be honored and respected, and steps to protect that special status must be taken. We are not dealing with a *person,* though, when we use embryos to advance stem cell research; we are dealing with genetic materials that, even in utero, have a long way to go before they become a person, an unlikely prospect, at that. *That is, in the end we are dealing with a thing, not a person*; that is what the classification of embryos as "simply water" entails. We surely *are* allowed—indeed, commanded—to use *things* to find ways to cure diseases. Moreover, in our case I do not see a serious danger of a slippery slope in the status of human genetic materials, for the use to which these embryos would be put is nothing less than another holy cause—namely, curing people of serious diseases. Thus, I do not consider the deliberate creation of embryos for purposes of stem cell research to demean the birth process in any way.

Removing a Cell from an Embryo

Obtaining stem cells by removing a cell from an embryo poses no problems for the Jewish tradition whatsoever. As the embryo itself outside the womb is at most "simply water," one cell from the embryo is certainly not equivalent to a human being. Moreover, in this procedure the embryo from which the cell was taken can still develop normally in a woman's womb, as the cell is removed before differentiation prevents substitution of other cells in the embryo for the removed one.

Cloning

Obtaining stem cells through cloning is now low in researchers' priorities because we are just at the beginning of research on cloning in general. Moreover, Congress is currently engaged in a dispute about whether to fund or even to permit therapeutic cloning—creating embryos or tissues to treat disease from existing, fully nucleated cells. We clearly do not want to support reproductive cloning, at least at this stage of development of the technique, for it is neither safe nor effective. It is one thing to kill or discard all 272 attempts to clone a sheep before Dolly was created as the first cloned sheep; it would be quite another thing to create and kill multiple human beings with major birth defects. In therapeutic cloning, though, we are dealing with cells or, at most, organs, and those we may discard, if necessary, in the process of perfecting the technique. When we can clone cells more effectively and safely, in fact, that method for obtaining stem cells should jump to the top of the list as our source of stem cells for cures, since it produces tissues from patients themselves and thus does not pose the problems of recipient rejection.

The Justice Issue

The Jewish tradition sees the provision of health care as a communal responsibility (Dorff 1998; Dorff and Mackler 2000), and so justice arguments have a special resonance for Jews. That is, when and if stem cell technology becomes available, poor people as well as the middle class and the rich should be able to benefit from it. That is especially true, since much of the basic science in this area was funded by public funds. At the same time, the Jewish tradition does not demand socialism, and for many good reasons, we, in the United States, have adopted a modified, capitalistic system of economics. The trick, then, will be to balance access to applications of the new technology with the legitimate right of a private company to make a profit on its efforts to develop and market applications of stem cell research. Within these constraints, though, rabbinic rulings written by rabbis in all American Jewish denominations not only permit embryonic stem cell research, but urge that we aggressively pursue this promising approach to curing and preventing many devastating diseases as partners of God in the ongoing task of healing.

Notes

The Mishnah was edited by Rabbi Judah, President of the Sanhedrin, in approximately 200 CE Pages in parentheses following the chapter and paragraph number indicate where the passage appears in the Babylonian Talmud. Other texts cited include: the Babylonian Talmud, edited by Ravina and Rav Ashi c. 500 CE; Maimonides' *Mishneh Torah*, completed in 1177; Joseph Karo's *Shulhan Arukh*, completed in 1565.

References

Altman, Lawrence K. 2001. Volunteer in asthma study dies after inhaling drug. *New York Times*, June 15.

Dorff, Elliot N. 1996. Artificial insemination, egg donation, and adoption. *Conservative Judaism* 49:3–60.

Dorff, Elliot N. 1998. *Matters of life and death: A Jewish approach to modern medical ethics.* Philadelphia: Jewish Publication Society.

Dorff, Elliot N., and Aaron L. Mackler. 2000. Responsibilities for the provision of health care. In *Life and death responsibilities in Jewish biomedical ethics,* ed. Aaron L. Mackler, 479–505. New York: The Jewish Theological Seminary of America.

Jakobovits, Immanuel. 1959. *Jewish medical ethics*. New York: Bloch.

Karo, Joseph. 1565. *Shulhan Arukh.*

Kolata, Gina. 2001. Researchers say embryos in labs aren't available: Few couples have agreed to allow frozen specimens to be used in experiments. *New York Times*, August 26.

Lemonick, Michael D. 2002. Risking business? Do infertility treatments damage babies' genes? Doctors used to think not. Now they are not so sure. *Time* 159 (March 18):68–69.

Mackler, Aaron L., ed. 2000. *Life and death responsibilities in Jewish biomedical ethics*. New York: The Jewish Theological Seminary of America.

Maimonides. 1177. *Mishneh Torah.*

Office of Technology Assessment, US Congress. 1988. *Infertility: Medical and social choices*, OTA-BA-358. Washington, DC: US Government Printing Office.

Rabbi Judah. 200 C E. *Mishnah.*

Ravina and Rav Ashi, ed. c. 500 CE. *Babylonian Talmud.*

Spirtas, Robert et al. 1993. Fertility drugs and ovarian cancer: Red alert or red herring? *Fertility and Sterility* 59:291–293.

Wade, Nicholas. 1999. Patient dies while undergoing gene therapy. *New York Times,* Sept. 29.

5

Islamic Perspectives on the Ethics of Stem Cell Research

Abdulaziz Sachedina

The ethical assessment of research that uses pluripotent stem cells derived from human embryos and aborted fetal tissue gives rise to serious ethical-legal question concerning embryonic sanctity and the legality of abortion in Islamic tradition. Although the rulings given by Muslim jurists in this regard are inferentially deduced on the basis of the precedents in the criminology connected with harming or destroying a fetus, the moral status of the fetal viability and embryonic sanctity remains unclear. There is no clear definition of "embryo" or "fetus" in the Muslim juridical tradition. Hence, its ethical-legal status remains unresolved. From the prevalent rulings about the graded penalty for inducing abortion, it appears that personhood of the fetus becomes established much later in the nine-month gestation period. This ambiguity has led to lax attitudes when assessing fetal viability in new bioethical rulings. The ethical dilemma concerning embryonic stem cells in biomedical research and all its ramifications for the dignity of the embryo has not been taken up in the current debates.

Embryonic Sanctity

On August 7, 2005, Christian groups in the U. S. announced the "Snow Flake Embryo" adoption program as part of their campaign to oppose stem cell research that uses in vitro fertilization (IVF) clinic surplus embryos to derive stem cells. The group believes that the only natural way to give moral weight to the lives of frozen embryos is to adopt and implant them, and carry the fetus through a full pregnancy. They are not opposed to assisted reproduction technology that uses these frozen embryos to help couples to have their first or second child. They have problems with the modern science reducing potential human life to "surplus" unwanted embryos that can be destroyed for research. From a strictly religious point of view, there cannot be anything like "surplus" or "unwanted" embryos, since such a devaluation of an embryo would be counted as an affront to God's claim on life. In the context of Islamic or any other religiously informed bioethics, as I shall discuss in this chapter, as long as the embryo is defined ontologically as possessing the potential ability to become a human being, there will be moral qualms and religious opposition in supporting the use of frozen embryos for research purposes. The litmus test will be the determination of whether the embryo at that very early stage of its development (five to eight days after fertilization when it is totipotent) is a being with rights.

Stem cells are defined by their potential for differentiation as totipotent, pluripotent, and unipotent, and by their source as embryonic or adult. Although all types of stem cells are currently being studied for their potential therapeutic benefits, most scientists agree that embryonic stem cells from the inner cell layer of the blastocyst, which have the potential to generate all types of cells in the human body, offer the greatest prospect for the study and treatment of incurable human diseases. There is little controversy over the morality of using adult stem cells, because they can be derived from living donors of bone marrow and other tissues. However, adult stem cells, which are undifferentiated and unipotent cells, have limited potential to reproduce themselves in culture and to differentiate into cell types besides those tissues and organs from which they were isolated. Consequently, the scientific community has concentrated its research on embryonic stem cells. Embryonic stem cells are usually harvested from donated frozen embryos that were produced for the purpose of assisted reproduction. The religious-legal acceptance of the technology in the Muslim world is corroborated by the mushrooming of fertilization clinics in all major cities. In spite of this endorsement of IVF technology, the problem is the total lack of ethical discussion in the Muslim sources regarding the beginning of life and morally questionable attitude towards clinical abortion. The revelation-based principle of sanctity of life would appear to rule out the termination of fetal life through clinically induced abortions in the early stages. And yet, both in the liberal opinions on abortion and legal permission to use "surplus" frozen embryos there is total disregard for the embryonic inviolability. By concentrating on the legal implications of the feticide and totally neglecting the moral philosophical dimensions of human embryology, Muslim jurists have limited the extension of the principle of the sanctity of life to the embryo that is in the womb. That sanctity principle is not extended to the embryos that are not implanted and that are frozen.[1]

One of the intriguing questions connected with embryonic sanctity in the Islamic revealed texts deals with the beginning of life. The question of the moral standing of embryonic and fetal life remains unresolved in Islamic jurisprudence because of the lack of a precise definition of life and of the beginning of life that involves religious, ethical, legal and social considerations. Although the jurists do not dispute the biological fact of life and the sanctity a fetus enjoys because of that, they differ as to the stage of fetal development at which the fetus enjoys absolute inviolability (*dhimma ṣāliḥa*) and possesses full rights as an independent person.[2]

Hence, there are disagreements about the moment of conception and the time when the ensoulment—the infusion of the soul into the body of the fetus, thus conferring moral status on the fetus—occurs, and whether the viability of the fetus is marked when it is capable of living as a newborn outside the womb. By definition, since the fetus (*janīn*) is 'concealed' (*istajann*) in the mother's womb until it is born,[3] it has no independent claim to life. In juridical terminology, the fetus is defined as an entity that in one sense does not acquire a personhood (*nafs*) directly so as to benefit from rights.[4] Furthermore, in Islamic jurisprudence the abortion rulings are not

[1]See, for instance, *Masā'il fī al-talqīḥ al-ṣanā'ī*, in *Masā'il wa Rudūd*, compiled by Muḥammad Jawād Raḍī al-Shihābī and edited by 'Abd al-Wāḥid Muḥammad al-Najjār (Qumm: Dār al-Hādī, 1412 AH) i. 99. The opinion is ascribed to prominent Shī'ite jurist Abū al-Qāsim al-Khū'ī.

[2]In the section on "The Crime against the Fetus," 'Abd al-Raḥmān al-Jazarī, *Kitāb al-fiqh 'alā al-Madhāhib al-'arba'a, Kitāb al-ḥudūd* (Beirut: Dār al-Kutub al-'Ilmiyya, 1392 AH), v. p.372ff.) takes up detailed comparative rulings on the status of the fetus and culpable actions leading to its abortion among four Sunni schools.

[3]*Lisān al-'arab* and other lexicons (*al- Miṣbāḥ al-Munīr, al-Mu'jam al-Wajīz*)

[4]This is the Ḥanafī definition of janīn as mentioned by Ibn 'Ābidīn, *Hāshiya*, vi. 587. See also: *al-Baḥr al-Rā'iq*, viii. 389; *Badā'i' al-Sanā'i'*, vii. 325.

framed in terms of a resolution to a conflict of rights between the pregnant woman and her fetus. According to the Ḥanafī Sunni scholars, for instance, as long as the fetus remains in utero it does not have independent and absolute inviolability because it is regarded as a part of the mother's body. However, as soon as it becomes separated from the uterus with the capability of surviving outside the womb then it is regarded as a person *(nafs)* possessing inviolability and rights like liberty, inheritance, proper lineage, and so on.[5]

In this sense, the fetus in the womb has a relative claim to life and to rights based upon its eventual personhood in that it is a potential human being while in utero. The closer to birth the fetus is, the closer to personhood it can be considered and the greater justification it has to be accorded rights.[6] Such an estimation of the personhood of the fetus is behind the contemporary liberal juridical opinions among the Ḥanafī Sunni scholars who do not regard abortion as forbidden if the mother's life is in danger at any stage of gestation, including the last days before the child is born.[7] This conditional permission linked to the impending danger to the mother's life is often overlooked when clinical abortions are readily performed in the Muslim world with no impunity. To add to this unscrupulous attitude towards the pre-ensoulment embryo, the ruling that permits abortion for the reason of poverty[8] actually leads to the abuse of abortion as a method of population control. Certainly, no school of Islamic jurisprudence allows abortion to function as a method of population control.[9]

The problem is that it is not until recently that abortion began to be treated independently under its own rubric in the juridical formulations. Like a number of topics that involved some kind of criminal act and the ensuing penalty in medical jurisprudence, abortion as an unlawful act found its place in Islamic penal system (*jināyāt*).

---

[5]Al-Jazarī, *al-Fiqh ʿalā al-madhāhib al-arbaʿa*, v. 372.

[6]Ibn ʿĀbidīn, *Ḥāshiya*, v. 517; *al-Baḥr al-Rāʾiq*, viii. 389; *Badāʾiʿ al- Sanāʾiʿ*, vii. 325.

[7]*Al-Mawsūʿa al-fiqhiyya al-kuwaytiyya*, xvi. 279.

[8]Ibid.

[9]A number of articles that include opinions and rulings issued by leading Muslim scholars, both Sunni and Shīʿite, have been published during the last three decades that permit safe methods of birth control but reject abortion as one of the legitimate ways of encouraging or enforcing population control. See, for instance, Jād al-Ḥaqq, Tanẓīm al-nasl wa fawāʾid al-bunūk al-muḥaddada, *Al-Fatāwā al-islāmiyya*, ix. p. 3110; Muḥamad ʿAlī al-Tashkīrī, Raʾy fī tanẓīm al-ʿāʾila wa taḥdīd al-nasl, *Majjala al- Tawḥīd*, xxxix. 76-85.

Consequently, in the juridical tradition one finds little or no discussion of the ethical dimensions of embryonic personhood. The fundamental assessment of an embryo in the Shariʿa is based upon the Qurʾanic reference and its elaboration in the Tradition that speak of a progressive acquisition of a human status without any concern for moral issues connected with the independent status of a fetus as a moral entity. Although there are a number of recent studies devoted to fresh rulings that deal with legality of abortion because of adultery *(zina)* and rape *(ightiṣāb)*, there is hardly any serious debate among Muslims about the ethical issues connected with a pre-implantation embryo and/or fetus as a person with its own rights and needs for protection.[10]

It is significant to note that the issue of intentional abortion does not come up in the Qurʾan. All the standard juridical references to the Qurʾanic passages actually deal with homicide *(qatl al-nafs)*, rather than abortion of the fetus through a miscarriage before it completes its nine-month gestation. In fact, there is no definition of the embryo as a living entity right from the zygotic stage anywhere in the Tradition. In their assessment of the tort committed against the fetus, the jurists have regarded implantation of the coagulated drop— zygote—in the uterus (*istiqrār al-nuṭfa fī al-raḥm*) as the determining stage of fetal life when any infliction of harm to it requires compensation.

The verses quoted for assessment of the compensation due to the fetus or any other party who participated in this wrong act treat the fetal development as a growing entity that resembles another organ of the body. To be sure, the verses do cover the stages of gestation from fertilization to personhood. But they do not in any way explain the nature of the zygotic stage, whether it holds life or carries pluripotent stem cells with a potential to generate all the cells of the human body. According to some recent rulings on allowing abortion in pregnancies that have resulted due to forced rape, like the ones that took

[10]In recent years a number of articles have appeared in Arabic and Persian that discuss abortion in the context of modern medicine. Unlike articles in Western languages on the subject of abortion in Islamic tradition, these are written by Muslim scholars of Islamic law, whose thorough grounding in juridical sources and methodology make these studies an important contribution to our understanding of the issue in jurisprudence. However, there is little attention paid to the ethical issues connected with the rightness or the wrongness of abortion with due analysis of personhood and rights that accrue to a fetus. See, for instance, a number of articles on the subject in the last five years in *Majalla al- sharīʿa wa dirāsāt al-islāmiyya*, published by the Kuwait University.

place in the Balkans in the 1990s, it appears that the embryo is treated as an entity that does not have all the factors that biologically lead to its ability to grow into a human being.[11] In fact, an interesting discussion providing guidelines to determine whether intentional abortion inflicted at the stage of coagulated drop (*al-dam al-malqā* = zygote) constitutes a tort underscores a completely different understanding of the crime against the embryo. According to majority of the jurists, if the aborted matter dissolves in hot water then it cannot be regarded as aggression toward the embryo.[12] In other words, it is only when an embryo has coagulated and lodged itself in the uterus and has grown into a clot and tissue that the crime needs to be assessed for compensation.

More pertinently, the passages that speak about the embryonic development of a fetus do not spell out the modern-day differentiation between a biological animation and moral-legal existence to argue for the embryo's integrity. Verses 12-14 of Chapter 23 describe the stages of the biological development of the embryo:

> We created man of an extraction of clay, then we set him, a coagulated drop (*nuṭfa*) in a safe lodging, then We created of the coagulated drop a leech-like clot (*ʿalaqa*), then We created of the clot a morsel of tissue (*muḍgha*), then We created of the tissue bones, then we covered the bones in flesh; thereafter We produced it as another creature. So blessed be God, the Best of creators.[13]

Some important conclusions have been drawn from the three primary stages of embryonic development to a human person in the Qur'an: First, perceivable human life is possible at a later stage of biological development of the embryo when God says: "thereafter We

[11]Saʿd al-Dīn Masʿad Hilālī, Ijhāḍ janīn al-ightiṣāb fī ḍawʿ aḥkām al-sharīʿa al-islāmiyya: dirāsa fiqhiyya muqārana, in *Majalla al-sharīʿa wa dirāsāt al- islāmiyya*, Vol. 15, Number 41 (1421/2000), pp. 282-315, deals with the new situation that arose when the Serbs in the Balkans used rape as a weapon against Muslim women.
[12]Ibn Rushd, *Bidāya*, ii. 416; Ibn Qudāma, *Mughnī*, ix. 539, 556-7; Ibn Ḥazm, *Muḥallā*, vii. 30.
[13]See Keith Moore, "A Scientist's Interpretation of References to Embryology in the Qur'an". *The Journal of Islam Medical Association* 18: 15-16. Moore has interpreted *nutfa* = drop as the zygote, the union of the sperm and the ovum; ʿ*alaqa* = lit. "leech" refers to a bloody form; *mudgha* = lit. "chewed substance" refers to a stage when fetus acquires flesh and solidity; and the fourth stage refers to the time when all organs attain their full perfection.

produced him as another creature."[14] Second, since all the factors that make up a human being are not present, it is possible to make a distinction between biological animation and moral-legal personhood, and, as indicated by the consensus among the Sunni jurists over the interpretation of the above verses, to place the latter stage after the first trimester of pregnancy. This consensus is based on the traditions that provide further elaboration of the Qur'anic embryology.

The single most important tradition that has provided religious grounds for legal estimation of a fetus inviolability has been reported in both the Sunni and Shīʿite compilations. In the version preserved in Bukhārī's (d. 870) compilation, which in the Sunni estimation is the most authentic collection, the Prophet is reported to have said as following:

> Each one of you in creation amasses in his mother's womb [in the form of a coagulated drop *(nuṭfa)*] for forty days; then he becomes a blood clot *(ʿalaqa)* for the same period; then he becomes a lump of flesh *(muḍgha)* for the same period; then the angel is sent with a mandate [to write down] four things [for the child]: his sustenance, his term of life, his deeds, and whether he will be miserable *(shaqī)* or happy *(saʿīd)*.[15]

There is no mention of the breathing of the spirit *(rūḥ)* in this version. However, in another equally authoritative version related in Muslim's (d. 875) collection the last part of the tradition includes an additional sentence which clearly states: "Then the angel is sent to breathe into him the spirit *(al-rūḥ)*."[16] In another variant reported in the same compilation, the angel is present from the time of implantation, some forty-five days after the conception, when the embryo lodges itself in the uterus. This report also mentions the determination of the sex of the child whether it will be male or female.[17]

These traditions provide the ontological interpretation of biological data as to when the embryo attains a status of another human in-

---

[14]Qurṭubī, *Jāmʿi*, xii. 6; Rāzī, Tafsīr, xxiii. 85; Ṭabarsī, *Majmaʿ*, vii. 101; Ṭabāṭabāʾī, *Mizān*, xv. 20-24

[15]Bukhārī, *Ṣaḥīḥ, kitāb al-qadar, ḥadīth* 1 and 2, i. 211. For variants see Muslim, *Ṣaḥīḥ, kitāb al-qadar, ḥadīth* 2643 and 2645 (Nawawī's commentary on *Ṣaḥīḥ*, Vol. 16:191). For Shiʿite version see: Ḥurr al-ʿĀmilī, *Wasāʾil, kitāb al-diyāt, ḥadīth* 35652.

[16]Muslim, *Ṣaḥīḥ, Kitāb al-qadar*, iv. 2036, *Ḥadīth* No. 2643.

[17]Ibid., iv. 2038, *Ḥadīth* No. 2644; also, Aḥmad b. Ḥanbal, *Musnad*, iv. 7.

dividual. From different versions of this tradition it is possible to speak about the stage of recording human destiny by an angel who is sent by God to breathe the spirit as occurring either on the fortieth, forty-second, or forty-fifth night or after 120 days. The jurists have identified this stage as the moment of ensoulment when the fetus attains ontological unity and identity as a human person. It is important to note that the moral-legal implication of the ensoulment is not the subject of these traditions.

With their limited knowledge about human embryology, ancient Muslim jurists did not emphasize the distinction between two periods of pregnancy to deduce decisions about the culpability and accruing penalty in the matter of induced abortion. In fact, the inference regarding the first trimester is particularly absent in Bukhārī's above-cited version. Others, like Ibn al-Qayyim al-Jawziya (d. 1350), have argued that the tradition suggests that all the three early stages from coagulated drop to clot to lump of flesh are covered in the first forty days because the tradition clearly states: "Each one of you in creation amasses *(yajmaʿu)* in his mother's womb for forty days."[18] Moreover, none of the authenticated traditions refer to the coagulated drop *(nuṭfa)* stage as a separate gestational stage. According to some jurists, even the phrase that literally means "like that" does not suggest that the reference is to "the same period," that is, forty days. The tradition simply suggests that before the angel is sent to write the child's destiny there is no spirit in the fetus whether in drop, clot, or lump form. For these jurists the ensoulment occurs at the end of the first forty days and not after that as asserted by others.[19]

However, it is possible to extrapolate the beginning of life from the time of conception in some Shīʿite traditions. Thus, for instance, the following tradition in which a disciple of the seventh Shīʿite Imam Mūsā al-Kāẓim (d. 800) asks for a solution to the case involving an induced abortion:

> [Isḥāq b. ʿAmmār reports:] I asked Abū al- Ḥasan [al- Kāẓim]: [What is your opinion about] a woman who fearing pregnancy

---

[18]Muḥammad b. Abū Bakr, known as Ibn al-Qayyim al-Jawziya, *al- Tibyān fī aqsām al-qurʾān* (Beirut: Muʾassassa al-Risāla, 1994), p. 337; also, Ibn Ḥajar al-ʿAsqalānī, *Fatḥ al-bārī*, xi. 481.
[19]Sharaf al-Quḍāt, *Matā tunfakh al-rūḥ fī al-janīn?*, (Dār al-Furqān, 1990), p. 45.

> takes a medicine and aborts what she has conceived? He responded: No, [it is not right.] At that I said: But it is a coagulated drop *(nuṭfa)*. He said: The beginning of human creation is a coagulated drop.[20]

This tradition explicitly mentions the beginning of human creation at the zygotic stage and rules the abortion at that stage illicit. A similar implication can be inferred from another tradition ascribed to ʿAli b. Abū Ṭālib (d. 660) in which he specifies the five stages of fetal development in the matter of the amount of compensation that must be paid when the abortion is induced:

> He specified five stages for man's semen [in order to fix the compensation]: when it is fetus before the ensoulment 100 dinars, this is because God created a human being from an extraction of clay. This is the coagulated drop. Hence, this is a part [of human creation]. Then it is a blood clot. This is another part. . . . [21]

In both these traditions, and in the light of the above-cited verse, it is possible to argue that the zygotic stage is regarded as the beginning of life and hence abortion at that stage carries the prescribed penalty. However, there is another tradition that specifically takes up the issue related to the timing of ensoulment. Saʿīd b. al-Musayyib (d. 715) is reported to have asked the fourth Shīʿite Imam ʿAlī b. al-Ḥusayn (d. 713) about ensoulment in the following case:

> I asked [the Imam]: [In your opinion] do the changes from one state to another that take place [in the fetus] during the gestation occur with or without the spirit *(rūḥ)*? He said: The changes occur through the spirit, with the exception of the preexistent life that is transferred in the loins of men and the wombs of women. If the fetus had no [independent] spirit other than the life that was there [because of the parent's existence], it could not have changed from one state to another in the womb. [The existence of the spirit and ensoulment are proven by the fact that had it not been for the presence of the spirit] the killer

[20]Ḥurr al-ʿĀmilī, *Wasāʾil al- shīʿa*, xix. 15.
[21]Kulaynī, *Furūʿ al-Kāfī*, vii. 342; Ibn Bābūya, *Man lā yaḥḍur*, iv. 54.

would not have been required to pay the blood money *(dīya)* at that [early] stage [of fetal development].[22]

This last case provides a clearer understanding about the beginning of life and ensoulment in Shīʿite tradition. Based on these accounts, a number of Shīʿite jurists have argued for embryonic inviolability much before the 120 days' cut-off period.

In any event, Sunni jurists are agreed upon the prohibition against aborting after the ensoulment. Al-Qurṭubi (d. 1273) has stated this most clearly in his juridical exegesis of the Qur'an: "There is no disagreement among the scholars that the ensoulment occurs after 120 days. This is after completing four months of gestation and having entered the fifth."[23] More importantly, the jurists regarded the deliberate termination of pregnancy at any stage of the embryonic development to be sinful. However, after the first trimester, when the majority of the jurists afforded the fetus ontological status of an individual, abortion was absolutely prohibited unless it conflicted with saving the mother's life.[24] Variant readings and interpretations notwithstanding, among the Sunni scholars there seems to be a juristic agreement that based on the above-quoted traditions in the two highly respected compilations of Bukhārī and Muslim, the fetus attains personhood after 120 days.[25] Nevertheless, the differences of opinion about the absolute inviolability of the fetus had to wait for the biomedical advancements in modern times, when biological data on the embryonic journey to a full human at times contradicted the traditional account of when conceivable life was scientifically discernible. But, even then, the jurists remained oblivious to the moral dimension of embryonic inviolability.

---

[22]Kulaynī, *Furūʿ al-Kāfī*, vii. 347. The use of *diya* as a form of penalty prescribed in this case indicates that the crime is seen as homicide. This signification of *diya* is supported by the lexical sense of the word. See, for instance, major classical Arabic lexicons under WDY: *Tāj al-ʿarūs*, x. 386; Ibn Athīr, *Nihāya*, iv. 202; *Lisān al-ʿArab*, xv. 383.

[23]*Al-jāmiʿ li-aḥkām al-qur'ān*, viii. 12. Al-Nawawī in his commentary on *Saḥīḥ Muslim*, xvi. 191 and Ibn Ḥajar in his *Fatḥ al-bārī*, being a commentary on *Ṣaḥīḥ al-bukhārī*, xi. 481, mention the consensus of the jurists in this regard.

[24]Nawawī after commenting on the traditions in Muslim's compilation makes this observation: "The jurists are in agreement that the ensoulment takes place only after four months [of the pregnancy]" (xvi. 190-91).

[25]Ibn ʿĀbidīn, *Ḥāshiya*, ii. 370; Ibn al-Humām, *Fatḥ al-qadīr*, ii. 495; Qurṭubī, *Jāmi'*, xii. 8; Nawawī, *Sharḥ*, xvi. 191; Ibn Qayyim al-Jawziyya, *Tibyān*, 337-8; al-Ramlī, *Nihāyat al-muḥtāj*, viii. 416; al-Mirdāwī, *al- Inṣāf*, vii. 386; Ibn Ḥajar, *Fatḥ al-bārī*, xi. 481 and 484.

Moral Problems Related to Technically Assisted Reproduction

While more or less in agreement about abortion before the first trimester, the jurists have disagreed about the legal status of the fetus in according it rights that are categorically and unqualifiedly due to an individual. The rulings about the legal status of the fetus are mostly inferred from the estimation of the three primary stages of embryonic development in the Qur'an rather than from the ontological assessment of the nature of the human person. Accordingly, there is no discussion about the dignity of the person, which serves as the foundation upon which all human rights rest. Just as there is no clear understanding of the zygotic stage of the fertilized ovum in Islamic sources, there is also a conspicuous absence of any ontological definition of a human person (*nafs*) in any estimation of the fetus as the unity of body and spirit. In jurisprudence, there is an assumption that the term *nafs* self-evidently stands for personhood (*ādamī*), whose life must be protected through a detailed penal code rather than a theory of inherent dignity of the fetus. Any argument to assert the fetal inviolability at all stages of its journey towards a full human being would have required the jurists to seriously engage the Qur'an in deriving an ethical framework to define human personhood in order to affirm the inherent dignity of the pre-ensoulment fetus. The juridical trend is simply to deny personhood to the pre-ensoulment fetus because it does not have full moral status in the absence of differentiated organs to indicate human shape and, as a consequence, to permit abortion at this stage as well as the derivation of embryonic stem cells from cyropreserved embryos.

The other reason for this relaxed attitude towards pre-ensoulment fetal personhood is the religious tradition that holds that no pre-ensoulment fetus will be resurrected on the Day of Judgment (*al-qiyāma*). In other words, anyone who has not been infused with the soul is not to be resurrected. Hence, it is not forbidden to abort the fetus at this stage because it is not a being as yet.[26] The problem with this argument is that from the point of view of the Qur'anic doctrine that speaks about God's omnipotence and omniscience, how can human beings prejudge that the fetus will not be ensouled and will not reach personhood to be capable of being resurrected if allowed

---

[26]*Al-Furū'*, i. 281

to continue in its gestational journey? Certainly, to abort or destroy the fetus at this stage is an act of aggression toward it, and unless there is a valid reason to do so it is forbidden to abort it.

In the 1970s in vitro fertilization (IVF) to treat human infertility marked the beginning of the revolution in making possible what is impossible in nature. "In vitro" (lit. 'in glass') refers to the Petri dish in which the sperm and eggs are mixed. This technique was originally developed to get around the woman's damaged or absent fallopian tubes, which connect the ovaries to the uterus. In 1978 the technique was successfully used to fertilize the eggs and implant the resultant products to treat infertility.

The major ethical concern with IVF is about a woman's egg being fertilized outside the body and then being injected into the fallopian tubes of the mother or surrogate mother. Although in the classical juridical corpus the jurists had accepted the possibility that a woman could asexually introduce the sperm that she considered to be her husband's into her uterus on her own, an asexual pregnancy without penetration through assisted reproductive technology had no precedent in the Shariʿa. Neither was surrogate motherhood known to the jurists. With the provision of polygamy, the immediate solution to infertility was always a second wife, sometimes with the encouragement and approval of the first wife, and at other times with the disapproval or even divorce of the first wife. But the new possibility that the second wife could now gestate an embryo that carried the gametes of the husband and the first wife (who, due to a medical condition, could not carry it to its full term) through IVF procedure required precise determination whether the procreation had occurred within the same family unit. The legal methodology provided an important rule that required the jurists to avert probable harm (*dafʿ al-ḍarar al-muḥtmal*) before any consideration of benefit that accrued to the agent through such a medical intervention.

Notwithstanding the unknowns in the traditional law, the benefits of IVF in treating infertility were obvious, as long as such fertilization was achieved within the legitimate boundaries of marriage. However, as was customary in Islamic juridical deliberations, little attention was paid to the moral and social implications of the procedure for the nature of the child's identity and relationship to the family, on the one hand, and the status of multiple human embryos that were produced in the Petri dish and then implanted to increase the possibility of pregnancy, on the other. In the case of multiple pregnancies the procedure requires the abortion of additional embryos to avoid endangering the mother's health and to improve the chances of survival for one embryo. Besides two to three embryos that were injected for gestation, there were surplus embryos that were frozen for use in further, future attempts. What is the status of these inseminated and frozen embryos? Could they be used later in further attempts at having a first child or for additional children? Who owns them if the couple later divorce or if one of them dies? Could they be simply discarded as "unwanted" embryos? Could they be used to derive stem cells for research and therapeutic purposes? Currently derivation of stem cells is justified on the basis of all therapeutic ramifications that are speculatively assessed in order to advance potential cures for many people with chronic, debilitating, fatal, or degenerative disorders.

Undeniably, IVF had a limited goal of correcting a natural condition to allow a would-be mother to carry a fertilized embryo to its full term of gestation. Keeping in mind the plurality in the Muslim legal opinions on new bioethical issues, in spite of the fact that there were some dissenting voices among both the Sunni and the Shīʿite scholars, the majority of them came to endorse the IVF technology, with the stipulation that the procedure itself should not lead to any sinful act contrary to the rulings about man-woman relationships in the Shariʿa. Hence, even when some scholars had reservations about the procedures of producing gametes asexually, including the morally questionable act of masturbation to derive the sperm, IVF became a routine medical practice in the Muslim world to help women who could afford expensive reproductive technology to conceive. In addition, in order to avoid surgery, doctors can now be guided by ultrasound to the ovaries to retrieve eggs vaginally, but this somewhat invasive procedure raises questions about a third party male physician (other than the woman's father, husband, or brother) having access to the private parts of the woman. This is morally problematic from the Islamic code of modesty for women. There was no immediate solution to this latter problem since there were not enough physicians, male or female, to perform the IVF procedures without invading the privacy that was protected by Islamic code of modesty.

In majority of the cases of infertility, if the family was well-to-do, the treatment was sought abroad, where such Islamic sensibilities about male-female relationships were altogether absent. Nevertheless, the issue, however academic in nature, was general enough to require a sensible and immediate solution in view of the shortage of women specialists in all areas of medicine. I have specified the problem as academic because for most people, medical care was so difficult to obtain that it hardly mattered whether the patient was seen by a male or female doctor. Before the spread of modern mass education, infertility was a serious problem in the family and in the society at large. Married couples who could not have a child left no stone unturned to become parents, which included even financial hardship. Consequently, the legal hair-splitting that was part of the seminary culture and was meant for consumption among the traditional scholars of Islamic law rarely filtered down to the ordinary folks looking for practical solutions, except in the form of ruling that either permitted or forbade a procedure.

For those jurists who wanted to provide moral-legal justification for the use of surplus embryos as the source for derivation of embryonic stem cells for research, juridical solutions were not hard to deduce when legal principles like public good *(maṣlaḥa)* that promotes what is beneficial, and necessity (*ḍarūra*) that overrules prohibition, could provide religious-legal justification and legitimization. After all, as some prominent jurists had already ruled, the sanctity of life principle does not apply to the embryos that are outside the womb. Consequently, the use of frozen embryos to isolate stem cells was justified, especially in view of enhancing the possibility of discovering cures for incurable diseases. The future use of frozen embryos for posthumous transfer of an intra-fallopian gamete to a widow was another problem that required meticulous understanding of the status of the frozen embryo to make sure it could be treated as a property that belonged to the legally married couple. If it were established that the frozen embryo is in the usual sense a property then it was subject to the laws governing ownership and transfer of property that belonged to the biological father and mother. However, is it really a property? There was no doubt that if the couple was alive, both had a right to determine the use of their embryos. What if one of them died? Then the complexity of the problem came to the fore as the jurists began to question the widow's right to use a frozen embryo that was the result of her husband's gamete being fertilized with hers, when legally, because of the death of her husband, the contract that wedded her to him became invalid. She was no more his wife, and hence, the newborn could not use his name as part of her/his identity.

The Problem of Endorsing Stem Cell Research

Although assisted reproductive technology has been endorsed by the Muslim jurists, one of the controversial issues in this technology is the asexual production of embryos through somatic cell nuclear transfer (SCNT), which involves the introduction of the nuclear material of a somatic cell into an enucleated oocyte. The use of donor eggs or sperm is out of the question in Islamic tradition, since the preservation of the child's lineage to its biological parents through marriage is obligatory. The ethical issues in assisted reproduction are associated with the SCNT technique through which the embryo is created. While the moral status of the embryo remains at the center of the controversy connected with the permission to use the frozen embryos, the problem of producing embryos with SCNT technique raises serious questions about the commodification of early forms of human life.

Retrieval of stem cells necessitates the embryo's destruction and its production just for that reason gives rise to incompatible notions of embryonic sanctity and the respect and rights owed to preimplantation embryos at the blastocyst stage. SCNT-derived stem cells may lead to the acceleration of research in reproductive genetics with direct impact on interhuman relationships that occur from a naturally occurring pregnancy, uninterrupted by science, through its natural course of development within a marriage. As discussed earlier in the context of embryonic sanctity, there is enough evidence in the revealed texts of Islam to argue for the moral status and rights of the embryo at the zygotic stage. Hence, its destruction for the derivation of stem cells cannot be ethically justifiable. And, although there is an absolute and collective moral duty in Islam for the physicians and scientists to undertake biomedical research that may result in beneficial treatments for a number of incurable diseases that afflict humanity today, there is an equally valid concern whether the potential benefits of the research involving embryos can be translatable into therapy. This requirement puts the burden of proof on public and private agencies in the Muslim world to provide evidence that stem cell research adheres to the standards of religious-ethical and scientific oversight.

In addition, the hesitation in endorsing the cloning method of creating pluripotent stem cells is the remote potential (at this time, at least) for the asexual reproduction of a human being outside the boundaries of a marriage. One of the essential principles of Islamic bioethics is "No harm, no harassment," which underlies all ethical decisions that might cause harm and lead to harassment of an individual who must be protected from such probable harm in future (Sachedina, 2006). There is no greater harm in Muslim societies than to cause an ambiguity in an individual's lineage. One of the essential purposes of the Shariʿa is the protection of the lineage of the offspring through marriage. Islam regards relationships growing out of one's natural connection to one's kin, in addition to relations that are negotiated in the social context on the basis of interpersonal justice, as fundamental to human religious and spiritual life.

Accordingly, Muslim scholars and their governments need to assess the risks and the benefits of stem cell research in the light of Islamic values related to the dignity of the embryonic being. Thus far the ethical-religious assessment of research uses of pluripotent stem cells derived from human embryos in Islam has been inferentially deduced from the rulings that deal with the fetal viability and embryo sanctity in the classical and modern juristic decisions. The jurists treat a second source of cells derived from fetal tissue following abortion as analogically similar to cadaver donation for organ transplantation to save other lives, and hence, permissible. As discussed above, the moral considerations and concerns in Islam have been connected with the fetus and its development to a particular point when it attains human personhood with full moral and legal status. Based on theological and ethical considerations derived from the Qurʾanic passages that describe the embryonic journey to personhood developmentally, and the rulings that treat ensoulment and personhood almost synonymously occurring over time rather than at the time of conception, it is correct to suggest that majority of the Sunni and Shīʿite jurists will have little problem in endorsing regulated research on embryonic stem cells that promises potential therapeutic value, provided therapeutic benefits are not simply speculative.

The inception of embryonic life is an important moral and social question in the Muslim community. Anyone who has followed Muslim debates over this question notices that the answer to it has differed with the different ages and in proportion to the scientific information available to the jurists. Accordingly, each period of Islamic jurisprudence has come up with its ruling (*fatwā*) consistent with the findings of science and technology available at that time. The search for a satisfactory answer as to when an embryo attains legal rights that must be protected has continued to this day.

As for the experimentation on human embryos in the Muslim world, at this time there is very little hard core evidence available even to regional government agencies. But with the growth of IVF reproductive technology and the so-called "surplus" embryos readily available for experimentation in the field of biomedical research and biotechnology, it is not farfetched to assert that in the Muslim world the integrity and the life of the early human embryo is in serious danger. Stem cell research in the field of developmental and cell biology is promising critical therapeutic benefits, and it is hard to imagine that Muslim scientists are not interested in harvesting stem cells from human embryos and fetuses in the hope of solving the problem of incurable diseases. The ethical dilemma posed by stem cell research on embryos created either by in vitro fertilization or nuclear transfer cloning is a serious one that requires a careful assessment of technically assisted reproduction that seems to be treating the embryo's life as a means to an end. Whether created for reproduction purposes or research, there is no legal or moral basis to deny the dignity of the so-called "spare" embryos. The loosely applied principle of public good (*maṣlaḥa*) seems to have provided unproblematic justification for any procedure that actually requires more scrupulous ethical analysis because it is a potential life that is involved. A more relevant principle in this case states in no uncertain terms that "averting corruption has preponderance over advancing public good."

However, in the juridical tradition, on the basis of the penalties prescribed for the feticide, there is an agreement that the fetus before ensoulment cannot be accorded the status of a full person. Also no funeral rites are to be performed for a fetus before the first trimester.[27] Overall the law stipulates the fetus' right to life, even when a universally recognized definition of an embryo that specifies personhood was lacking. Accordingly, in the penal system it requires postponement of a pregnant woman's execution until after she has delivered her baby and made provisions for its care after her death. Besides maternal safety, the majority of the Sunni and some Shīʿite

[27]Qurṭbī, *Tafsīr*, xii. 6; Ibn Qudāma, *Mughnī*, ii. 398; Ibn Qayyim al-Jawziyya, *Tibyān*, p. 351; Ṭūsī, *Nihāya*, i. 50.

scholars rule that abortion may take place during the first trimester if the woman's pregnancy threatens the well-being of an already existing infant. However, in the case of a pregnancy that threatens a woman's life, the jurists have prioritized saving the life of the ensouled fetus over saving the mother's life.[28] These were the rulings formulated in the classical age.

The phenomenal advancement of biomedical technology in saving the life of a woman faced with complicated pregnancy as well as in determining the beginning of a conceivable life in a fetus has led to modification of some classical formulations about embryonic inviolability. A number of juristic principles, including the "contrariety between two harms" (*taʿāruḍ al-ḍararayn*) and "protection against distress and constriction" (*ʿusr wa ḥaraj*) have been evoked to rule on the priority of saving the mother's life on whose well-being depends the life of the fetus.

The differences of opinion regarding the appropriate penalty for the act of feticide and the lack of agreement among the jurists regarding its assessment as homicide shows that the fetus was not accorded personhood when ensouled. In the case of a man who causes a woman to miscarry (or aborts her) and if the fetus is unformed, his crime carries the penalty of *al-ghurra* (monetary compensation). However, if the expelled fetus is formed then the penalty is full blood money (*diya kāmila*). Notwithstanding the estimation of the earlier stages of fetal development as less than human person, this latter opinion treats feticide at that advanced stage as more of a crime than the death of just an ensouled fetus. When that happens an additional penalty of *al-kaffāra* ("expiation") may be levied, requiring the aggressor to fast for two consecutive months.[29]

Under Islamic penal code the penalty for self-induced abortion or abortion inflicted by others, whether intentionally or accidentally, is treated in some detail.[30] There, the abortion of the fetus at any time is regarded as a crime because it is prematurely removed from the womb where it was created to complete the gestation until it was ready to sustain life outside the uterus. In fact, according to Ibn ʿĀbidīn, contrary to the modern liberal rulings, the permission for

[28]Ibn 'Ābidīn, *Hāshiya*, i. 602, vi. 591; Ghazālī, *Iḥyā'*, ii. 53; Ibn Rajab, *Jāmi' al-ʿulūm*, 46.

[29]Ibn Rushd, *Bidāya*, ii. 416; Ibn Qudāma, *Mughnī*, ix. 539, 556-7; Ibn Ḥazm, *Muḥallā*, vii. 30; Ṭūsī, *Nihāya*, ii. 803.

[30]See detailed comparative rulings on the subject among four Sunni schools in al-Jazarī, *Kitāb al-fiqh'alā al-Madhāhib al-'arba'a*, v. 372ff.

abortion to save the mother's life is not unconditional, because the assessment of the situation is a conjecture. His ruling makes this clear:

> If the fetus is alive, and [even] if there is fear for the life of the mother as long as the fetus remains alive, it is not permissible to dismember it. The reason is that the mother's death on account of the fetus is conjectural. As such, it is not permissible to kill a human being (*ādamī*) on the basis of a conjecture.[31]

The question of fetal rights assumes importance in relation to determining a fetus' personhood. The penal system imposes monetary fines progressively in relation to the age of the fetus on anyone involved in inducing abortion with the intention of terminating the pregnancy, including physician, father, or mother. However, if the wife decides to abort with the permission of her husband, then, according to some Ḥanafīs, there is no need to pay the monetary compensation to anyone.[32] The fully formed fetus is treated like an independent human being with rights to inherit and to be compensated for any damage done to it. This is so even when in another situation the law regards the fetus as an integral part of the mother's body, identical to an organ. It is for this latter reason that it permits donating fetal tissue, treated like other bodily organs for medical research, including the derivation of stem cells. However, in the context of IVF reproductive technology, the jurists seem to maintain a moral distinction between the embryo that is already implanted and developing in the uterus and "surplus" or "spare" embryos. Whereas the implanted embryo enjoys the fetal rights, including the right to life, surplus embryos are not treated as aborted since these existed outside the body of a woman and never reached the stage of ensoulment. Hence, there is no prescribed penalty for discarding these pre-implanted embryos. In fact, there is permission for using them to derive stem cells.

Such a devaluation of pre-implanted human embryos in the IVF clinics gives rise to their exploitation because they have become an important source for potential therapeutic and commercial value. It might even lead to the commercialization of human embryos specifically fertilized as the source of therapeutic products. Muslim jurists

[31]Ibn ʿĀbidīn, *Ḥāshiya*, i. 602.
[32]Ibn ʿĀbidīn, *Ḥāshiya*, vi. 591.

have not considered all the negative facets of their ill-conceived ruling allowing both unregulated in vitro fertilization and the discarding of unused embryos, as if potential human life could be treated like a commodity. The problem is essentially related to Islamic jurisprudence, which ignores any moral analysis of the case and simply deduces its rulings from legal principles that are fundamentally ethical in their purport, but never acknowledged as such. If the contemporary Sunni Muslim jurists were adequately trained in Islamic theological ethics where the objective nature of human action is analyzed in terms of moral good and evil, then they would have considered the moral status of the embryo meticulously. The advancement of medical technology makes it imperative to reconsider the moral status of the fetus in the light of fresh interpretations of religious, cultural and social beliefs.

A human embryo is a potential human life. It has moral-legal status and deserves respect from the time it is conceived. If that were not the case, why would Islamic penal code impose fines for induced abortion from the early stage of progressive fetal development? It certainly cannot be simply used as a product or as a means to an end. Hence, any ruling that permits creation of human embryos to use them as the source of therapeutic products and, in doing so, destroy them is an affront to the divine purposes of the Shariʿa. The Qurʾanic description of the human embryonic journey to acquire human status underscores God's special endowment on humanity by providing it meaning in life. At no point does the Qurʾan or the Tradition suggest that using human embryos to benefit human society is permissible. Those who support such permission base their opinion on the argument that regards destruction of pre-implanted, unensouled embryos that existed outside the uterus as a minimum sacrifice of "left over" embryos for the greater good of the entire society.

The major problem, as I see it, pertains to the denial of the moral status of a fetus outside the mother's womb because, according to this assessment, fetal life even when the fetus is implanted in the womb is still a remote possibility. The progressive development and viability of the embryo indicate and express a moral and legal progression of rights and dignity. And, even if full personhood and ensuing complete rights are achieved only after birth, by its own terms the Shariʿa takes into account the various stages of biological development to assess the level of damage done to the fetus, indirectly contributing to a progressively growing respect, dignity, and rights of the fetus. At no point does the tradition differentiate between pre-implanted embryos and those that are already implanted in the uterus. All the laws in the criminology know of only the in uterine embryo.

Nonetheless, the reference in the Qur'anic embryology about God breathing His "spirit in him" (Q. 23:5) raises the question about when that happens. From a purely scientific viewpoint the debate is irrelevant since whether or not a fetus acquires a soul is a matter in the realm of theology and metaphysics. But in Islamic jurisprudence the matter of ensoulment is of utmost importance because it determines the stage at which the fetus acquires legal personhood and provides a dividing line in a pregnancy after which abortion is not permissible unless carrying the fetus through complete gestation is a threat to a woman's life.

The question of ensoulment is critical to determine the legal status of a fetus in Islamic criminology. The problematic in jurisprudence stems from the terminology used to express different stages of the fetal development from pre-implanted fertilized ovum, embryo, embryo-fetus in utero, and newborn that is strictly derived from the references to embryonic development in the revealed texts. These texts, as discussed earlier, do not speak about the ensoulment until much later. More importantly, these texts do not refer to aborted fetuses. Let us examine juridical terminology for abortion in some detail to get to the root of the problem in according legal-moral status to the fetus.

Legal-moral status of an entity is described in terms of its relation to other moral agents and the obligations and relationships that other moral agents have toward this entity. If fetal tissue is treated as organic non-human life form, then it might be permissible to treat it as an organ in a woman's uterus, and, when aborted, lawful to use it for research with compelling reasons and just ends. On the other hand, if human embryonic cells are considered human entities then our moral relationship and obligation toward them shifts sharply. In Islam, as pointed out earlier, a fetus is guaranteed legal rights by the Shariʿa, and the evidence is abundant in many sections of Islamic jurisprudence. If the moral status of the fetus is conceded from the moment of conception, then it puts constraints in medical research that wishes to use the "spare" IVF embryos to derive, for instance, stem cells, because of our obligation to another entity that is not capable of protecting itself and consenting to its use for research purpose. In the Shīʿite moral tradition, like the Catholic formulation, eradication of a fetus is a sin. But unlike Catholicism, not all Shīʿite jurists regard the moment of conception as the moment of ensoulment. Some maintain that ensoulment occurs at the time of implantation of the fertilized egg in the uterine wall (that is, after 11-14 days); whereas others, taking the hint from the Prophetic tradition, rule that it occurs at the end of first trimester (120 days).

That the fetus has a moral-legal status in the legal corpus cannot be questioned. Hence, for instance, in the matter of inheritance related to the fetus, the jurists are in agreement that if a man dies and a pregnant wife survives him, the right of the fetus is secure, and the inheritance cannot be disposed of before the share of the fetus is set aside. Moreover, if the woman delivers more than one baby, the other heirs have to pay back the share of the other babies. In the case where a woman aborts a fetus, at any stage of its life, and if it betrays any sign of life, such as a cough, sneeze, or finger movement, the fetus is entitled to inherit any legitimate relative who has died after its conception. If the fetus does not survive, its legal heirs inherit its share.

In ruling on the wrongness of abortion, the jurists have ruled that since the Shariʿa guarantees the sanctity of fetal life it protects a fetus against deliberate abortion without legitimate excuse. As noted above, in the case of a pregnant woman who is sentenced to death, execution is postponed until after delivery, and according to some jurists, until the mother completes nursing the child.[33] The fetal sanctity is further maintained in the Islamic penal code that does not regard the embryo in its early stages as simply a dormant mass. The treatment of both induced or accidental abortions, in the form of interdiction, admonition, or penalties, indicate that the Shariʿa recognizes human dignity at conception. It safeguards the life of a fetus, as well as its rights, while it is still in the womb.

However, with all the social changes and scientific findings today, we find ourselves before an open window, with hands almost reaching in to tamper with the lives of fetuses. Permission and prohibition seem to be obscured by the interpretations of modern science concerning fetal growth, stages, and movement, as well as the inception of embryonic life. Modern methods of fetal diagnosis, such as fetoscopy, ultrasound, and other means for examining a fetus and monitoring its growth inside the uterus were not available to ancient

[33]Ibn Qudāma, *al-Mughnī*, viii. 170

scholars. These scientific methods allow us now to see the embryo inside the mother's womb from the earliest moments and to follow its growth, hour after hour and day after day, until it fully grows into a human being.

On the basis of the textual evidence examined, it is possible to maintain the following conclusions about the stem cell research in Islamic societies: 1. The silence of the Qur'an over a criterion for moral status of the fetus allows the jurists to make a distinction between a biological and moral person, placing the latter stage after, at least, the first trimester in pregnancy, and extrapolating a number of rulings that deny dignity to pre-implanted "surplus" embryos in IVF clinics and allow their use for stem cell research. 2. Since the Tradition regards perceivable human life possible at the *later* stage in biological development of the embryo, and since there is hardly any discussion of the early stages of fetal development on the basis of which to assess moral culpability if the embryo were destroyed, Muslim jurists tend to ignore the ethical dilemma concerning their use to derive stem cells.

References

Sachedina, Abdulaziz. 2006. "No harm, no harassment": Major principles of healthcare ethics in Islam. Pages 265–289 in David E. Guinn, ed., *Handbook of Bioethics and Religion*. New York: Oxford University Press.

PART 2

Information Technology: Inequality, Identity, and Invasions of Privacy

6

Our Information Future: From Open Source Software to the Digital Divide

Travis Kriplean and Daniel Lee Kleinman

Introduction

It is often said that the computer and the Internet have ushered in the "information age." In this world, we are inundated with information, access is easy, transmission is rapid, production of content is decentralized, and barriers to entry are low. Search engine and data-mining techniques have been designed to crawl through masses of data. The amount of time and effort expended to find answers has been vastly reduced. And an astounding amount of content is now published online through Web sites, blogs, social networking sites, and wikis.

What do all of these developments mean? What implications will they have? Have these phenomena led to substantial changes in the organization of society, for how power relations play out, and how people relate to one another? Or are the many claims about the information age exaggerated?

In our brief overview chapter, we cannot hope to answer all of these questions. Instead, we take up four central developments in our new knowledge economy. In our first section, we discuss open collaborative production, focusing on the open source software movement. Here, we raise questions about how our traditional notions of property, technology development, and knowledge production comport with the information age. Next, we consider what new information technologies mean for politics, locally, nationally, and globally. Many assert that recent technological developments will mean a new political world. We explore this claim. Third, we discuss the social implications of search engines. Mechanisms of search are central to our use of the World Wide Web, and few of us probably consider how such technologies shape the information to which we have access and how accessible the information we produce is. Finally, we discuss the so-called digital divide. We explore how deeply separated the information haves and have-nots are and consider what reducing the digital divide would mean.

Open Collaborative Development

A second generation of Web applications aimed at supporting sustained and open collaborative production, distribution, and discussion is the latest development to garner attention. These projects are driven by communities that explore what can be done collectively with information technologies; examples include blogs, self-published journal-like Web sites that link to one another, and wikis, collaboratively created Web sites that make it easy for any reader to quickly change the content of the site (O'Reilly 2005). Much of the spirit and process behind these technologies can be traced to an older movement called Free and Open Source Software (FOSS), which introduced a novel legal mechanism for ensuring that people can never be excluded from freely using software.[1] Advocates of this "commons-based peer production" approach have been challenging the conventional wisdom regarding ownership of content and its production with respect to new content distribution, moving from packaged and closed to freely shared content, from certified experts to amateur enthusiasts, from content produced behind closed doors to open

[1]In its early years, "free software" didn't appeal to business minds. In 1997, a group of free software advocates got together to address this issue. They defined the term "open source" to recast free software in a more corporate-friendly tone. Since then, the Free Software and Open Source movements have been distinct, with the latter being more "pragmatically" oriented to the technical merits of open production and the former more ideologically oriented to the freedoms it affords (Perens 1999; Raymond 1999; Stallman 2002c).

processes visible to all (Benkler 2002; Raymond 1998b). Any time there are challenges to conventional wisdom and accepted practices there are also controversies, in this case over the quality, security, legality, and accountability of content created by peer producers. In this section we link advocates of "commons-based peer production" to a movement organized around its vision of community and property in the digital age, describe the basics of online collaboration, and discuss how these new ways of collaborating may challenge established institutions. We focus on FOSS and Wikipedia, an online encyclopedia that allows anyone to edit articles.

Software has two forms: "source code," which contain commands written in a computer language by software developers, and the executable application, which is generated by a computer based from the source code and can be directly understood by the computer. FOSS is a movement based on the philosophy that users should have the freedom to: 1. run the program for any purpose, 2. study how the program works and adapt it to individual user needs, 3. redistribute copies, and 4. improve the program and release the changes back to the community (Stallman 2002a). To be considered FOSS, the software has to be copyrighted under a license upholding these tenets. Prominent examples of FOSS software include the Linux operating system, the Apache Web server, and the Mozilla Firefox Web browser. FOSS has recently garnered substantial support in the IT sector; some companies like IBM and Novell have even built their businesses around it.

While a substantial amount of noncopyrighted, public domain source code has always been floating around academia and industry (Campbell-Kelly 2003), FOSS copyright licenses were a reaction to the traditional proprietary model of software development. This traditional approach is based on selling a "shrink-wrapped" executable application without releasing the source code, creating "lock-in" through dependency on the product, and then making money by extending the features and fixing bugs in the software. FOSS supporters object to these practices on technical, economic, and moral grounds. They argue that in the digital age, where spatial distance is less important than in earlier eras and digital copies are free to produce, profit is better realized through servicing content (e.g., bug fixing, advertising) than through controlling distribution. They argue that the exclusionary aspect of proprietary economics is inefficient because much work is duplicated across organizations (Graham 2005; Wheeler 2005).

Most significantly, FOSS challenges traditional views of ownership and property (Weber 2004). As originally instituted, the US intellectual property system aims to balance individual rights and the public good. It sets up a collection of exclusive rights that grant a temporary monopoly to authors, artists, and inventors. After a certain period of time, anyone has the right to distribute the material (copyright) or make the product (patent) without paying royalties. FOSS advocates believe that exclusive control of content is unnecessarily constraining; instead, a vibrant "knowledge commons"—the free and open world of ideas that anyone can draw upon—should be nurtured because creativity is about building upon ideas and inventions that have come before (Anderson 2004; Lessig 2004; Moglen 2003; Stallman 2002b). To FOSS supporters, the rights of the copyright/patent holder have become overemphasized to the detriment of the public good (Lessig 2004; Vaidhyanathan 2001). This perspective is codified in FOSS copyright licenses, often called "copyleft" because they are firmly rooted in intellectual property law but are antithetical to the economic model of exclusionary distribution.

The FOSS copyright licenses and underlying ideology have provided a framework for organizing production based on the volunteer efforts of many people. For software projects, volunteers from around the world help write the code, find and report bugs, and suggest features (Karels 2003; Mockus, et al. 2002). Contributors give a variety of reasons for participating: it's fun, it's social, one learns good coding techniques, it has social significance, code should be open, and a problem needed a solution. None of the reasons appear to dominate (Ye and Kishida 2003). Projects are organized through online forums where people can communicate and debate the technical merits of alternative architectures. Although there are some very large projects, most don't gain more than one or two developers and most aren't under active development for long. Wikipedia has similar communication forums and nearly half a million volunteers have contributed to its articles.

While one might have difficulty imagining how these communities produce working software or an accurate encyclopedia with such an open production structure, many of these volunteer projects are competing successfully with commercial alternatives. FOSS advocates argue that the success is due to the ease with which people can contribute to these projects. Coding isn't necessary; even contributing a bug report helps the effort. With low barriers to participation, many more people can contribute to a single project (Anderson 2004; Graham 2005). Moreover, there is less overhead in finding the communities that one would like to participate in (Anderson 2004). The contributions are said to be of high quality because people can choose to work on projects they enjoy (Graham 2005) or need (Raymond 1998b). And finally, because many people look at the content, someone will likely notice an error and either fix it or alert others—"Given enough eyeballs, all bugs are shallow" (Raymond 1998b).

Perhaps counterintuitively, these communities are usually highly organized, with explicit governmental structures set up for each project. For example, the Linux operating system is a "benevolent dictatorship" with Linus Torvalds deciding which code changes are incorporated into the Linux codebase. Since the Linux codebase is millions of lines long, Torvalds delegates responsibility to "trusted lieutenants" who control sections of the source code after earning trust through good contributions.

Although projects may be autocratically controlled, the copyleft licenses guarantee that the content remains free. Ownership is more like control of a brand (e.g., Wikipedia). Anyone can take the content and start another project independent of the original. For instance, Wikipedia articles can be freely used elsewhere, as long as that project is copyrighted under the same license and acknowledgment given. This is called the "right to fork" (see Raymond 1998a).

Controversies have erupted over these projects as they have grown in popularity. At the root of these conflicts is the relationship between expertise and quality. The traditional mode of quality content production is through certified experts—programmers, academics, journalists—working behind closed doors (e.g., corporations, universities, the syndicated press). It is often assumed that amateur volunteers cannot meet the same standards as certified experts. Unfortunately, measuring quality differences is extremely difficult—definitions of quality are elusive, fair comparisons are lacking, and the studies are too easily politicized—but the data that do exist should lead us to question our assumptions. On the matter of software, Linux and Firefox are widely used, suggesting consumer satisfaction. As for the encyclopedic content of Wikipedia, a recent study by *Nature*, a prominent scientific journal, produced results that elicited quite a stir. *Nature* gathered a group of experts to blindly evaluate the accuracy of articles in Wikipedia and *Encyclopedia Britannica* and found that Wikipedia had only a marginally larger number of inaccuracies (Giles 2005). Although the study had some major flaws, as noted in the fiery Britannica rebuttal (2006), the study validated the claim that Wikipedia at least approaches encyclopedia quality.

The transparency of production in these open systems is a cause of concern for many people. This is especially true for Wikipedia. As a major institution in defining truth, encyclopedias are held to high cultural expectations of accuracy, objectivity, and neutrality; as one journalist pointed out, if Wikipedia had been called a "Big Bag O'Trivia," the online encyclopedia wouldn't be causing such a stir (Orlowski 2005). The transparency of Wikipedia—where the discussion pages for each article reveal the conflicts and "edit wars" taking place over entries—is said to signify its untrustworthiness, especially for more controversial topics, where it is not clear that Wikipedia articles achieve encyclopedic standards (Scott 2006). Moreover, the open process may be used as a political platform hidden in the guise of truth. In one instance, a political controversy erupted when US senators' aides (from both major parties) were discovered to be editing the Wikipedia biographies of members of Congress (Davis 2006). Wikipedia is undoubtedly a great resource, but we should be aware of how it is produced.

The issue of accountability in FOSS is another major source of dispute. Although transparency might lead to integrity in process, accountability is traditionally based on holding parties responsible for problems through certification and ownership. If a company's product fails in a critical task, a doctor makes a life-threatening error, or a newspaper misreports, legal recourse is possible. However, if an open source program crashes or a Wikipedia article is incorrect, who can be blamed and called upon to fix the problem? This issue became abundantly clear with a recent wrangle involving John Seigenthaler, a former aide to John F. Kennedy, who had libelous information about him included in his "false" Wikipedia biography (Seigenthaler 2005). While some argued that, since he had the ability to change the material himself, responsibility shifted to Seigenthaler, it is ridiculous to claim that Siegenthaler should be constantly vigilant in policing all open content.

FOSS advocates believe that these new models for organizing high-quality content creation, ownership, and distribution are technically, morally, and economically superior to mass media methods of supporting production and innovation. Such challenges to the traditional relationship between experts and amateurs, centralized control and distribution of content, and the transparency of production have led to significant quarrels over the legal, economic, and cultural establishments that evolved together throughout the mass media era. Only time will tell whether recent FOSS-related developments mark a critical transition in our society or will simply serve to bolster existing social institutions.

Will the Internet Transform Politics?

The rhetoric about the transformation of our society facilitated by digital technologies carries into the political world where many academics, businesspeople, and analysts suggest that the Internet is ushering citizens worldwide into a new political era and changing the nature of civic life. Some claim the Internet is "transforming" American politics (Nagourney 2006). Others suggest the blogosphere, the entire community of blogs, amounts to nothing less than "the reinvention of the public square" (Bai 2006). Still others contend that the Internet has fundamentally altered the character of political organizing (Boyd 2003).

We need to evaluate carefully such sweeping claims. When we talk about the impact the Internet is having on democracy, are we referring to a kind of public discussion or deliberation, or are we talking about the process of decision-making on policy matters? Is the Internet altering traditional politics or making a new kind of politics? To explore these issues, we look first at the Internet's mobilization role in American politics in terms of fund-raising and building new social movements. We next consider how Internet technologies are being used to connect politicians and citizens on a more routine basis. And finally, we consider the prospects of the Internet to enrich civic life through the promotion of sustained debate and discussion.

As an organizing and a fund-raising tool, the Internet has been touted as revolutionary to politics, especially in the United States. The case in point is Howard Dean's 2004 presidential campaign. Although John Kerry was the original Democratic front-runner, Dean took an early lead, raising large sums of money quickly and in small amounts from large numbers of supporters. The Dean team used an aggressive Internet-based initiative involving e-mails, blogs, and Meetup.org, a Web site for bringing supporters together in local areas across the country, to mobilize volunteers and raise money. Dean may have been hurt, however, by his lack of a well-developed grassroots structure on the ground in places like Iowa. Without such organization, Dean could not withstand the scrutiny he faced as the Democratic front-runner, and reservations about his electability mounted among party elites. Thus, while the Internet serves to aid unknowns with popular ideas and help organize fund-raising efforts, it has yet to significantly alter the anatomy of American political campaigns or render traditional means of winning an election obsolete.

Another area where activists and analysts point to the transformative potential of the Internet is in building and sustaining social movements. According to some, the Internet "has the capacity to alter the structure of opportunities for communication and information in civic society" (Norris 2001, 191) and thus facilitate the building of alternative movements. In February of 2003, the Internet was used to organize, quickly and efficiently, hundreds of peace demonstrations across the globe (Boyd 2003). In New York alone, some 400,000 people came together. A month later, the US organization MoveOn.org, a group with only five paid staff people at the time, laid the groundwork for protests at 6,000 sites in 130 countries with more than a million people in total (Boyd 2003, 14). While the Internet has been successful at rapidly mobilizing concerned citizens, it is uncertain whether these efforts will lead to citizens remaining engaged over long periods of time (Kleinman 2005, 46). Moreover, while Internet technologies offer opportunities for keeping people in touch and bringing together people with similar concerns, whether the Internet will significantly change how citizens do the hard work of building durable organizations is still an open question.

Some argue that Internet technologies can connect politicians and citizens by enabling governments and politicians to post a wide array of information about programs and policies. These same proponents contend that the use of the Internet in this manner can increase the transparency of government action (Perez 2004, 88; Norris 2001; Warschauer 2003), and insofar as information is power, easy access to government and political party materials can enhance public debate and even alter outcomes (Froomkin 2004). Other research suggests that, while government Web sites increase transparency, they may not promote greater citizen-government interaction (Norris 2001, 122).

The Web sites of political parties offer a similar picture. A worldwide study of political party Web sites does not suggest widespread opportunity for "bottom-up" influence on party positioning (Norris 2001, 150) or the leveling of the playing field for minor political parties; however, simply making more information available more easily may increase political party responsiveness (Norris 2001, 169). Moreover, even if the Internet increases information availability, that doesn't mean it will be used; the "political culture must want it," and that is by no means certain (Agre 2002).

Unfortunately, information technologies also permit new techniques of surveillance and propaganda (Agre 2002). For example, some political consulting firms offer politicians detailed information about what to say and how to say it by analyzing massive amounts of stored demographic and policy preference data about millions of Americans. For example, one firm created a "message tester." After distributing free WebTVs, with the only stipulation being that the viewers respond to periodic surveys, this firm was able to measure reactions to the "subtle variations" in political messages on subpopulations by broadcasting ads to these devices before the message was publicly aired (Howard 2003).

The most talked-about Internet tool for connecting politicians and citizens is "blogs," self-published journal-like Web sites. These Weblogs allow politicians to outline their positions in detail and to respond to critics in short order. Furthermore, they permit regular and substantive interaction between candidates and citizens. Analyst David Perlmutter describes the blogs of most US politicians (if they have them) as thin, although some, like 2008 presidential hopeful John Edwards, are prolific bloggers, providing detailed discussions and analysis of contemporary issues. Perlmutter, however, suggests this may be because Edwards has "no place else to go: no longer in the Senate himself, he is not as much in the public eye as many of his likely opponents" (Perlmutter 2006). But 2008 presidential candidate Hillary Clinton's blog comments are "safely worded and read more like press releases" than real interventions into ongoing political discussion (Perlmutter 2006). Participation in the blogosphere takes time and relatively spontaneous writing involves risk. For established politicians, time may be better spent giving a speech garnering $2,000 per dinner plate than blogging. In the former case, the audience can be controlled, public statements can be carefully crafted, and the money is assured. But if more people come to expect dynamic blogging from political leaders, social pressure may force a greater number of politicians to participate in the blogosphere to compete with early adaptors.

Blogs, however, are designed for any interested Internet user, not just politicians. The essential characteristics of a blog are a series of posted messages that anyone can comment on and a list of blogs that the author reads; the interlinked community of blogs is called the "blogosphere." Blogging has recently grown in popularity: by late 2004, 7% of US Internet users had created a blog, 12% had posted comments on others' blogs, and 27% had read blogs (Rainie 2005b). A substantial number of bloggers have turned out to be aspiring commentators and reporters desiring to contribute to ongoing political discussion. Some political blogs have risen above others and hold mass readership. For example, DailyKos, a politically liberal blog, was visited by nearly 5 million different readers in one month in the summer of 2005, and is now attracting some 600,000 visitors daily (Bai 2006; Perlmutter 2006).

Bloggers have played a role in shaping the political landscape. For example, political parties, recognizing the increasing importance of bloggers, certified a host of bloggers to cover their nominating conventions, even setting up Web sites to aggregate the blog posts (Adamic and Glance 2005). Although a minority of Americans read blogs (about 9% of Internet users said they sometimes read political blogs during the 2004 presidential campaign [Rainie 2005b]), the blogosphere's political influence extends to the general political landscape through their interaction with mainstream media (Farrell and Drezner 2004). Because of bloggers' ability to identify and frame breaking news, many mainstream media sources keep a close eye on the best-known political blogs (Adamic and Glance 2005, 2). This position of bloggers as political watchdogs has led to some of the more prominent stories in the recent past. For example, conservative blog writers were among the first to challenge the memo about President Bush's National Guard service first reported on CBS, while liberal bloggers brought attention to Trent Lott's statements about Strom Thurmond and framed them as an endorsement of racial segregation (Lessig 2004).

Some community blogs are more participatory than others. Many see these venues as having the greatest potential for promoting serious, sustained, and thorough political discussion. A prominent example is Slashdot.org, which focuses on technological news and attracts about a third of a million people a day (Lampe and Resnick 2004). Anyone can suggest a topic for discussion. The editors of the site post a subset of these suggestions to which any reader can respond. Slashdot relies on a democratic process for organizing and managing online debate. Slashdot readers use "moderation points" to affect the rating and categorization of comments as they follow the debate. This rating system is used by other readers to filter out low-rated comments. Moderation points are kept scarce so that users only use them when they feel strongly about a comment. Slashdot creates incentives for people to continually participate by allowing users to gain or lose "karma" points based on their comments and moderation of comments. Despite some problems, there is evidence that ratings generally follow community consensus (Lampe and Resnick 2004, 3, 4, 6). Sophisticated democratic mechanisms for moderating discussion may improve debate. But this kind of technology is not intended as a decision tool, and it is not entirely clear how citizens could move from discussion to decision were a more direct form of democratic decision-making the objective.

Other challenges remain before it can be claimed that the blogosphere has radically democratized politics and civic life. First, only a few blogs exercise political clout through wide readership. For instance, a survey asking journalists to list the top three blogs they read revealed that 10 blogs represented over half of the citations (Farrell and Drezner 2004), although there is evidence of rapid turnover in the authoritative blogs (Marlow 2004). It is also unclear whether blogs encourage readers to "view life from the perspective of others" (Froomkin 2004, 11) or whether these "cybercafes" are merely echo chambers for the like of mind, leading to "cyberbalkanization" (Putnam 2000; Sunstein 2001). On the one hand, a recent survey shows that people who go online tend to be exposed to a greater number of arguments on issues and politicians' stances than those who do not (Horrigan, et al. 2004). On the other hand, the way people hear about alternative arguments matters: another study measured the degree of interaction between liberal and conservative blogs in the months preceding the 2004 presidential election and found that there was minimal overlap in the types of news stories discussed, and relatively few links going from liberal to conservative blogs (and vice versa) (Adamic and Glance 2005). Finally, like other Internet technologies, blogs seem to energize those already engaged, and it is doubtful they can reach those for whom politics holds little obvious interest (Nagourney 2006, A17).

Technologies tend to amplify the existing balance of social forces because the powerful elements of society are able to exercise asymmetrical force over the use and adoption of technology through established institutions, including the media, school systems, and the courts (Agre 2002). The Internet is only one part of a larger political ecology, and any transformative role that the Internet takes will be through people who occupy powerful positions in professions, the mass media, legislatures, and interest groups. While there are a host of interesting developments on the political front made possible by the Internet, whether politics will be "transformed" is a different question. To date, there seems to be some evidence that these technologies have energized the public sphere and made some marginal contributions to shaping public debate and participation. But the constraints on people using the Internet to transform politics are significant, and whether people will use the Internet to bring us to a significantly different political environment remains to be seen.

Internet Companies and the Search

Arthur C. Clarke once stated that "any science or technology which is sufficiently advanced is indistinguishable from magic" (1973). When we enter a complicated query into a search engine and are served a ranked ordering of the most "relevant" pages milliseconds later, search engines certainly appear magical. Our dependence on search engines has grown quickly—by mid-2005, over 60 million Americans were using search engines daily, with Google's computers alone answering around 5 billion searches a month from 90 million unique visitors (Rainie 2005a). At the same time, brand names like Yahoo! and Google have taken on cultural significance beyond their technical features, as reflected in the popular neologism "to Google," a verb meaning the act of seeking and finding information online. The search bar is actually a rather profound mark of our time—through this single portal, we can find the most "relevant" authorities for any question or topic.

Since finding credible online information is a central problem of the information age, the technical means by which quality and relevancy are established carry political implications (Hargittai 2000; Introna and Nissenbaum 2000). Internet companies are taking on more power in our society because they define the algorithms which locate and rank answers to any question; have become primary gatekeepers for a wealth of information; and aggregate data about people's use of their popular services (e.g., search, e-mail, calendar). In this section, we look at each in turn.

There are three basic elements to a search engine—crawling, indexing, and searching. Before a search engine can answer a user's query, it must have a snapshot of what exists on the Web. A search company uses a computer program called a "spider" (or "crawler") to discover the structure of the Web. The spider is given an initial set of Web sites to visit. When it visits a Web site, the spider stores a copy of the whole site, makes a note of all the Web site's hyperlinks, and then follows one of them. Search companies have millions of spiders continually crawling the web and storing new and updated Web sites. The second part of a search engine is "indexing." Indexing determines how the Web site should be stored so that it can be used later in response to search queries.

With the crawlers continually indexing Web content, the search engine is in a position to answer search queries through the use of a search algorithm, an automated method for finding a ranked ordering of the most relevant, "authoritative" Web sites for a given search query. Internet companies incorporate hundreds of factors in these algorithms. For example, the number of times a Web site mentions a term in the search query is an important metric. A less obvious but important idea is "link analysis"—because Web sites hyperlink to one another, these links give information about the Web sites themselves (Brin and Page 1998; Kleinberg and Lawrence 2001; Kleinberg 1999). Imagine the Web as a social network—the "popularity" of a Web site is proportional to the number and popularity of the sites that hyperlink to it. This popularity is then used in determining if a site is authoritative.

The technical process raises questions about power. Unless a Web page is crawled and indexed, it essentially doesn't exist, because searching is one of the only methods people use to find content (Introna and Nissenbaum 2000). The method by which spiders discover new sites thus determines whether that site will be discoverable by anyone who doesn't know the URL beforehand. Moreover, for a site to be accurately portrayed in search results, Web site designers need to follow style and content guidelines lest the algorithm used to index sites misunderstand their site.

Representation in a search engine's index, however, is not enough—ranking matters. Most people do not look past the first page of search results (Huberman, et al. 1998; Joachims, et al. 2005). One example of the political implications of this ranking system is "Google bombing." Google bombing occurs when many people link to a Web site from their own Web site, surrounding the link with a particular phrase. If the Google bombing is successful, the link analysis used by search engines causes the targeted Web site to be ranked highly in response to a search for that text. Examples include "miserable failure" returning George W. Bush's White House biography (Jimmy Carter's is No. 2) and John Kerry's Web site for "waffles" (McNichol 2004; Memmott 2004). Likewise, a back and forth struggle for the top hit for "Jew" has been occurring between Wikipedia and the anti-Semitic site JewWatch (Tatum 2005). A second example of the social ramifications of ranking is that Google returns Wikipedia as the number one search result for most proper nouns, a sort of algorithmically encoded allegiance to FOSS.

Ranking also has media implications. Many newspaper, magazine, and TV outlets, who gain over 30% of their readers online, are submitting to the mechanical examination of the spiders by removing puns and cultural references in order to be as literal as possible and thus garner higher relevancy rankings (Lohr 2006). The technical aspects of search engines can thus systematically lead to social exclusion, provide new venues of political representation, and reshape cultural style and content.

It is not just the technical aspects of search engines from which political implications arise. Internet companies also take on the authoritative role of gatekeeper by mediating between creators and consumers of cultural products (Hargittai 2000). Perhaps the ultimate political power of a gatekeeper is to filter, to make it appear that content doesn't exist when it does. Substantial public concern over search filtering wasn't raised until Google concluded a recent deal with China to offer services there. Critics worry that Google's grab for the Chinese market will stifle free speech and aid in China's efforts at censorship. However, it is doubtful that free speech in China will be affected by Google's presence because China has for some time employed a hardware infrastructure to filter material it deems objectionable (Open Net Initiative 2005; Thompson 2006). Google argues that its involvement in China will have a positive effect, since, unlike other Chinese search engines, Google alerts users when results have been filtered, thus making users aware of the absence of information. Unfortunately, this alert is unconvincingly in the fine print.

Internet companies also take on the role of data aggregator. Digital media have improved to the point of being capable of storing large amounts of searchable data cheaply, efficiently, precisely, and automatically. People are generating more and more digital traces as technologies move further into our lives (e.g., credit card companies record purchases, Internet service providers record data about traffic). Internet companies are a prime example, potentially recording all behavior, content, and communication that are mediated through the e-mail, search, social networking, calendar, and blog services they offer. In some sense, a "database of intentions" is being amassed as people pour their hopes, dreams, and desires into search portals (Battelle 2005). In this environment, data mining, a set of algorithmic techniques for discovering patterns in large amounts of data (Frawley, et al. 1992), can then be used to provide targeted advertising based on personal behavior stored in the databases. Governments have been eager to obtain this information, sometimes casting a wide net, aiming to use data-mining techniques to uncover suspicious activities or find highly specific information about particular individuals. The US Justice Department's 2005 subpoena of Google, for example, demanded a week's worth of searches and a random sample of a million of the Web sites Google has indexed (*Gonzales v. Google Inc.* 2006; Richtel and Hafner 2006; Claburn 2006). The Department of Justice wished to use this information to show that harmful content was still easily accessible by minors.

Charging Google with bias against unpopular sites and the creation of a "Googlearchy" (Hindman, et al. 2003) might misrepresent its place as a general search engine (Brooks 2004). But there are dangers when there is such a strong asymmetry between understanding a tool and the trust that we have in the institutions that produce it. The algorithms underlying search engines matter; algorithms can systematically exclude certain classes of content and bias attention toward the already link-rich, while also providing new methods of political expression. Aside from the power of defining such algorithms, Internet companies are also taking on authoritative roles as gatekeepers and data aggregators, roles which have been important in a host of controversies ranging from filtering of search results, to imprisonment of dissidents, to use in federal investigations.

Attempting to Bridge the Digital Divide

In this section, we explore the so-called "digital divide," the gaps in access between those who have and those who lack computers and Internet access. We consider the meaning and consequences of this high-technology inequality.

Globally, the digital divide is huge. There are approximately 459 computers per 1,000 people in the United States as compared to seven per 1,000 in sub-Saharan Africa, where 98% of students go through school without even seeing a computer (The Codebreakers 2006; Norris 2001, 50). In terms of the Internet, about 84% of people online live in highly industrialized countries; moreover, over 50% of people in the United States use the World Wide Web, but only 0.1% of Nigerians do (Norris 2001, 15, 45).

The digital divide does not just refer to a rift between developing and industrialized countries, but also to divisions within industrialized countries. In the United States, the term "digital divide" was popularized in the mid-1990s by the Clinton administration. Worried that inequality of access to computers and the Internet would leave many Americans without the benefits of the information economy, President Clinton advocated a series of programs to make access to crucial information technologies universal in the United States. The administration's concern was highlighted by a set of US Commerce Department reports, which found great disparities in ICT access and use across class, gender, race, and rural and urban households. With the advent of high-speed Internet, new divides were created between those who had high-speed access and those who did not.[2]

Many believe that if we do not close the digital divide, the information "have nots" will be left behind in the wave of prosperity that the new knowledge economy will bring. Nicholas Negroponte has proposed one solution. "We really believe we can really make literally hundreds of millions of these machines around the world," asserts Negroponte (quoted in Twist 2005). Negroponte, a prominent technologist from MIT, speaks about the $100 laptops that his research group has designed and worked with industry to build. Negroponte claims that these laptops, which run on the Linux operat-

---

[2]Recent efforts to track the digital divide in America suggest that gender gaps in Internet access have mostly disappeared (Hargittai 2003), while racial gaps have narrowed (Fox 2005; Horrigan 2006). The divide persists, however, with regard to age, rural v. urban, economic status, and education (Fox 2005; Horrigan 2006; Madden 2006).

ing system, have low power consumption, and are human-powered by crank, will solve significant problems for both the global digital divide between developing and industrialized countries and for educating people in the remotest parts of the world. Organized in mid-2005 through the nonprofit association One Laptop Per Child (OLPC), this effort has received support from the United Nations. As of early 2006, OLPC was moving to deploy its machines to China, India, Brazil, Argentina, Egypt, Nigeria, and Thailand, where each government committed to buying 1 million machines.

Yet, as innovative as recent initiatives are, these programs typically fail to consider that the digital divide reflects historically rooted systems of stratification and inequality. In the United States, in 2001, average household income was slightly higher than $58,208, but the 20% of the population with the lowest income made between zero and just under $18,000 (OMB Watch 2003). Across the globe, 6 billion people live on less than $2 per day (Collins and Yeskel 2000, 61). Under these circumstances, improving access to ICTs may create new opportunities for previously excluded social groups, but such access is not likely, by itself, to bring social advantage.

Access is merely a first step to "digital equality" (DiMaggio and Hargittai 2001). As digital divide expert Mark Warschauer argues (see his chapter in this volume), without attention to the human and social systems in which we live, simply offering the disenfranchised entrée to the information society will do little good (2003, 6). Being a full-fledged member of the "information society" requires not just access to technology, but an array of skills for effectively using them. Basic literacy is clearly a prerequisite for effectively bridging the digital divide. Without "the broader skills of composition, research, [and] analysis," access to ICTs is meaningless (Warschauer 2003, 112; see also 144). Thus, before improving access to computers and the Internet, our society may be better served by increasing access to high-quality basic primary and secondary education (Kleinman 2005, 38-42).

Of course, even with the skills that one can acquire with basic education, access to technology will not allow most people to reap great benefits from the information age. Consider the many different skills that factor into the topics we have discussed so far. First, there is a wide variation in expertise for finding, analyzing, and critically interpreting information using search technologies and other Web tools (Hargittai 2002). Second, despite claims of the transformative effects on public deliberation, one study found that nearly

40% of study participants were unable to find a Web site that compared the viewpoints of presidential candidates on abortion (Hargittai 2002). Other studies have indicated that only 38% of Internet users know what a blog is (Rainie 2005b). This variation in skill and knowledge may point to a "democratic divide" between those who can use information technology to "engage, mobilize, and participate in public life" and those who cannot (Norris 2001, 4, 13). Third, the ability to create Web sites or participate in online communities (e.g., the blogosphere or Wikipedia) also varies widely. Even in the United States, only 14% of Internet users have created Web sites, 8% have started blogs, and 26% have shared content in some way or another (Horrigan 2006). If the Internet is to "democratize" society, more than just a fraction of citizens will need to have the literacy to give voice to their opinions and debate with others in these digital mediums.

Outside the highly industrialized countries, such knowledge itself (along with computers and Internet access) would make little difference, since, as Warschauer shows, the vast majority of digital content does not currently meet the needs of the diverse peoples around the globe (2003, 80). In 2000, more than two-thirds of all international Web sites were in English (Warschauer 2003, 96), and most of the world's citizens are not native English speakers. Locally useful information on education, family, finance, health, and housing is sorely lacking for many potential Internet users (Warschauer 2003, 88).

There have been a number of interesting efforts to address local needs. One such project is the Digital Study Hall (Wang, et al. 2005). This program is specifically tailored to address the primary education gap between middle-class urban Indian schools and rural Indian schools through a low-cost content sharing network based on existing TV and postal service infrastructures. Lectures recorded by high-quality teachers in urban schools are made available to teachers in rural India who can use the lectures as a platform for teaching the subject matter. Through involvement in the program, rural children may be able to receive higher quality primary education, rural teachers may be able to develop greater mastery of their subject material, and ties between rural and urban villages may be established.

Such innovations are important steps in a positive direction; however, if digital technologies are to enhance local economies and not amount to technological colonialism, people on the "wrong side" of the divide must have a measure of control over the technology. In this context, while often the focus is placed on the provision of ICTs, maintenance is often overlooked. Dependence on large, Western IT companies for both hardware and software will likely simply reinforce inequality. A strategy for invigorating local autonomy may be through the use of FOSS (The Codebreakers 2006). While it isn't clear whether FOSS is less expensive to use than proprietary software, the cost is shifted to service rather than software. Through the training of local software experts, money can be kept inside the local economy while a software infrastructure capable of being adapted to local needs is developed (The Codebreakers 2006). The costs of learning to code in order to fix and adapt software are significant, however; regions that take this route must build a knowledge infrastructure and will likely need help in doing so (Benner 2006).

Increasing access to ICTs, by itself, will not bridge the digital divide. Programs for deploying digital technologies need to be part of a broader strategy that promotes the development of economic infrastructure, digital literacy, and social capital. This is not to disparage programs like OLPC; we need to be cautious, however, about the homogeneity and scale of the massive technological provisioning projects being touted as solutions to the digital divide, and pay attention to projects that are highly tailored to local conditions and are attentive to matters of sustainability and autonomy.

Conclusion

Information scientist Philip Agre contends that "Technologies often come wrapped in stories about politics. These stories may not explain the motives of the technologists, but they do often explain the social energy that propels the technology into the larger world" (2003, 39). The stories we've described in this chapter are political—FOSS as the foundation for a movement espousing new conceptualizations of creativity and property; the Internet as a platform upon which radical transformations to the political sphere may take place; the search engine as the answer to our information overload; the potential of cheap ICTs to bridge inequalities. These are tales of losers and beneficiaries, of hype and reality, and of technical and social remedies. They are about choices. And the choice of which particular path to take reflects value judgments. Decisions about social preferences are fundamentally political.

The revolution in information technology in which we find ourselves offers much promise. But most often these opportunities are obscured by exaggerated claims that focus more on the technologies themselves than on the social circumstances in which they are embedded. Compounding this, it seems that every technological innovation is hailed as the next revolution. Often, these political stories take on a technologically reductionist tone that prescribes technologies to cure social woes. Our belief in the virtue of the "cutting edge" has become even more pronounced in the digital age, which has left little room for discussion, evaluation, and criticism. When we hear these political stories, we must ask ourselves if the adoption of new technology will improve our lives, rather than accepting new technologies without question. If we see the potential for progress, there are other important questions to answer regarding how the adoption will take place, who will control it, and who will benefit from it.

In this chapter, we have highlighted some important topics surrounding new information technologies. We have considered some of the complicated matters of political and social power that are shaping the uses and development of new technologies, as well as attendant inequalities. But these brief descriptions have only been deep enough to promote further discussion and exploration with family, coworkers, and others. The remaining chapters in this section bring a wider set of considerations to the many possibilities and problems raised in our increasingly information intensive world.

References

Adamic, L. A., and N. Glance. 2005. *The political blogosphere and the 2004 U.S. election: Divided they blog*. Chicago, IL: ACM Press.

Agre, P. E. 2002. Real-time politics: The Internet and the political process. *The Information Society* 18:311.

Agre, P. E. 2003. P2p and the promise of Internet equality. *Communications of the ACM* 46:39–43.

Anderson, C. 2004. The long tail. *Wired Magazine* 12:170–177.

Bai, M. 2006. Can bloggers get real? *New York Times Magazine,* May 28.

Battelle, J. 2005. *The search: How Google and its rivals rewrote the rules of business and transformed our culture.* New York: Portfolio.

Benkler, Y. 2002. Coase's penguin, or, Linux and "the nature of the firm." *The Yale Law Journal*, 112:369–446.

Benner, C. 2006. Beyond the Open Source hype. http://www.foreignpolicy.com/story/cms.php?story_id=3471 (accessed May 26, 2006).

Boyd, A. 2003. The web rewires the movement. *The Nation* 277:13.

Brin, S., and L. Page. 1998. The anatomy of a large-scale hypertextual web search engine. *Computer Networks & ISDN Systems* 30:107.

Brooks, T. A. 2004. The nature of meaning in the age of Google. *Information Research-An International Electronic Journal* 9. http://informationr.net/ir/9-3/paper180.html.

Campbell-Kelly, M. 2003. *From airline reservations to Sonic the Hedgehog: A history of the software industry.* Cambridge, MA: MIT Press.

Claburn, T. 2006. Justice department subpoenas reach far beyond Google. *InformationWeek,* March 29.

Clarke, A. C. 1973. *Profiles of the future; an inquiry into the limits of the possible,* rev. ed. New York: Harper & Row.

Collins, C. and F. Yeskel. 2000. *Economic apartheid in America: A primer on economic inequality and insecurity*. New York: The New Press.

Davis, M. 2006. Congress 'made Wikipedia edits.' *BBC News,* February 9.

DiMaggio, P., and E. Hargittai. 2001. From the 'digital divide' to 'digital inequality': Studying Internet use as penetration increases. Princeton: Center for Arts and Cultural Policy Studies, Woodrow Wilson School, Princeton University.

Encyclopedia Britannica. 2006. Fatally flawed: Refuting the recent study on encyclopedic accuracy by the journal *Nature*. http://corporate.britannica.com/britannica_nature_response.pdf.

Farrell, H., and D. Drezner. 2004. The power and politics of blogs. Paper presented at the Annual Meeting of the American Political Science Association, September 2-5 in Chicago, IL.

Fox, S. 2005. Digital Divisions. Washington, DC: Pew Internet and American Life Project. http://www.pewinternet.org/pdfs/PIP_Digital_Divisions_Oct_5_2005.pdf.

Frawley, W. J. et al. 1992. Knowledge discovery in databases: An overview. *AI Magazine* 13:57–70.

Froomkin, M. 2004. Technologies for democracy. In *Democracy online: The prospects for political renewal through the internet,* ed. P. M. Shane, 3–20. New York: Routledge.

Giles, J. 2005. Internet encyclopedias go head to head. *Nature* 438:900–901.

Gonzales v. Google, Inc. 2006. CV 06-8006MISC JW (N.D. Cal. 2006). Ruling available online at http://www.google.com/press/images/ruling_20060317.pdf.

Graham, P. 2005. What business can learn from open source. http://www.paulgraham.com/opensource.html.

Hargittai, E. 2000. Open portals or closed gates? Channeling content on the world wide web. *Poetics* 27:233–253.

Hargittai, E. 2002. Second-level digital divide: Differences in people's online skills. *First Monday* 7(4). http://www.firstmonday.org/issues/issue7_4/hargittai/index.html.

Hargittai, E. 2003. The digital divide and what to do about it. In *New economy handbook*, ed. D. Jones, 822–841. San Diego, CA: Academic Press.

Hindman, M. et al. 2003. Googlearchy: How a few heavily-linked sites dominate politics on the web. Paper presented at the Annual Meeting of the Midwest Political Science Association, Chicago, IL, April 3–6.

Horrigan, J. 2006. Home broadband adoption 2006. Washington, DC: Pew Internet & American Life Project. http://www.pewinternet.org/pdfs/PIP_Broadband_trends2006.pdf.

Horrigan, J. et al. 2004. The internet and democratic debate. Washington, DC: Pew Internet & American Life Project. http://www.pewinternet.org/pdfs/PIP_Political_Info_Report.pdf.

Howard, P. 2003. Digitizing the social contract: Producing American political culture in the age of new media. *The Communication Review* 6:213–245.

Huberman, B. A. et al. 1998. Strong regularities in world wide web surfing. *Science* 280:95.

Introna, L., and H. Nissenbaum. 2000. Shaping the web: Why the politics of search engines matters. *The Information Society* 16:169–185.

Joachims, T. et al. 2005. *Accurately interpreting clickthrough data as implicit feedback.* Salvador, Brazil: ACM Press.

Karels, M. J. 2003. Commercializing open source software. *ACM Queue* 1(5): 46–55.

Kelly, K. 2006. Scan this book! *New York Times Magazine*, May 14.

Kleinberg, J., and S. Lawrence. 2001. The structure of the web. *Science* 294:1849.

Kleinberg, J. M. 1999. Authoritative sources in a hyperlinked environment. *Association for Computing Machinery* 46(5):604.

Kleinman, D. L. 2005. *Science and technology in society: From biotechnology to the Internet.* Malden, MA: Blackwell.

Lampe, C., and P. Resnick. 2004. Slash(Dot) and burn: Distributed moderation in a large online conversation space. In *Proceedings of the ACM CHI Conference on Human Factors in Computing Systems,* Vienna, Austria, April 24–29, ed. E. Dykstra-Erickson and M. Tschegli, 543–550. New York: ACM Press.

Lessig, L. 2004. *Free culture: How big media uses technology and the law to lock down culture and control creativity*. New York: Penguin Press.

Lohr, S. 2006. This boring headline is written for Google. *New York Times*, April 9.

Madden, M. 2006. Internet penetration and impact. Washington, DC: Pew Internet and American Life Project. http://www.pewinternet.org/pdfs/PIP_Internet_Impact.pdf.

Marlow, C. 2004. Audience, structure, and authority in the weblog community. Paper presented at the International Communication Association Conference, New Orleans, LA., May 27–31.

McNichol, T. O. M. 2004. Your message here. *New York Times*, Jan. 22.

Memmott, M. 2004. Kerry gets served up with 'waffles'; web pranks twist results of searches. *USA Today*, April 12.

Mockus, A. et al. 2002. Two case studies of open source software development: Apache and Mozilla. *ACM Transactions on Software Engineering and Methodology*, 11(3): 309–346.

Moglen, E. 2003. The dotcommunist manifesto. http://www.emoglen.law.columbia.edu/publications/dcm.html.

Nagourney, A. 2006. Politics is facing sweeping changes via the internet. *New York Times*, April 2.

Norris, P. 2001. *Digital divide: Civic engagement, information poverty, and the Internet worldwide*. Cambridge and New York: Cambridge University Press.

OMB Watch. 2003. Income watch: The rich are getting richer and getting bigger tax breaks. *OMB Watch* 4(13). http://www.ombwatch.org/article/articleview/1615/1/178/?TopicID=2.

Open Net Initiative. 2005. Internet Filtering in China in 2004–2005: A Country Study. http://www.opennetinitiative.net/studies/china/.

O'Reilly, T. 2005. What is Web 2.0: Design patterns and business models for the next generation of software. O'Reilly Media, Inc. http://www.oreillynet.com/pub/a/oreilly/tim/news/2005/09/30/what-is-web-20.html.

Orlowski, A. 2005. There's no Wikipedia entry for "moral responsibility." *The Register*, London, England: Dec. 12.

Perens, B. 1999. It's time to talk about free software again. http://lists.debian.org/debian-devel/1999/02/msg01641.html.

Perez, O. 2004. Global governance and electronic democracy: E-Politics as a multidimensional experience. In *Democracy online: The prospects for political renewal through the Internet,* ed. P.M. Shane, 83–94. New York: Routledge.

Perlmutter, D. D. 2006. Political blogs: The new Iowa? *Chronicle of Higher Education* 52:B.6.

Putnam, R. D. 2000. *Bowling alone: The collapse and revival of American community.* New York: Simon & Schuster.

Rainie, L. 2005a. Search engine use shoots up in the past year and edges towards email as the primary Internet application. Washington, DC: Pew Internet and American Life Project. http://www.pewinternet.org/PPF/r/167/report_display.asp.

Rainie, L. 2005b. The state of blogging. Washington, DC: Pew Internet and American Life Project. http://www.pewinternet.org/pdfs/PIP_blogging_data.pdf.

Raymond, E. S. 1998a. Homesteading the Noosphere. *First Monday* 3(10). http://www.firstmonday.org/issues/issue3_10/raymond/index.html.

Raymond, E. S. 1998b. The cathedral and the bazaar. *First Monday* 3(3). http://www.firstmonday.org/issues/issue3_3/raymond/index.html.

Raymond, E. S. 1999. Shut up and show them the code. *Linux Today* 15:51.

Richtel, M., and K. Hafner. 2006. Google resists U.S. subpoena of search data. *New York Times*, Jan. 20.

Scott, J. 2006. The great failure of Wikipedia. http://www.archive.org/details/20060408-jscott-wikipedia (accessed May 10, 2006).

Seigenthaler, J. 2005. A false Wikipedia "biography." *USA Today*, Nov. 29.

Stallman, R. 2002a. The free software definition. In *Free software, free society: Selected essays of Richard M. Stallman*, ed. J. Gay, 41–45. Boston, MA: Free Software Foundation.

Stallman, R. 2002b. The gnu manifesto. In *Free software, free society: Selected essays of Richard M. Stallman*, ed. J. Gay, 31–41. Free Software Foundation.

Stallman, R. 2002c. Why free software is better than open source. In *Free software, free society: Selected essays of Richard M. Stallman*, ed. J. Gay, 55–61. Free Software Foundation.

Sunstein, C. R. 2001. *Republic.com.* Princeton, NJ: Princeton University Press.

Tatum, C. 2005. Deconstructing Google bombs: A breach of symbolic power or just a goofy trick? *First Monday* 10(10). http://www.firstmonday.org/issues/issue10_10/tatum/index.html

The Codebreakers. 2006. First broadcast by BBC World TV, May 10.

Thompson, C. 2006. Google's China problem (and China's Google problem). *New York Times Magazine*, April 23:64–76.

Twist, J. 2005. UN debut for $100 laptop for poor. *BBC News*, Nov. 17. http://news.bbc.co.uk/1/hi/technology/4445060.stm.

Vaidhyanathan, S. 2001. *Copyrights and copywrongs: The rise of intellectual property and how it threatens creativity.* New York: New York University Press.

Warschauer, M. 2003. *Technology and social inclusion: Rethinking the digital divide*. Cambridge, MA: MIT Press.

Wang, R. et al. 2005. The digital study hall. Technical report TR-723-05, Computer Science Department, Princeton University.

Weber, S. 2004. *The success of open source*. Cambridge, MA: Harvard University Press.

Wheeler, D. A. 2005. Why open source software/free software (Oss/Fs, Floss, or Foss)? Look at the numbers! http://www.dwheeler.com/oss_fs_why.html.

Wyatt, E. 2005. Google adds library texts to search database. *New York Times,* Nov. 3.

Wyatt, E., and J. Markoff. 2004. Google is adding major libraries to its database. *New York Times*, Dec. 14.

Ye, Y., and K. Kishida. 2003. Toward an understanding of the motivation of open source software developers. In *Proceedings of the 25th International Conference on Software Engineering*, Portland, Oregon, May 3–10, 419–429.

7

Whither the Digital Divide?

Mark Warschauer

The powerful role of information and communication technology (ICT) in social and economic development has led many to ponder the consequences of unequal access to it. In the mid-1990s, journalists, scholars, and political leaders started discussing the significance of what they termed a *digital divide* between the information haves and have-nots. In this chapter, I review the issue of a digital divide and efforts to address technological inequality.

Unequal Access

The idea of a digital divide first arose in the United States. Indeed, early studies of US Internet access revealed sharp disparities among users according to race, income, and educational level (National Telecommunications and Information Administration 1999). As computers and the Internet become more widely available at lower cost, however, and access rates among high user groups level off, access rates among minority groups, such as African-Americans, are catching up (Marriott 2006).

Internationally, gaps in Internet access remain more persistent. According to one recent analysis, the percentage of the population who use the Internet in each major world region ranges from 68.6% in North America to only 2.6% in Africa (Miniwatts Marketing Research 2006; see Table 7.1). Data suggest that eventually this gap will decrease, as in the United States, since access rates are growing fastest in the regions of the world with least access. Given the very low access rates in some countries and regions, however, vast gaps in ac-

Table 7.1 World Internet Usage

Region	Percent of Population Using the Internet	Usage Growth 2000–2005
Africa	2.6	423.9%
Asia	9.9	218.7%
Europe	36.1	177.5%
Middle East	9.6	454.2%
North America	68.6	110.3%
Latin America/Caribbean	14.4	342.5%
Oceania/Australia	52.6	134.6%
World Total	15.7	183.4%

Source: Miniwatts Marketing Research (2006)

cess rates might exist for decades, serving to keep large sections of the world's population isolated from the so-called information superhighway.

Statistical measures of Internet usage alone obscure several important questions. First, is lack of ICT access a cause or effect of poverty or other measures of social exclusion? Second, what does it really mean to have access to ICTs? And third, what is the best approach to dealing with unequal access?

Cause or Effect?

There is no doubt that unequal access to computers and the Internet is, to a large extent, an *effect* of poverty. It is not surprising that only 0.6% of the population of Malawi uses the Internet when the average annual income level in the country is barely the cost of a personal computer (Internet World Stats 2006). However, lack of access to ICTs is also believed to be a *causal* factor in impoverishment.

Consider the global relationship between gross domestic product, global exports, and Internet usage. According to data from the United Nations Development Programme, the fifth of the world's people living in the highest-income countries control 86% of global Gross Domestic Product (GDP) and 82% of global exports, whereas Internet users in those countries constitute 93.3% of the world total. In contrast, the 20% of the world's population living in the poorest nations control only 1% of global GDP and 1% of global exports, and

Internet users in those countries comprise 0.2% of global Internet users (Warschauer 2003a; see Table 7.2).

Analysis by Castells (1996) helps explain why GDP, global exports, and Internet usage are so tightly linked. As he points out, exports from the impoverished sub-Saharan countries in Africa tend to be predominately low-value primary commodities, the market value of which has fallen steadily over the last several decades. The exports of the wealthy countries are based on high-technology and high-knowledge goods and services whose corresponding market value has risen since the onset of the computer and Internet era. From 1960 to 1997, the share of world trade composed of high- and medium-technology goods rose from 33% to 54%, while the share of world trade composed of primary products fell from 45% to 24% (World Bank 1999). These data suggest that the richest countries of the world were able to make use of ICT to expand their control of world wealth, whereas the lack of ability to access and use ICT in economic production further weakened those nations that were already poor.

A similar dynamic is at play in ICT use and wealth among individuals in the United States. Prior to the computer era, there were large numbers of high-paying jobs in manufacturing, construction, or mining. However, from the early 1940s to 2000 the percentage of workers employed in manufacturing declined from 32% to 13% (Forbes 2004) and the percentage of the US workforce that is unionized fell at a similar rate (Joyce 2005). During the same period, there was a dramatic rise in the power, pay, and prestige of what Robert Reich (1991) has called *symbolic analysts*. These are professionals such as scientists, engineers, consultants, systems analysts, and designers who "solve, identify, and broker problems by manipulating symbols" (1991, 178). Symbolic analysts rely on ICT to network and communicate with other people, seek out and analyze information, and produce high-quality written and multimedia documents. In

Table 7.2 Shares of Global GDP, Exports, and Internet Users among World's Population

	Poorest 20%	Middle 60%	Richest 20%
GDP	1%	23%	86%
Exports	1%	17%	82%
Internet users	0.2%	6.5%	93.3%

other words, those who have the greatest access to and mastery of ICT have increased their socioeconomic position.

Elements of Access

What, then, does it mean to have access to ICT? If we consider access to mean the mere presence of a computer and an Internet connection, then virtually 100% of the people in the United States have access, if not at home or work, then through a public library. However, a number of analysts (Carvin 2000; Hargittai 2003; Warschauer 2003) have suggested that a broader view of access is required if we are to understand what enables people to deploy ICT in personally or socially meaningful ways. In my book on the digital divide (Warschauer 2003b), I analyzed these factors as falling into four general areas: *physical resources*, *digital resources*, *human resources*, and *social resources*.

Physical resources refers to the requisite device (desktop, laptop, handheld computer, Internet-connected personal digital assistant, or mobile phone) and Internet connection (via phone line or wired or wireless broadband link). A good deal of variation exists in terms of quality of computers and of Internet connections, so the physical resources themselves enable diverse types of ICT usage.

Digital resources refers to the content that is available online. According to research by Zook (2001), 65% of the world's Internet domains are located in the United States, Great Britain, or Germany, and the majority of the domains in these and other countries are concentrated in major urban areas. This is important in considering the digital divide on an international level, because the information needs of the rural population in Asia, Africa, or Latin America might not be well met by Internet sources from New York, London, or Berlin. Content diversity is also related to language diversity, with a disproportionate amount of Web content in English (Paolillo 2005). According to one recent estimate, there are hundreds of times as much content available in English per native speaker of that language as is available in Arabic, for example (Warschauer 2003a). The transition from the first-generation World Wide Web, which facilitated browsing but left writing and publishing difficult, to what has been called *Web 2.0*, which more easily allows writing, content creation, and publishing through blogs, wikis, and other means, can be helpful in reducing the vast divide in digital resources, if worldwide initiatives are able to mobilize local communities to produce their own Web content.

Human resources refers to the knowledge and skills required for meaningful use of computers and the Internet, which include both the traditional literacies of reading and writing, as well as a set of new digital literacies. The latter include *computer literacy* (the fluency, comfort, and skill in using a computer and its programs); *information literacy* (the ability to find, critique, evaluate, and deploy information from online and other sources); *multimedia literacy* (the ability to produce and publish quality work that draws on texts, images, sounds, and video); and *computer-mediated communication* literacy (the interpretation, writing, and thinking skills necessary to communicate effectively in, or help arrange and manage, various types of synchronous and asynchronous online interaction; for further details, see Warschauer 2003a).

Social resources refers to the social relations, social structures, and social capital that exist to support effective use of ICT in families, communities, and institutions. Two decades of research in schools, governments, businesses, and other institutions has revealed how successful incorporation of ICT inevitably depends on multifaceted and ongoing reform of social relations and incentives rather than merely on a one-time infusion of equipment (see overviews in Kling 2000; Warschauer 2003a).

The interaction between these sets of resources helps explain why problems of technology and inequality are so daunting, and why notions of a binary divide based purely on whether or not somebody has a computer are too simplistic. Rather, there are many degrees of access to ICT, depending on a complex combination of physical, digital, human, and social resources available.

Examples from Education

To illustrate the above points, I will draw examples from the deployment of technology in K-12 schools in the United States and Egypt. In the United States, educators and policy makers have long viewed public schools as a key venue for providing children with more equitable access to new technologies and the ability to use them. A number of federal and state programs have thus been set up to provide funding for hardware, software, and professional development related to educational technology. Due in part to these programs, as

Table 7.3 Ratio of Public School Students to Instructional Computers with Internet Access

Percent of Students Eligible for Free or Reduced-Price Lunch	1999	2000	2001	2002	2003	2005
Less than 35 percent	7.6	6.0	4.9	4.6	4.2	3.8
75 percent or more	16.8	9.1	6.8	5.5	5.1	4.0
Differential of high-poverty schools compared to low-poverty schools	+121.1%	+51.7%	+38.8%	+19.6%	+21.4%	+5.3%

Source: Wells, et al. (2006)

well as to the falling cost of computer and telecommunications equipment, the overall ratio of students to Internet-connected instructional computers in public schools fell from a national average of 9.1 students per computer in 1999 to 3.8 in 2005 (Wells, et al. 2006). Throughout this six-year period, the student-computer ratio was larger in what are considered *high-poverty schools* (with 75% or more of their students eligible for the federal government's free or reduced price lunch program) than in *low-poverty schools* (with 35% or fewer of their students eligible for the lunch program). However, the differential of student computer ratios between high-poverty schools and low-poverty schools fell from 121% in 1999 to about 5% in 2003 (see Table 7.3). A declining gap in student-computer ratios also emerges when analyzing schools with high and low numbers of minority students.

Yet, as the equipment gap steadily shrinks between low- and high-socioeconomic status (SES) schools, there still exist substantial differences in how technology is used to support learning. A number of studies have found that teachers of high-SES students use new technology more frequently than do teachers of low-SES students (Schofield and Davidson 2004; Warschauer and Grimes 2005). In addition, high-SES students more frequently use technology in school for tasks that promote higher-order thinking, such as simulations and project-work, whereas low-SES students more frequently use technology for remedial drills (Becker 2000; Wenglinsky 1998).

I have conducted comparative case studies of educational technology implementation in high- and low-SES K-12 schools in California, Maine, and Hawaii (Warschauer, et al. 2004; Warschauer 2000;

Warschauer 2006) drawing on observations, interviews, examination of teacher and student artifacts, and comparison of survey and test score data. The studies helped illuminate the complex array of reasons why technology is used unequally by low- and high-SES students.

First, there is an important interaction between school and home access to technology. Teachers in low-SES schools are keenly aware that a number of their students lack home access to a computer and the Internet. They thus spend a good deal of time teaching their students how to use software. In contrast, teachers in high-SES schools correctly assume that virtually all their students have computers and the Internet at home, and most have support from their families in how to use them. This allows them to focus student use of computers in classroom on learning objectives, rather than software use. In addition, they are much more ready to assign research-based computer homework than are teachers are low-SES students, some of whom fear their students will not be able to complete it.

Second, the low-SES and high-SES schools in our studies had very different support structures for integration of technology. Low-SES schools in the United States have higher teacher and administrator turnover; a higher number of teachers without appropriate credentials; and less experienced teachers and staff than do high-SES schools; and fewer and less educated parent volunteers. Teachers in low-SES schools more frequently complained of lack of support structures to make scheduling arrangements, keep equipment functioning, or provide pedagogical support than did teachers in high-SES schools.

Third, there are important differences in the capabilities of students. Learners in high-SES schools, who have greater degrees of language and literacy skills, background content knowledge, and prior experience with technology, can much more easily make use of new technology to achieve sophisticated learning objectives. In contrast, low-SES schools contain a disproportionate number of English-language learners, students with limited literacy, and students who lack basic keyboarding or other technological skills. It thus becomes much more challenging to introduce technology to these students while also focusing on academic objectives.

And fourth, schools in high- and low-SES communities often have very different visions and goals, with former schools more sharply focused on preparing their students for university and the latter often explicitly or implicitly more directed to vocational training. Technology diffusion may help schools better achieve their goals, but is unlikely to change the underlying vision and goals of a school.

Egypt represents a very different context for introducing technology into schools. There, the government seeks to develop a more modern technological society and recognizes that ICT has a valuable role to play in that. The government and Ministry of Education thus created a large and ambitious Technology Development Center (TDC) with the mission of injecting more advanced technology in public schools. Initiatives of the TDC included placing multimedia rooms with two to three Internet-connected high-end computers in all high school and middle schools in Egypt, plus building a national network of 27 expensive videoconference centers.

I investigated the use of these facilities from 1998-2001 and found that their actual contribution to educational improvement was minimal (Warschauer 2003b). The stated goals of the TDC—to foster more interactive hands-on learning with technology—contradicted the social context of schooling in Egypt, which is known for poorly trained teachers, large class sizes, and a focus on rote knowledge. As a result, the multimedia rooms were seldom used and in many schools the computers remained locked up. The videoconference centers were used for high-profile international events that similarly had little impact on education. As a result, the considerable expenditures on technology brought few demonstrable results, and diverted funding from other educational needs, such as opening up more rural schools for the large numbers of girls in Egypt's countryside who lack educational opportunities. Research on the use of technology in education internationally, as well as in a wide range of social and economic contexts in the United States and around the world, suggests that similar dynamics are at work, with the usefulness of technology for addressing social problems highly dependent on other contextual factors (Warschauer 2003a). What strategies, then, are successful for using technology to promote greater social equality and inclusion?

Closing the Gap

First, analyses of targeted problems must begin with the examination of social structures, social problems, social organization, and social relations, rather than with an accounting of computer equipment and Internet lines. An accounting of equipment is part of the overall analysis, but a fairly small part; if interventions are designed to address social problems, they must be planned by focusing on the overall structures and relationships that give rise to those problems.

Once social problems or goals are identified, programs should be based on a systemic approach that recognizes the primacy of social structure and promotes the capacity of individuals or organizations for ongoing social change through innovation of those structures using technology. Corea discusses this strategy in depth, pointing out that information technology implementations often create only superficial change, with organizations returning to their ingrained ways once the new systems have been "absorbed into the previous web of calcified inefficiencies" (2000, 9). Rather than just foisting technologies haphazardly on people, a better solution is to foster the "long-term nurturing of behaviors intrinsically motivated to engage with such technologies" with the goal of achieving "an 'innovating' rather than a 'borrowing' strategy of growth as a means to reduce technological disparities" (Corea 2000, 9). This can bring about a "catching up process" through development of capacity "in the generation and improvement of technologies, rather than in the simple use of them" (Perez and Soete, quoted in Corea 2000, 9). All of this requires changes in the social environment to facilitate "the learning of new behaviors that propagate continuous improvements in conditions of living" (Corea 2000, 9). This process of innovation might take many forms. Rural teachers might learn how to create their own technology-based materials based on local conditions rather than only using commercial software developed for other contexts. A crafts cooperative might learn how to develop and manage its own Web site rather than just posting its announcements on somebody else's. Nongovernmental organizations might learn to establish and run their own networks of telecenters rather than just attending cybercafes.

In promoting such efforts and programs, it is essential to understand and exploit possible catalytic effects of ICT. Many important changes in social relations may come from the human interaction that surrounds the technological process, rather than from the operation of computers or use of the Internet. For example, a new computer laboratory in a low-income neighborhood may also become a meeting hub for at-risk youth and college-student mentors. Or the involvement of community members in planning the laboratory may bring together new coalitions that can also work for other types of community improvement. The social importance of ICT in the information economy and society means that ICT initiatives often have powerful leveraging potential that can be used to support broader strategies for social inclusion.

The role of leadership, vision, and local "champions" (McConnell 2000) is crucial to the success of ICT projects for social inclusion. A common mistake made in ICT development projects is to make primary use of computer experts rather than of the best community leaders, educators, managers, and organizers. Those, however, who are capable of managing complex social projects to foster innovation and creative and social transformation will likely be able to learn to integrate technology in this task. On the other hand, those with technological skills, but who lack understanding of the complex human issues at hand or the leadership ability to address them, will usually prove less effective.

Finally, the process of organizing, designing, implementing, and evaluating ICT projects must itself be open to innovation and flexibility. Good big things come from good small things, and room for innovation, creativity, and local initiative is critical to give the space for good small things to emerge. Flexible pilot programs are thus an important part of the development process. Scalability is of course an important aspect, and the potential for scaling up has to be part of the formative and summative evaluation of pilot programs. But lockstep, centrally organized large-scale initiatives with no room for local experimentation and innovation do not meet the needs of a rapidly changing information society, economy, or educational system.

Conclusion

ICT is an amplifier of other social and economic factors and processes. It thus is rightly seen as having the potential to help individuals, groups, and even nations leapfrog over developmental stages. Yet, at the same time, infusions of ICT can also amplify existing inequalities, as the effective use of ICT requires other human and social resources and can magnify differences in their distribution.

The notion of a digital divide has focused the attention of the public and policy makers on the important intersection between technology and inequality. Research on technology and social change has indicated how complex this intersection is, and has suggested some possible directions to take in addressing the problem. By focusing on the diverse range of resources that enable meaningful use of technology, and seeking long-term solutions that strengthen marginalized groups' agency, we can best make sure that ICT is used to further a process of social reform, equity, and inclusion.

References

Becker, H. J. 2000. Who's wired and who's not? *The Future of Children* 10(2): 44–75.

Carvin, A. 2000. Beyond access: Understanding the digital divide. Paper presented at the New York University Third Act Conference, September 1, 2000.

Castells, M. 1996. *The rise of the network society.* Malden, MA: Blackwell.

Corea, S. 2000. Cultivating technological innovation for development. *Electronic Journal on Information Systems in Developing Countries* 2(2): 1–15.

Forbes, K. J. 2004. U.S. manufacturing: Challenges and recommendations. http://www.whitehouse.gov/cea/forbes_nabe_usmanufacturing_3-26-042.pdf (accessed February 2, 2006).

Hargittai, E. 2003. The digital divide and what to do about it. In *The new economy handbook*, ed. D. Jones, 822–841. San Diego, CA: Academic Press.

Internet World Stats. 2006. Africa Internet usage and population statistics. http://www.internetworldstats.com/africa.htm (accessed December 5, 2006).

Joyce, F. 2005. Fate of the union. http://www.alternet.org/story/21312/ (accessed February 2, 2006).

Kling, R. 2000. Learning about information technologies and social change: The contribution of social informatics. *The Information Society* 16(3): 1–36.

Marriott, M. 2006. Digital divide closing as blacks turn to Internet. *New York Times,* March 31, 2006. http://www.nytimes.com/2006/03/31/us/31divide.html (accessed June 1, 2006).

McConnell, S. 2000. A champion in our midst: Lessons learned from the impact of NGOs' use of the Internet. *Electronic Journal on Information Systems in Developing Countries* 2(5): 1–15.

Miniwatts Marketing Research. 2006. Internet usage statistics-the big picture. http://www.internetworldstats.com/stats.htm (accessed June 1, 2006).

National Telecommunications and Information Administration. 1999. *Falling through the net: Defining the digital divide.* Washington, DC: NTIA.

Paolillo, J. 2005. Language diversity on the Internet. In UNESCO Institute for Statistics, *Measuring linguistic diversity on the Internet*, 43–89. Paris: UNESCO.

Reich, R. 1991. *The work of nations: Preparing ourselves for 21st century capitalism.* New York: Knopf.

Schofield, J. W., and A. L. Davidson. 2004. Achieving equality of student Internet access within school. In *The social psychology of group identity and social conflict,* ed. A. Eagly et al., 97–109. Washington, DC: APA Books.

Warschauer, M. 2000. Technology and school reform: A view from both sides of the track. *Education Policy Analysis Archives* 8(4). http://www.epaa.asu.edu/epaa/v8n4.html.

Warschauer, M. 2003a. Demystifying the digital divide. *Scientific American* 289(2): 42–47.

Warschauer, M. 2003b. *Technology and social inclusion: Rethinking the digital divide.* Cambridge: MIT Press.

Warschauer, M. 2003c. The allures and illusions of modernity: Technology and educational reform in Egypt. *Educational Policy Analysis Archives* 11(38). Available online at http://www.epaa.asu.edu/epaa/v11n38/v11n38.pdf.

Warschauer, M. 2006. *Laptops and literacy: Learning in the wireless classroom.* New York: Teachers College Press.

Warschauer, M., and D. Grimes. 2005. First-year evaluation report: Fullerton School District laptop program. http://www.gse.uci.edu/markw/fsd-laptop-year1-eval.pdf (accessed February 2, 2006).

Warschauer, M. et al. 2004. Technology and equity in schooling: Deconstructing the digital divide. *Educational Policy* 18(4): 562–588.

Wenglinsky, H. 1998. Does it compute?: The relationship between educational technology and student achievement in mathematics. ftp://ftp.ets.org/pub/res/technolog.pdf (accessed February 2, 2006).

Wells, J. et al. 2006. *Internet access in U.S. public schools and classrooms:1994–2005.* Washington: National Center for Educational Statistics.

World Bank. 1999. World development report 1998/1999: Knowledge and development. Washington, DC: World Bank.

Zook, M. A. 2001. Old hierarchies or new networks of centrality?: The global geography of the Internet content market. *American Behavioral Scientist* 44(10): 1679–1696.

8

The Balance of Privacy and Security[1]

Eugene H. Spafford and Annie I. Antón

Introduction

In early 2006, reports surfaced in the media of large-scale, clandestine eavesdropping on telephone communications being performed by the US National Security Agency for purposes of counterterrorism. This followed on earlier controversial efforts such as the Defense Advanced Research Projects Agency (DARPA) Total Information Awareness Program (Anderson 2004) and the Transportation Security Administration's Secure Flight Initiative (Secure Flight Working Group 2005). All of these programs collected and analyzed vast amounts of data on large segments of the population with the stated purpose of trying to identify terrorists. This has resulted in public outcry because many view this as an invasion of privacy and one that might not accomplish its purported goal (Stirland 2005). Moreover, the availability of these data raised concerns about the potential for other unauthorized or unwarranted uses thereof. This chapter examines the tension between Internet privacy and the legitimate needs of law enforcement.

[1]Full report submitted for publication (CERIAS Technical Report TR 2006-36). If this paper is to be cited, contact the authors for current status.

Much of what we do in today's society involves information exchange and surveillance—for example, government, commerce, health care, education, and entertainment. For people to be comfortable and willing to take full advantage of IT (Information Technology) services, however, we need to safeguard their privacy and identity. Privacy is important because it helps us maintain our individuality, autonomy, and freedom of choice. Individuality is what distinguishes us from others. Autonomy enables us to freely and independently perform everyday tasks and activities without feeling that we must act differently because of unwanted external influence or control by others. Some would argue that surveillance in certain contexts can erode one's autonomy and liberty. Security technologies such as surveillance are making it easier to invade people's privacy. In the past, these technologies were discussed within the context of spying on the enemy or deterring crime. The ubiquity of security technologies today, however, has brought the ability to engage in spy-like activities in relation to the masses. Technology has made it easy, simple, and cheap to collect, store, and search data. Consequently, we observe an escalating tension between those who are eager to use security technologies and those who value privacy as an inherent right and distrust those who monitor their activities. Whatever the context, once privacy is lost, it can seldom be fully recovered.

A Basic Conflict

The Internet is a powerful mechanism for distributing ideas, images, sound, and other intellectual content with low cost and great rapidity. It provides a means for collaborative development of artifacts, either in real time or incrementally, and enables "time-shifting" so that the audience can choose the time and place to peruse posted content. These characteristics provide the means for a "marketplace of ideas" (as articulated by Holmes and Brandeis in 1919 in the dissenting opinion in *Abrams v. United States*) that makes it possible for anyone who can afford time at an Internet café to publish, without having to be a traditional publisher.

Within a large enough population there will likely be those with interests that do not meet the definition of "lawful" or even "sane" as articulated by the majority. Hence, the potential emergence of communication streams devoted to criminal enterprises, prurient (or deviant) interests, radical or revolutionary political expression, bigotry and hate speech, and more. It may well be that some of the individuals pursuing these streams are mentally disturbed and sociopathic in nature. Unfortunately, the same Internet that allows them to exchange ideological texts also allows them to formulate joint plans, obtain technical information, and order dangerous materials. Given the fragility of many of our social and technical systems, and the widespread availability of technologies that may be used offensively, it is in the interest of law enforcement and government, as representatives of the people, to discover and prevent offenses before they occur.

As individuals, we decide what is offensive, but as a society we define offenses. At what point does something actually become "an offense"? Individuals who are unhappy with the government should be free to express displeasure with their officials and disdain for their actions. Invective may even help diffuse anger over official actions (and misdeeds), and act as a cathartic. The United States has long embraced the ability to speak freely as a cherished right of the public, and the government is prohibited from exercising prior restraint. Even the act of officially observing such speech, or forcing the identification of the speaker (or author) may restrain such speech, and has been prohibited by courts in some venues.

There are other reasons why an author or speaker may wish anonymity beyond simply voicing political discontent. Discussing information about matters that may be viewed as highly private, such as issues about medical conditions, religion, or sexuality, can be hampered by observers matching the questions with the questioners or responders. Individuals raising issues of public safety ("whistle-blowers") may fear for their financial or personal safety and seek to report anonymously in a public forum. Those who have suffered physical, mental, or sexual abuse may wish to share their stories as both therapy and guidance for others without the chance of damaging their images as held by those who know them.

In these and other cases, private conversations preserve something of value to the participants. Truly anonymous posting helps to encourage openness and honesty by participants without fear of retribution. However, we cannot tell, a priori, if those conversations are socially benign or of a more egregious nature. Thus, we cannot determine in advance whether a conversation should have the strongest possible privacy protections, or some more penetrable form of protection. When made as a default choice of privacy (non)protection, the choice colors all subsequent communications, for good or ill.

In addition to the choices made for the privacy of personal communication, we also see choices made for the privacy protection of messages by organizations and governments. Those organizations are also capable of misdeeds that can be concealed by undue secrecy. Because of their additional resources, misdeeds by government and organizations may be broader in scope and more difficult to redress. Exposure of the operations and behavior of these entities provides a check on their potential misbehavior. Complete transparency is not desirable, however, because those organizations maintain information on individuals that must be kept private, and they handle sensitive information that is necessary both for competitive advantage and security. The use of computing for data storage and processing, for communication, and for process control greatly complicates the ability of these organizations to protect their information from prying eyes, but also provides new opportunities for visibility into their operations.

Security and privacy are not opposites, but their interests often conflict. The increasing use of information technology may amplify those conflicts. In the following sections we will outline some of the conflicts that have arisen in recent years regarding the use of Internet communications.

What Is Privacy?

There is no absolute definition of privacy, as it means different things to different people. Within the context of information collection, storage and use, we define privacy as the expectation that information about you, as an individual or as a member of a group, and that is not generally known, will not be disclosed. This can include your activities, your name, your affiliations, and other information about you. In today's world of ubiquitous technology, this information spans anything from your telephone records to your medical history. It is important to emphasize that the definition of privacy depends on what you do not want revealed about you to others, and that this is subjective. Some people are "open books," whereas others are very secretive individuals. In American society, the tradition has been to support whatever view of privacy different individuals may have. Exceptions to this come into play for purposes of public health, public safety, and law enforcement. However, despite the apparently clear language of law, this is interpreted subjectively. An example in 2006 was the controversy of the US government wiretapping the phones of suspected terrorists (Cauley 2006). Individual beliefs concerning whether government should be able to compel us to provide information about ourselves differs from person to person.

There are valid social reasons why one's privacy may be intruded upon, such as when a child is suspected of being a victim of abuse, or to investigate or prevent criminal activity. Generally speaking, people's perceptions of privacy and privacy invasions are relative to materials and timing. For example, some people view any unsolicited e-mail as an intrusion, whereas others are happy to have 24-hour Webcams in every room of their houses. For purposes of this discussion, we thus consider any information that you do not want revealed about yourself to be a privacy invasion because, for example, if revealed, it may cause you to change your behavior. If an information disclosure is not damaging to you, but it still causes you to change your behavior, it is an intrusion; for example, purchasing items in cash so that they cannot be tied to you (e.g., magazines, alcohol, etc.) or to ensure that others will not treat you differently is a means of protecting your privacy.

Identification

A time-honored method of protecting individual privacy is the use of anonymity. Throughout recorded history, materials have been written under pseudonyms to protect the author, as was done with *The Federalist Papers* in the early American colonies (Madison, Hamilton, and Jay 1987). To this day, the authorship of some of those papers—viewed as foundations of US political thought—is still uncertain. Anonymity in political expression is still a major concern of self-preservation: political dissidents in many countries around the world today are subject to arrest, torture, and possibly execution if their identities are discovered.

Pseudonyms offer another way to protect one's privacy; pseudonyms may promote the consideration of the ideas expressed independent of any existing reputation or image of the author. For example, Mary Anne Evans used the pseudonym George Eliot to hide that she was female because she wanted her work to be taken seriously and because she wanted to guard her private life from the public. Similarly, Charles Dodgson used the pseudonym Lewis Carroll to hide that his fiction was written by a mathematician. In online games and discussion groups, individuals often choose personae that portray different ages, genders, or ethnicities so as to relate in ways independent of their actual selves. Adolescents, in particular, may use these pseudonymous identities as a way of helping them to refine their own real identities and interests.

Although anonymity is often used to conceal one's identity for legitimate reasons, anonymity can also be misused. Libelous and hurtful statements made anonymously cannot easily be refuted, nor can the authors be held accountable for the harm caused by their falsehoods. Impersonation of others to commit fraud is a significant form of criminal activity. Stalking and solicitation of minors by sexual predators are two crimes that occur all too frequently online, and which are often difficult to investigate and resolve when pseudonyms or anonymous communication is involved. Communication within loose organizations of criminals trading in stolen intellectual property or access codes is difficult or impossible to investigate without knowing the underlying identities of participants. For anonymous participation to succeed in those cases where it is most needed, such as under threat of financial or political retribution, the anonymity needs to be so strongly protected as to be effectively inviolable. However, because we cannot know a priori whether such communication is lawful or harmful, there is a basic conflict—do we support mechanisms that allow for true anonymity to anyone seeking it, or do we force all participants to have some form of (eventually) verifiable identity?

Log Analysis and Data Mining

The pattern of someone's behavior can reveal a great deal about what they are doing. For example, given the information that someone has used a search engine to search for a term such as "Fanconi's syndrome," followed by visits to Web sites on bone marrow registries and information on "antithymocyte globulin" may well lead to a likely conclusion that the person's child has been diagnosed with aplastic anemia. This inference, even if incorrect, could be viewed as an invasion of privacy by the person doing the searches. Other inferences can be drawn based on items purchased online, patterns of e-mail received, and even the times at which the accesses occur. The results of such inferences could be used for directed marketing, public health studies, or to reduce the coverage available under a health insurance policy. They might also be used for targeted fraud ("phishing") or extortion. If the subject making the searches is a researcher working for a large pharmaceutical firm, observation of the searches performed might provide clues about current research interests and drug development. When employed by a competitor, this form of surveillance can be used to "scoop" development on a new product. The financial incentives for such inferences can be quite significant in a number of industries, and may drive some agents to employ illegal means of observation as well.

Law enforcement has proposed the use of similar techniques to identify criminals. By observing who has a history of seeking information on hydroponics and purchasing high-intensity daylight spectrum lights, drug enforcement agents hope to find people with clandestine marijuana-growing operations. By observing who has a history of research into explosives, purchases of nitrogen fertilizer, and online searches of building plans, some federal agents hope to identify terrorists before they strike. In both cases there is a presumption of guilt based on purely circumstantial evidence, and the searches and inferences are likely to be viewed as an invasion of privacy by anyone incorrectly targeted for law enforcement activity.

Some policy makers want to require Internet service providers (ISPs) and online services to keep extensive logs and records of user behavior so that criminals such as child pornographers and identity thieves might be traced and prosecuted. Keeping such extensive records, however, would potentially allow others to gain access, for reasons of civil litigation, directed marketing, or criminal activity. The potential for accidental disclosure of these stored records is also a concern. For instance, the Privacy Rights Clearinghouse announced in December 2006 that over 100 million personal data records had been disclosed or compromised in a two-year period—records that were presumably better protected than ISP logs because of their more sensitive nature.[2]

Long Term Storage and Processing

Historically, information that was stored for long periods of time was generally information that had been analyzed and synthesized, or had proven value. Data without known value was often not collected or saved because of the cost of storage media and space. Also, the process of manually indexing large data stores so as to have ac-

[2] <http://www.privacyrights.org/ar/ChronDataBreaches.htm>

cess to specific items at some future time was itself time-consuming and difficult. As such, the whole process of collecting and storing data had some built-in limits.

Currently, modern computer storage of information is relatively inexpensive, and indexing and sophisticated searching are performed by advanced algorithms and fast computers. There are currently few disincentives to storing significant quantities of data, including personal data. At the same time, advances in data-mining and matching technologies provide for massive data searching and inference formation. It is possible to derive previously unknown correlations and patterns from massive collections of data. Depending on the agency performing the searches, these correlations may provide new insights into a range of issues such as disease transmission and epidemiology, tendencies to vote for particular political candidates, the likelihood of buying certain sets of products, or social networks of criminals.

Clearly, there are strong reasons that such data would be collected, saved, and sold to new aggregators. Governments would want the data so as to anticipate citizen needs and provide needed services, law enforcement would want the data to identify malfeasors, and commercial entities would want to use the information for purposes of advanced marketing and pricing. Historians and other social scientists would also find detailed data to be of value in understanding events and trends. However, the collection and use of such information facilitates intrusions into the privacy of individuals and groups. Not only is it possible to discover information in the data (and its relationships) about individuals, but erroneous information also presents possibilities for intrusions. Transcription errors, old (stale) data, and simple errors in algorithms used to process the data may result in both type I (false positive) and type II (false negative) errors. Actions taken on erroneous conclusions may result in everything from mistargeted advertisements to false accusations of criminal behavior—all of which would be considered to be (at least) privacy violations. These are often all the more surprising to the subjects when they discover the extent to which data about them has been saved.

Unfortunately, the propensity for errors and inaccuracies in the information that data brokers collect is exacerbating the problem. Studies have revealed that data broker files are riddled with errors and that there is no easy way to fix these errors (Sullivan 2005). Moreover, a recent study examined the quality of data provided by two data brokers, ChoicePoint and Acxiom (Pierce and Ackerman

2005). Of the credit and background reports given out by ChoicePoint, 100% had at least one error in them. The error rates for basic biographic data (including information people had to submit to receive their reports) fared almost as badly: Acxiom had an error rate of 67% and ChoicePoint had an error rate of 73%. The majority of participants in the study had at least one significant error in their reported biographical data from each data broker.

Resolution?

It is clear that there are a number of privacy-related conflicts relating to the collection, storage, analysis, and use of information in computing systems and networks. Most of these conflicts are rooted in fundamental differences of opinion about the role of privacy and the grounds under which the veil of privacy may be lifted. Because there is such a range of circumstances under which these determinations may be made, and because many cannot be determined in advance (as illustrated in the examples above), we are left with an environment where deployment of technology and establishment of policy must be guided by principles that are designed to protect privacy yet allow for legitimate uses of personal information by others.

In early 2006, in response to a steady stream of privacy breaches and news accounts of significant government programs to surveil and analyze citizen information, the US Public Policy Committee of the Association for Computing Machinery (ACM) created a set of privacy recommendations. The ACM[3] is the oldest scientific and professional organization for computing professionals, with over 80,000 members worldwide. The privacy recommendations were created based on professional experience, previous recommendations such as the Fair Information Practices (Ware, et al. 1973), and current law in many countries around the world (e.g., the Canadian Privacy Act[4]). The committee sought to codify the best practices that would maximize transparency and accountability, reduce error and surprise, minimize the potential for misuse and data exposure, and yet also allow reasonable continued and future use of aggregated data by business and government. The 24 recommendations are shown in Box 8.1, grouped by major concept.

---

[3]<http://www.acm.org/usacm/>

[4]<http://www.privcom.gc.ca/legislation/02_07_01_01_e.asp>

The ACM Privacy Recommendations represent an attempt to balance enhanced personal privacy with legitimate use. The following are examples, one from each major grouping:

Minimization, #2: *Store information for only as long as it is needed for the stated purposes.* This principle recognizes the need for organizations to store personal information for various uses. However, to reduce the time during which that information may be exposed, and to reduce the potential amount of information that may be exposed, the principle states that information should only be kept until its defined uses are met. Too often, organizations collect information and then keep it indefinitely because "it might be useful later." This leads to information drift (becoming less accurate) and increases the potential for disclosure.

This principle implies an assumption that the organization has thought through and formulated an explicit privacy policy and a definite set of use cases for the collected data—two exercises that are conducted by too few organizations.

Consent, #7: *Whether opt-in or opt-out, require informed consent by the individual before using personal information for any purposes not stated in the privacy policy that was in force at the time of collection of that information.* When individuals provide information (or they are informed that it is collected), it should be with a set of stated purposes and policies about how it is used. The individual can choose to consent or not under those circumstances, or to file objections with the organization through appropriate channels. If an organization later decides to alter those parameters, the individual should be fully informed and given the same chance to consent to participate as when the information was originally collected. This allows organizations to find new uses for already-collected information, without going to the effort of recollecting and validating that information. It does not, however, allow unrestricted use of the data. Note that the concept of "informed consent" is commonly used in scientific experiments and medicine to ethically determine participation.

Openness, #13: *Communicate these policies to individuals whose data is being collected, unless legally exempted from doing so.* To support informed consent, as per #7, individuals must actually be informed about the uses of their personal information. Other principles in this group stress that the provided information be explicit, bounded, and understandable. However, this principle also recognizes that some law enforcement and government databases may con-

Box 8.1. USACM PRIVACY RECOMMENDATIONS

Minimization

1. Collect and use only the personal information that is strictly required for the purposes stated in the privacy policy.
2. Store information for only as long as it is needed for the stated purposes.
3. If the information is collected for statistical purposes, delete the personal information after the statistics have been calculated and verified.
4. Implement systematic mechanisms to evaluate, reduce, and destroy unneeded and stale personal information on a regular basis, rather than retaining it indefinitely.
5. Before deployment of new activities and technologies that might impact personal privacy, carefully evaluate them for their necessity, effectiveness, and proportionality: the least privacy-invasive alternatives should always be sought.

Consent

6. Unless legally exempt, require each individual's explicit, informed consent to collect or share his or her personal information (opt-in); or clearly provide a readily-accessible mechanism for individuals to cause prompt cessation of the sharing of their personal information, including when appropriate, the deletion of that information (opt-out). (NB: The advantages and disadvantages of these two approaches will depend on the particular application and relevant regulations.)
7. Whether opt-in or opt-out, require informed consent by the individual before using personal information for any purposes not stated in the privacy policy that was in force at the time of collection of that information.

Openness

8. Whenever any personal information is collected, explicitly state the precise purpose for the collection and all the ways that the information might be used, including any plans to share it with other parties.
9. Be explicit about the default usage of information: whether it will only be used by explicit request (opt-in), or if it will be used until a request is made to discontinue that use (opt-out).
10. Explicitly state how long this information will be stored and used, consistent with the "Minimization" principle.
11. Make these privacy policy statements clear, concise, and conspicuous to those responsible for deciding whether and how to provide the data.

Box 8.1. USACM PRIVACY RECOMMENDATIONS (*CONT'D*)

12. Avoid arbitrary, frequent, or undisclosed modification of these policy statements.
13. Communicate these policies to individuals whose data is being collected, unless legally exempted from doing so.

Access

14. Establish and support an individual's right to inspect and make corrections to his or her stored personal information, unless legally exempted from doing so.
15. Provide mechanisms to allow individuals to determine with which parties their information has been shared, and for what purposes, unless legally exempted from doing so.
16. Provide clear, accessible details about how to contact someone appropriate to obtain additional information or to resolve problems relating to stored personal information.

Accuracy

17. Ensure that personal information is sufficiently accurate and up-to-date for the intended purposes.
18. Ensure that all corrections are propagated in a timely manner to all parties that have received or supplied the inaccurate data.

Security

19. Use appropriate physical, administrative, and technical measures to maintain all personal information securely and protect it against unauthorized and inappropriate access or modification.
20. Apply security measures to all potential storage and transmission of the data, including all electronic (portable storage, laptops, backup media), and physical (printouts, microfiche) copies.

Accountability

21. Promote accountability for how personal information is collected, maintained, and shared.
22. Enforce adherence to privacy policies through such methods as audit logs, internal reviews, independent audits, and sanctions for policy violations.
23. Maintain provenance—information regarding the sources and history of personal data—for at least as long as the data itself is stored.
24. Ensure that the parties most able to mitigate potential privacy risks and privacy violation incidents are trained, authorized, equipped, and motivated to do so.

tain information that individuals should not know about—either how it is used, or that they are profiled in those databases. Thus, for good reasons, law may exempt disclosure of those databases, but the requirement that those exemptions be defined in law presumes that some appropriate deliberation and oversight has occurred related to making that determination.

Access, #14: *Establish and support an individual's right to inspect and make corrections to his or her stored personal information, unless legally exempted from doing so.* One aspect of information collection that frightens or annoys many people is the prospect of inaccurate information being used to make decisions about them. This principle recognizes that concern, and encourages the data holder to provide meaningful mechanisms to allow that information to be examined, and, if appropriate, corrected by the individuals involved. We note that this might actually be to the benefit of the organization holding the data because it might lead to a more accurate, and therefore higher-quality and more cost-effective, data collection.

This principle is currently supported by too few organizations. For example, ChoicePoint used to state that it could not correct errors in its records, but that consumers must locate the original source from which ChoicePoint gathered the information and correct any mistakes there (Otto, et al. 2007). In contrast, the right to access and correct erroneous financial information has been a part of the Fair Credit Reporting Act since its passage in 1970. Recently, ChoicePoint has announced plans to allow individuals to review and correct their personal information via a single point of access (Husted 2005); this system, however, is not yet available to consumers.

As with principle #13, principle #14 recognizes that there may be databases that should not be made known to the subjects, or the extent of the information on them should not be made known during the lifetime of the information. As before, the requirement that the exemption be defined in law presumes that some appropriate deliberation and oversight has occurred related to making that determination.

Accuracy, #18. *Ensure that all corrections are propagated in a timely manner to all parties that have received or supplied the inaccurate data.* This matches, in part, with principle #14, and also with #17. It recognizes that data are collected for particular purposes and may be supplied to business partners. Furthermore, it recognizes that data are sometimes shared between organizations for valid purposes, and as allowed by law or policy. As examples, banks provide transaction information to the Internal Revenue Service, and airlines share some flight information with their associated frequent flier programs. When data are shared, this principle requires the holders of the data to propagate any valid corrections rather than make that task a burden on the individuals involved. Further, it is defined as a "push" operation rather than a "pull" operation, so that it is more likely to occur when the correction is made, rather than in response to a periodic bulk request.

Additionally, this principle implies that a data aggregator must keep a record of where data are shared, and from where they are received. Thus, if an individual discovers and corrects an error in the middle of a "chain" of data sharing, it will propagate to both the initial end sites—the individual does not need to discover each location, or the first location, where the data were collected as currently required by ChoicePoint. This is especially important if some of the data collections are exempt from disclosure or review by the individual; so long as the corrections will traverse the same path as the data that was provided, presumably all copies of the data will be corrected in a timely fashion.

Security, #20: *Apply security measures to all potential storage and transmission of the data, including all electronic (portable storage, laptops, backup media) and physical (printouts, microfiche) copies.* This principle addresses one of the most common and often overlooked routes of exposure of personal information—physical compromise of the data. Too often, IT professionals focus on access control and encrypted communication, but fail to realize that the data they are trying to protect exist in multiple forms in multiple locations, many of which are easily stolen. The holder of the data should be responsible for security of the information in the original database and in any additional copies, in whatever format.

Accountability, #22: *Enforce adherence to privacy policies through such methods as audit logs, internal reviews, independent audits, and sanctions for policy violations.* Excellent policies and training do little unless they are actually applied. It is not always possible to determine whether all policies and safeguards are working correctly 100% of the time simply by observing the system or asking the personnel involved. Mistakes may happen, errors in code occur, and personnel may be careless or untruthful for a variety of reasons. It is important for organizations to conduct meaningful, regular examinations to ensure that privacy policies are being followed, and that deficiencies are appropriately addressed. Furthermore, making these compliance activities and remediations public will likely increase the confidence and comfort of individuals whose information is held by the organization (assuming that the safeguards are appropriate and found to be working correctly).

So long as people view information as private, and so long as others find some value in the collection and analysis of that information, there are likely to be conflicts. Although the full set of recommendations does not obviate all possible privacy conflicts that may result over use of personal information online and in stored data, they are designed to help to reduce the chance of such conflicts when applied as a whole. What remains to be seen is how many organizations and governments embrace these recommendations.

Closing Discussion

> They that can give up essential liberty to obtain a little temporary safety deserve neither liberty nor safety.
>
> –Benjamin Franklin, Historical Review of Pennsylvania, 1759

The tension between law enforcement and privacy is unlikely to cease to exist. New laws and regulations governing information collection, use, sharing, and storage are sure to be introduced in the years to come. At best, we can hope that decisions about privacy policy will be carefully considered by those who are neither afraid of the results nor advocates for either extreme, and that those decisions are made in an open manner, subject to oversight and comment by an informed public. In addition, policies should be subject to reevaluation and reconsideration to ensure that, as technology changes, the existing policies still meet the intended goals.

Government is in a unique position of power over individuals, having the capability to deprive them of their freedoms and even lives. Government is likewise in the position of guarantor of citizen rights. The right to privacy should be strongly protected as a social good. Erosions of privacy can occur unintentionally, but by their nature are difficult to repair because, once privacy is lost, it is practically impossible to regain it. Moreover, unintended consequences of small disclosures may be amplified when combined with other disclosures and inferences.

Increasingly those who violate privacy policies, or who are responsible for privacy breaches, are being held accountable for these transgressions. As long as there are differences in motivation, values, and experiences among all stakeholders, we are unlikely to have a society in which everyone behaves according to the law. Thus, mechanisms such as tracing networks of offenders are needed to identify and prosecute those who violate the laws on which society is based. Furthermore, given the differences in societies and their foundational principles, it is likely that some individuals may be at risk of aggression and malice from those in other countries and social settings. Societies need to have mechanisms and procedures in place to defend themselves from such threats.

In the United States, we have found that there is strength in diversity, and great creativity in heterogeneity. American society has benefited by encouraging tolerance and allowing people the freedom to push boundaries, sometimes past the point of comfort for some other individuals and groups. Both history and philosophy encourage us to interfere as little as possible with the exercise of individuals' actions when they are not explicitly hurtful of others. This is consistent with Judge William Brandeis's analysis that "Privacy is the right to be let alone."

References

Anderson, S. R. 2004. Total information awareness and beyond: The dangers of using data mining technology to prevent terrorism. Bill of Rights Defense Committee. http://www.bordc.org/threats/data-mining.pdf.

Cauley, L. 2006. NSA has massive database of Americans' phone calls. *USA TODAY*, May 11. http://www.usatoday.com/news/washington/2006-05-10-nsa_x.htm.

Husted, B. 2005. Exec: ChoicePoint will be more open. *Atlanta Journal-Constitution*, April 1.

Madison, James, Alexander Hamilton and John Jay. 1987. The Federalist Papers, 1787–1788, ed. Isaac Kramnick. Harmondsworth: Penguin.

Otto, P. N., A.I. Antón, and O.L. Baumer 2007. The ChoicePoint dilemma: How data brokers should handle the privacy of personal information. *IEEE Security & Privacy*, in press.

Pierce, Deborah, and Linda Ackerman. 2005. Data aggregators: A study of data quality and responsiveness. http://www.privacyactivism.org/docs/DataAggregatorsStudy.html (accessed May 19, 2005).

Report of the Secure Flight Working Group: Presented to the Transportation Security Administration. 2005. http://www.schneier.com/secure-flight-report.pdf (accessed September 19, 2005).

Stirland, S. L. 2005. Privacy, security experts urge delay of passenger screening system. *National Journals Technology Daily*, September 22.

Sullivan, Bob. 2005. ChoicePoint files found riddled with errors. *MSNBC News*, March 8.

Ware, Willis et al. 1973. *Records, Computers and the Rights of Citizens.* Department of Health, Education and Welfare, US Government Printing Office.

9

Digital Search and Seizure: Updating Privacy Protections to Keep Pace with Technology

James X. Dempsey

Every day, Americans use the Internet and wireless services to access, transfer, and store vast amounts of private data. Financial statements, medical records, travel itineraries, and photos of our families—once kept on paper and secure in a home or office—are now stored on networks. Electronic mail, online reading and shopping habits, business transactions, Web surfing, and geo-location data create detailed personal profiles. More and more, our lives are conducted online, and more and more personal information is transmitted and stored electronically. We benefit from the convenience, efficiency, and access to information these technologies provide. They offer huge benefits for democratic participation and human development. Yet these advances also create new privacy challenges, for they make possible opportunities for more intrusive surveillance. This essay discusses the impact new digital technologies have on our ability to keep personal information private and explains how privacy law has not kept pace with technical innovation.

Privacy is an important constitutional value and a crucial component of the trust necessary for the flourishing of digital commerce and democracy. However, while technology has changed dramatically in the past twenty years, privacy law has not. Much of the recent debate about government surveillance has centered around the

PATRIOT Act, but the changes in law wrought by that legislation are minor in comparison to the changes being brought about by technological change. The Electronic Communications Privacy Act (ECPA) of 1986 set an important precedent by establishing privacy rules for major new technologies that emerged in the 1980s. But opportunities for surveillance have expanded in ways not contemplated when ECPA and other privacy laws were written. Constitutional interpretations issued by the Supreme Court before the Internet was invented, if read broadly, would leave much electronic data outside the coverage of the Fourth Amendment's privacy protections. Court cases grappling with new technology so far have been few in number and often inconclusive in their holdings, providing few clear limitations on government surveillance and insufficient guidance to service providers. Meanwhile, differences in data storage and transmission technologies that affect privacy protections are rarely obvious to consumers, who would probably be astonished at the lack of protection accorded their personal information.

The Storage Revolution

Internet Service Providers (ISPs) and other online service providers are increasingly offering their customers the ability to store, on the service providers' computers, quantities of data that were unimaginable twenty years ago. In April 2004, Google started to beta test its "Gmail" system, at first providing users with one gigabyte of free storage space. This represented 500 times the amount of storage in an equivalent MSN/Hotmail account at the time (Festa 2004). In response, Yahoo! announced that it would increase free customer storage space to 100 megabytes and that paid customers would receive two gigabytes (Festa 2004). MSN/Hotmail followed, declaring that it would upgrade the storage space of free accounts to 250 megabytes and paid accounts to two gigabytes. In April 2005, Google boosted the capacity of Gmail to two gigabytes and indicated that it would continue increasing capacity for the foreseeable future (Hicks 2005). This dramatic growth in storage capacity comes as more e-mail is being read via Web mail accounts. At the time when current e-mail privacy laws were written, many e-mail users accessed their e-mail by downloading it onto their personal computers. That process often resulted in the deletion of the e-mail from the computers of the service provider. Now, e-mail—including e-mail that has been read but still has value to the user—often sits on a third-party server accessible via the Web.

Such free or low-cost storage offers Internet users the convenience of accessing their e-mail, documents, and photographs from any Internet-connected computer in the world. However, it also has unintended consequences for personal privacy. Under legal principles described in more detail below, e-mail and other data stored on the computers of third parties receive less privacy protection than data stored in one's home or office.

Moreover, a remarkable development is now on the horizon: the routine storage of voice telephone calls. As Ohio State law professor Peter Swire has noted, this storage is likely to become far more common with the imminent growth of a new technology, Voice over Internet Protocol ("VoIP"). VoIP uses the packet-switching network of the Internet to connect telephone calls. Broadband access makes it reliable, and its lower cost, especially for long-distance calls, makes it highly attractive to both business and residential users (Swire 2004; Swire 2005). Professor Swire has noted that the VoIP revolution brings with it the likelihood that there will be systematic storage of telephone communications. Indeed, a growing range of hardware and software products is being offered to facilitate the recording of VoIP calls not only by large enterprises and call centers but also by small businesses and individuals. These vendors specifically note that routine call recording, once a practice limited to large customer service centers, is easier with VoIP and should become the norm for businesses if not for individuals (Keating 2006).

While these digital technologies offer a welcome set of new services, most users are not aware of the consequences that flow from the decision to remotely store their communications, personal information, and files. Unless the law catches up, loss of privacy may be a hidden and unintended price of these new services.

The Current Rules for Government Access to Stored Information

The Fourth Amendment to the US Constitution shields individuals from unreasonable government searches and seizures of their "persons, houses, papers, and effects." The Supreme Court has held that the Fourth Amendment protects not only a person's home or apartment and physical person, but also the content of his or her telephone calls. In *Katz v. United States*, the Supreme Court ruled that the Fourth Amendment protects "people not places" (389 U.S. 347, 351 [1967]). In the Court's analysis, it asks whether the individual has a "reasonable expectation of privacy"—a two-part inquiry that asks first whether the individual's conduct reflects "an actual (subjective) expectation of privacy" and, if the answer is yes, whether that expectation is "one that society [objectively] is prepared to recognize as reasonable" (*Katz*, 389 U.S. 362; Harlan, concurring [1967]). While the Court has never explicitly ruled on e-mail, it seems logical that the same Fourth Amendment protection would apply to e-mail in transit, though some analysis disagrees (Bellia 2004).

In a series of cases in the 1970s, however, the Supreme Court held that the Fourth Amendment does not apply to personal information contained in records held by third parties. Once an individual voluntarily discloses information to a business, the Court reasoned, the individual no longer has a reasonable expectation of privacy and the government can access the record without raising any constitutional privacy concerns. In *Couch v. United States*, the Court held that subpoenaing an accountant for records provided by a client for the purposes of preparing a tax return raised neither Fifth nor Fourth Amendment concerns (409 U.S. 322 [1972]). In *United States v. Miller*, the Court held that records of an individual's financial transactions held by his or her bank were outside the protection of the Fourth Amendment (425 U.S. 435 [1976]). Lastly, in *Smith v. Maryland*, the Court held that individuals have no legitimate expectation of privacy in the phone numbers they dial, and therefore the installation of a technical device (a pen register) that captured such numbers on the phone company's property did not constitute a search (442 U.S. 735 [1979]) (Mulligan 2004; Dempsey 1997).

Although these "business record" decisions predated the digital revolution, they are still cited to support the argument that the privacy of personal information and records voluntarily disclosed to businesses is not constitutionally protected. Under this theory, everything from medical records at hospitals and insurance companies, to copies of cancelled checks held by banks, to records of who calls whom compiled by telephone companies fall outside the Constitution and, unless protected by statute (which some business records are), can be freely disclosed by the business entity to the government and to others.

There are serious questions as to whether the business records doctrine is still constitutionally sound, given the huge amounts of transactional data generated by electronic systems today. Moreover, it has never been clear whether the business records doctrine properly applies to the content of stored communications (U.S. Department of Justice 2002; Solove 2002). The doctrine was developed when courts did not foresee the ability of a communications service provider to store the content of communications and documents that the subscriber never intended the service provider to read or use. Nor did courts anticipate the role of the Internet in decentralizing data storage outside the home or office. The doctrine does not take into account an alternative analogy, based on Fourth Amendment cases limiting government access to items held by a third party in physical storage, such as a storage locker. When an individual stores personal property with a third party, the owner of the property retains a privacy interest in the stored items, and a warrant would be required to search the storage space. Under that analogy, transactional information regarding the terms of storage might not be protected by the Fourth Amendment, but the stored items themselves—in this case, the contents of stored e-mail—should be covered (Bellia 2004; Mulligan 2004).

Perhaps because Congress did not foresee the ways in which storage of e-mail would change, the business records doctrine played an important role in shaping the statutory privacy protections currently applied to e-mail and other records when they are in storage with a service provider. In 1986, Congress adopted the Electronic Communications Privacy Act (ECPA) (Public Law 99-508, U.S. Statutes at Large 100 (1986):1848). ECPA set rules for real-time interception of electronic communications, requiring essentially the same special warrant for access to e-mail in transit that had been required for tapping voice communications. It also restricted law enforcement access in real time to transactional information about all forms of electronic communications with the Pen Register/Trap and Trace statute, and adopted rules on access to stored electronic communications and stored transactional records held by service providers.

The part of ECPA addressing stored data, known as the Stored Communications Act (SCA), set rules for the government to obtain the content of stored e-mails (and now voice mails), stored transactional information related to communications, such as the "To" and

"From" lines on e-mail, and subscriber identifying information about the users of electronic communications services (Kerr 2004). The rules are complex and draw many fine distinctions that no longer match patterns of Internet usage and about which users are probably completely unaware (Mulligan 2004).

ECPA's standards for government access to e-mail messages vary depending on whether the e-mail is "in transit" on its way from the sender to the recipient or resting in storage on the server of the recipient's ISP. If the e-mail is in transit, it is protected by the highest standard in the wiretap law (U.S. Code 18, §2518 [1986]). E-mail, voice mail, and other communications (such as VoIP communications) stored with a service provider are entitled to less protection, and the level of protection depends on how long the communication has been stored and possibly on whether it has been accessed by the recipient. In general, e-mail, voice mail, and VoIP communications stored with a service provider for 180 days or less are afforded Fourth Amendment protection (although not the higher protection of the wiretap laws) and can be disclosed to the government only pursuant to a warrant issued on the basis of probable cause (U.S. Code 18, §2703(a)). Communications stored on the server of an ISP or other service provider for more than 180 days can be disclosed pursuant to a court order or even a mere subpoena (U.S. Code 18, §2703(b) [1986]). And the Department of Justice maintains that even very recent communications stored on the computer of a service provider fall under the lower standard of protection as soon as they are listened to or read by the customer (U.S. Code 18, §2510(17); U.S. Code 18, §2703(b)). A federal appellate court in California rejected the government's position, finding in *Theofel v. Farey-Jones* that, within the 180-day period, opened and unopened messages enjoy uniform privacy protections. In June 2007, the federal appellate court in Cincinnati ruled that stored email is constitutionally protected.

The rules concerning the ability of private parties to get access to stored data are even less clear. The use of civil subpoenas to obtain information from ISPs has received some media attention, in part due to the recording industry's initiative to subpoena the ISP records of individuals who are suspected of sharing copyrighted music files. For years, however, civil subpoenas have been served on ISPs in civil disputes such as divorce or custody cases, employment litigation, defamation, and other cases between private parties. ECPA focuses on government surveillance concerns, and it offers no clear guidance on access to records by private litigants. ECPA generally prohibits disclosures of the contents of stored e-mail to private parties, with certain exceptions, none of which expressly authorizes disclosures pursuant to a civil subpoena (U.S. Code 18, §2702 [1986]). On the other hand, ECPA provides that ISPs can disclose any records pertaining to subscribers other than the content of communications to private parties without the subscriber's permission and without a subpoena (U.S. Code 18, §2702(c)(6) [1986]). (As a matter of policy, many ISPs do not disclose subscriber information without a subpoena.) In addition, there is no requirement in ECPA that either the service provider disclosing records or e-mail content to a private litigant or the private litigant obtaining them via subpoena give any notice to the person whose information is being sought (*Virginia Code Ann.* §8.01-407.1(A)(3)). In contrast, if the ISP is covered by the Cable Act, that law requires parties to civil suits to obtain a court order and requires the cable operator (offering ISP service in this instance) to provide notice to the subscriber (U.S. Code 47, §551(c)(2)(B) [1984]).

A June 2004 decision by a three-judge panel of the federal appeals court in Boston triggered a controversy that illustrated another way in which ECPA does not match user expectations. The case, *United States v. Councilman*, noted that an ISP could read and use for its own business purposes (but not disclose to others) the e-mails of subscribers held in storage on the service provider's computers. The court went one step further and held that e-mails could be read by service providers even when they were in the very brief temporary storage that occurs as an e-mail is being transmitted. A larger panel of the court reversed that decision, holding that the e-mail was in transit when it was intercepted and therefore could not be read and used by the service provider. But the second decision did not question the provision of ECPA that allows service providers to read their customers' e-mail for any purpose at all once it is in storage on the ISP's server. Many in the industry felt that, given the practices of legitimate ISPs, which do not read their customers' e-mails, the *Councilman* controversy was overblown. Nevertheless, the case drew attention to an overlooked gap in the law (Center for Democracy and Technology 2005). ECPA's failure to prohibit ISPs from reading their subscribers' e-mail is in contrast to the law governing telephone companies, which does prohibit them from listening to customer conversations except to ensure service quality, detect fraud, or otherwise provide service (U.S. Code 18, §2511(2)(a)(i) [1968]).

In the 108th Congress (2003-04), legislation was introduced to address some of these issues. The E-mail Privacy Act, sponsored in the House by Representative Inslee (D-WA), would have made it clear that law enforcement officials have to obtain a wiretap order to gain access to Internet communications at any point during their transmission between the sender and the point at which they are made available to the recipient. The Inslee bill also would have prevented ISPs from reading their customers' e-mail except in cases where it is necessary to provide service or with consent. With the same intent, Representative Nadler (D-NY) introduced the E-mail Privacy Protection Act. While both bills would have helped to close the loophole highlighted by the *Councilman* decision, they did not address other shortcomings of ECPA. Neither was enacted.

Location Tracking

Other technologies are rapidly being adopted that can track our physical movement, raising additional privacy concerns. Cell phones have always had an inherent capability to locate their users. The Federal Communication Commission's Enhanced 911 rules have made the location capabilities of cell phones more accurate, and wireless carriers and others who have been required to make enhancements to their networks to incorporate location capabilities for emergency response purposes have sought revenue-generating applications for the same technology. Computers are the next step. "With the transition of the notebook from just being a portable computer to being a mobile computing device, LAC [location aware computing] holds a lot of promise towards growing the value and excitement of the mobile notebook platform" (Bajikar 2003). The FCC has already extended its E911 rules to some VoIP services, and is considering requiring that a location capability be built into essentially all computers. Meanwhile, entrepreneurs are developing wireless platforms for location aware computing, marrying GPS and wireless technologies to create a wireless network that will serve as a backbone for location applications (David 2004). Other technologists are working on tracking systems for high-density urban areas and indoor settings, where GPS capability is limited (Spring 2004; Location privacy workshop 2004). "With numerous factors driving deployment of sensing technologies, location-aware computing may soon become a part of everyday life" (Hazas, et al. 2004).

Location information can reveal a person's acquaintances and physical destinations. Such data may imply—correctly or incorrectly—additional information about the individual. Informational privacy about one's movements in society implicates the constitutional right to travel and the freedom to associate. Without assurance that one's movements are not arbitrarily being watched and recorded by the government, full exercise of these liberties will be impaired.

The capability of current location technologies to record both present and past movements of individuals raises greater privacy concerns than older electronic beepers. Beepers emit an electronic signal that police can monitor with a receiver; the signal becomes stronger or weaker depending on how close the receiver is to the beeper. This basic technology is not conducive to long-term, remote surveillance. With newer technologies, however, tracking can be done automatically by a remote computer, making it possible for law enforcement to monitor the movement of many more people for longer periods of time. It is now possible to compile and retain comprehensive records of individuals' movements over a period of years. And the technology will continue to improve in the coming years, making it easier and easier to monitor individuals' precise locations over prolonged periods of time.

In some ways, location tracking is more intrusive than a traditional physical search. Whereas a search warrant restricts the physical areas police may enter to search, a record of a person's movements cannot be similarly limited so as to provide only location information that may relate to criminal activity. Police can monitor a person's movements continuously. Also, location tracking is often covert. In the execution of the traditional search warrant, an announcement of authority and purpose ("knock and notice") is required so that the person whose privacy is invaded can observe any violations in the scope or conduct of the search and seek to halt or remedy them. In contrast, an individual whose movements have been monitored by law enforcement agents might never be aware that he or she was a target of surveillance (Bray 2005).

No existing statute sets explicit standards for government location tracking. There is a federal statute on tracking devices, U.S. Code 18, §3117 [1986], but it does not provide a particular standard for approving use of a tracking device. However, in the Communications Assistance for Law Enforcement Act of 1994 (CALEA), Congress specified that law enforcement agencies could not use an order authorizing a pen register or trap and trace device to acquire location information (U.S. Code 47, §1002(a)(2) [1994]; U.S. Code 18, §3121-3127 [1986]). While it did not specify what authority could be used to acquire location information, Congress made it clear that the very low standard of the pen/trap law was inadequate. Congress reaffirmed that location tracking information is entitled to special treatment in 1999, when it specified that telecommunications carriers cannot disclose wireless location information for commercial purposes except with the prior express approval of the customer (U.S. Code 47, §222 [1996, 1999]). By requiring prior express authorization, Congress set a higher standard for location information than for other telephone transactional data.

How, then, should the courts and Congress develop an appropriate standard for access to location information? Until recently this issue was unresolved, and government agents were routinely obtaining orders for disclosure of location information on less than probable cause. In 2005, however, two federal magistrate judges ruled that a warrant is required for government agents to obtain location information from service providers.[1] Nine more magistrates and two district court judges followed suit in 2006, while judges sided with the government in a number of other cases. In those cases ruling that real-time access to location information requires a search warrant based on probable cause, the magistrates and judges noted that the tracking of a person's location as he or she moves into private spaces like a home or office intrudes on matters protected by the Fourth Amendment, and is not comparable to the kind of information voluntarily disclosed to a business in the course of a commercial transaction. Real-time location information, these courts have held, cannot be acquired by a mere pen register, issued without serious judicial scrutiny, nor is it equivalent to stored records. The analysis and reasoning of the cases requiring a warrant are persuasive, but it is likely that the issue will continue to draw conflicting opinions until the Supreme Court or Congress settles it.

---

[1]In re Application for Pen Register and Trap/Trace Device with Cell Site Location Authority, Magistrate No. H-05-557M (S.D. Tex., Oct. 14, 2005); In the Matter of Application of the United States for an Order Authorizing Pen Register and Trap and Trace Device and Release of Subscriber Information (E.D.N.Y. Oct. 24, 2005). Both opinions, and other supporting materials, are online at http://www.eff.org/legal/cases/USA_v_PenRegister/.

Spies in Our Computers: "Keystroke Loggers"

Even more intrusive than monitoring our physical location is the monitoring of every keystroke entered into a computer. The extent of this type of surveillance is unknown; however, regulation has been slow to catch up. Keystroke loggers are computer programs that record every letter and command typed—every keystroke—on a computer. Most keystroke loggers also record the name of the application with which the keystrokes are associated and the time and date the application was opened. These programs are commonly used for relatively benign purposes to detect sources of error in computer systems, as a means for companies to assess their employees' work habits, and as a parental control device to monitor children's Internet use. Loggers can also be used, however, to obtain passphrases to encryption programs, in order to decrypt files scrambled by the owner. In June 2003, the antivirus and antispam software vendor Sophos reported that the W32/Bugbear-B virus contains a keystroke logger that allows confidential information, including passwords and credit card numbers, to be stolen from infected computers. When users who have been hit by the virus log onto password-protected Web sites—such as online banks or e-commerce sites—their passwords and account details are secretly stored (Vamosi 2003). Keystroke loggers can be physically installed on a computer or they can be remotely installed without obtaining physical access to the computer. Once installed, keystroke loggers enable the recording of information entered into a computer within the sanctity of one's home and access to one's private matters, including letters, diary entries, e-mails, and financial information, whether conveyed to a third party or not.

Governmental agencies have disclosed little information about keystroke loggers, citing the desire to protect law enforcement and national security interests. One of the few publicly known instances of government use of keystroke loggers surfaced in late November 2001 with the first court decision addressing this new method of surveillance, *United States v. Scarfo*. In *Scarfo*, a keystroke logger was installed when the FBI broke into the target's office and obtained physical access to his computer. Following increased press attention in response to the *Scarfo* case, the FBI acknowledged that it had developed an even more sophisticated type of keystroke logger, code-named Magic Lantern (Bridis 2001). According to press reports, Magic Lantern can be remotely installed on a computer via e-mail containing a virus disguised as a harmless computer file, known as a "Trojan horse" program, or through other common vulnerabilities hackers use to break into computers. Keystrokes recorded by Magic Lantern can be stored to be seized later in a raid or can even be transmitted back to the FBI over the Internet (Vamosi 2001; Wired News 2001).

Depending on how it is used, the technology also could capture information from individuals other than the target. Many personal computers located in homes and businesses are used by multiple people, and keystroke loggers could be installed on public computers at libraries or Internet cafés. In 2003, an individual placed keystroke loggers on computers at various New York City Kinko's to obtain people's banking passwords (Jesdanun 2003). Until they closely examine the entire stream captured from a multiple-user computer, law enforcement officers have no way of determining which captured keystrokes the surveillance target entered. In other contexts where such overbroad collection is likely to occur, such as pay phones or phones in a home or business, the wiretap laws require trained personnel to minimize, in real time, the recording of innocent conversations, to protect against abuse.

The use of keystroke loggers raises privacy concerns not contemplated by the current legal standards courts apply to determine whether a search has been conducted in a lawful manner. So far, the only court to rule on government use of keystroke loggers has held that law enforcement officers do not have to obtain a wiretap order before installing a keystroke logger on a targeted computer. The opinion, *United States v. Scarfo*, was issued by a federal trial court in New Jersey in 2001. In *Scarfo*, in which the FBI had obtained a search warrant to enter the target's office and physically install the keystroke monitor, the court denied a challenge to the introduction of evidence. The court failed to recognize that keystroke logging surveillance involves a level of intrusiveness not addressed by an ordinary search warrant but rather is closer to, and in some ways even more intrusive than, a wiretap and therefore should be subject to enhanced protections.

Because the Fourth Amendment "protects people, not places," the Supreme Court made it clear in 1967 in *Katz v. United States* that a search occurs even if there is no physical trespass (389 U.S. 347, 361 [1967]). In 2001, relying on *Katz*, the Supreme Court in *Kyllo v. United States* concluded that where the government "uses a device that is not in general public use, to explore details of the home that would previously have been unknowable without physical intrusion, the surveillance is a 'search' and is presumptively unreasonable without a warrant" (533 U.S. 27, 40 [2001]). Given the strong expectation of privacy in the keystrokes entered into one's computer, the Fourth Amendment certainly applies to the use of keystroke loggers, whether the keystroke logger is installed on-site or remotely.

However, keystroke loggers are different in crucial ways from an ordinary search. They perform the electronic equivalent of general searches, yet in *Scarfo*, the court held that use of the keystroke logger did not constitute an unlawful general search even though it recorded every keystroke typed. Traditional search warrants are normally executed at a specified time and place and are limited to a search for specified items. Previously disposed of or destroyed items cannot be subject to seizure. Nor does a search warrant permit ongoing surveillance. In contrast, keystroke loggers can record even deleted communications. And they entail an ongoing search, over a period of days, weeks, or longer.

In response to the uniquely intrusive aspects of electronic surveillance, the Supreme Court in 1967 imposed additional Fourth Amendment requirements (*Berger v. New York*, 388 U.S. 41 [1967]). A year later, Congress implemented those requirements by enacting the federal wiretap law, which imposes special privacy protections on electronic surveillance to comply with the Fourth Amendment: wiretaps are available only for the most serious cases; authorization to conduct a tap is issued only when all other investigative techniques have failed; applications are subject to rigorous judicial scrutiny; wiretaps are conducted in such a manner as to minimize the interception of innocent conversations; and parties whose conversations are intercepted are entitled to obtain after-the-fact judicial review of the authorization and conduct of wiretaps (U.S. Code 18, §2518 [1968]).

According to the district court's holding in *Scarfo*, the more stringent protections of the federal wiretap laws were not applicable to the keystroke logger surveillance at issue because the logger did not record keystrokes entered into the computer while the modem was operational. The court's decision was based on the fact that the wiretap laws cover communications, not data stored on the hard drive of a personal computer. In *Scarfo*, the FBI had configured the keystroke logger not to record while the modem was in use, specifically to prevent the interception of any electronic communications.

The court in *Scarfo* erred in two respects. First, it failed to account for the recording of keystrokes that compose documents to be sent later as e-mail or as attachments to e-mail messages. There is no rational basis for according a higher level of protection to communications contained in an e-mail at the moment of transmission over the Internet than to the very same message while it is being composed. A person's expectation of privacy regarding that communication remains the same. If anything, the privacy interests implicated by accessing keystrokes that have yet to be conveyed to a third party should be higher.[2] Second, the *Scarfo* court disregarded the constitutional problem of allowing precisely the type of general search that the Supreme Court attempted to prevent when it imposed strict requirements on wiretaps and bugs, including minimization requirements. A keystroke logger effectively facilitates government access to the inner thoughts of its surveillance target, even those thoughts that have been abandoned (that is, deleted), and even those thoughts entirely irrelevant to the investigation.

The current statutory framework fails to provide concrete guidelines for the application of existing law to new surveillance technologies. Vital questions about the government's use of keystroke loggers remain unanswered. How often is keystroke-logging surveillance employed and under what circumstances? Are judges carefully overseeing how the devices are being used? Mere FBI assurance that it will comply with all existing privacy laws when employing keystroke loggers does not alleviate concerns about governmental intrusion on privacy interests, since existing standards do not explicitly address the intrusive potential of these new surveillance technologies.

Revitalizing the Fourth Amendment Right to Privacy

The issue here is not about trying to limit innovation. All of the technologies examined here have legitimate uses. They afford con-

[2]In October 2004, a federal judge in California dismissed criminal charges under the Wiretap Act against a company employee who installed a keystroke logger on another employee's computer. The court held that interception of a transmission between a keyboard and a computer's processing unit was not covered by the Wiretap Act because a personal computer is not a system affecting interstate commerce, as required by the statute, regardless of whether the computer is connected to the Internet or another network and regardless of whether the logger captures e-mails being composed on the computer. *United States v. Ropp*, Cr. No. 04-3000-GAF (C.D. Cal., Oct 8, 2004) (unpublished). In our view, the case contains a cramped interpretation of the statute. It is inconsistent with the premise in *Scarfo* that keystroke loggers should not be used to capture e-mails without a Title III order when a computer is connected to the Internet. It is also inconsistent with the longstanding understanding that computers connected to the Internet are in interstate commerce.

venience, support new lines of business, and even enhance security. Their deployment should be applauded. Nor is the issue about denying any authority to the government. Especially in the face of terrorism, the government needs the authority to monitor advanced communications technologies. The concern is with assuring that new surveillance capabilities are subject to appropriate checks and balances.

It is time for a broad-based dialogue about the ways in which technology is undermining traditional privacy expectations. Technology companies should be aware of the issues so they can design products and services in ways that promote privacy and user control. The courts should reexamine the assumptions on which Fourth Amendment interpretations have been based and should be more careful in approving government surveillance requests. And Congress should update statutory protections to ensure that the principles that govern traditional surveillance techniques continue to apply to new technologies. Just as, in 1968, Congress permitted the use of wiretaps only if approved by a judge, and just as, in 1986, Congress extended protections to e-mail and wireless communications, Congress should ensure that new surveillance technologies are subject to appropriate standards. Government officials often argue that changing technology requires new powers to combat sophisticated terrorists and other criminals. Sometimes that is true. But the American people also need new protections to ensure that technological changes do not result in an unjustified loss of privacy.

Acknowledgments

This essay is based on *Digital search and seizure: Updating privacy protections to keep pace with technology*, a report released by the Center for Democracy and Technology, which was co-written by Ari Schwartz, Lara Flint, Deirdre Mulligan, Gemma Suh, and Indrani Mondal. Ben Karpf provided additional research assistance.

References

Bajikar, Sundeep. 2003. New notebook capability: Location aware computing. http://www.intel.com/design/mobile/platform/downloads/lac_white_paper.pdf.

Bellia, Patricia L. 2004. Surveillance law through cyberlaw's lens. *George Washington Law Review* 72:1375.

Berlind, David. 2005. Keystroke loggers must send Microsoft back to firewall drawing board. *ZDNet*, July 1. http://techupdate/stories/main/microsoft_firewall.html.

Bray, Hiawatha. 2005. GPS spying may prove irresistible to police. http://www.boston.com/business/technology/articles/2005/01/17/gps_spying may prove irresistible to police.

Bridis, Ted. 2001. FBI is building a "Magic Lantern"; Software would allow agency to monitor computer use. *Washington Post*, November 23.

Center for Democracy and Technology. 2005. *Federal appeals court reaffirms email privacy protections*. CDT Policy Post 11.20. http://www.cdt.org/publications/policyposts/2005/20.

David, Mark. 2004. I was where woz was: Location-aware computing. http://www.macnewsworld.com/story/I-Was-Where-Woz-Was-Location-Aware-Computing-37878.html.

Dempsey, James X. 1997. Communications privacy in the digital age: Revitalizing the federal wiretap laws to enhance privacy. *Albany Law Journal of Science and Technology* 8:65.

FBI 'fesses up to net spy app. 2001. *Wired News*, December 12. http://www.wired.com/news/conflict/0,2100,49102,00.html.

Festa, Paul. 2004. Google to offer gigabyte of free email. CNET News.com April 1, 2004.

Hazas, Mike et al. 2004. Location-aware computing comes of age. *Computer* 37:95–97.

Hicks, Matthew. 2005. Google boosts Gmail storage to 2GB. *eWEEK*, April 1. http://www.eweek.com/article2/0,1895,1781392,00.asp.

Jesdanun, Anick. 2003. Kinko's spy case highlights risks of public Internet terminals. *Detroit News*, July 23. http://www.detnews.com/2003/technology/0307/23/technology-224836.htm.

Keating, Tom. 2006. VoIP call recording. http://blog.tmcnet.com/blog/tom-keating/voip/voip-call-recording.asp.

Kerr, Orin. 2004. A user's guide to the Stored Communications Act, and a legislator's guide to amending it. *George Washington Law Review* 72: 1208.

Kinko's keystroke caper underscores need for diligence. 2003. http://www.bankersonline.com/technology/techalert_072303.html. *Bankers Online* July 23.

Location privacy workshop–individual autonomy as a driver of design. 2004. http://www.spatial.maine.edu/LocationPrivacy/backgroundReadings.html and http://www.spatial.maine.edu/LocationPrivacy/program.html.

Mulligan, Deirdre K. 2004. Reasonable expectations in electronic communications: A critical perspective on the Electronic Communications Privacy Act. *George Washington Law Review* 72:1557.

Solove, Daniel J. 2002. Digital dossiers and the dissipation of Fourth Amendment privacy. *Southern California Law Review* 75:1083, 1135.

Spring, Tom. 2004. Location reigns supreme with future PCs; MIT conference looks at the future of location-based computing. *PC World*, October 1. http://www.pcworld.com/news/article/O,aid,118031,00.asp.

Swire, Peter P. 2004. Katz is dead, long live Katz. *Michigan Law Review* 102:904. http://papers.ssrn.com/sol3/papers.cfm?abstract_id=490623.

Swire, Peter P. 2005. Testimony before the Subcommittee on Crime, Terrorism, and Homeland Security of the Judiciary Committee of the U.S. House of Representatives, Oversight hearing on the implementation of the USA PATRIOT Act: Sections of the act that address crime, terrorism, and the age of technology, April 20. http://judiciary.house.gov/media/pdfs/Swire042105.pdf.

United States Department of Justice, Computer Crime and Property Section. 2002. Searching and seizing computers and obtaining electronic evidence in criminal investigations. *Search and Seizure Manual §III.A.* http://www.usdoj.gov/criminal/cybercrime/s&smanual2002.htm.

Vamosi, Robert. 2003. Bugbear.b is on the prowl. *20-Net*, June 5. http://techupdate.zdnet.com/techupdate/stories/main/0,14179,2913939,00.html.

Vamosi, Robert. 2001. Warning: We know what you're typing (and so does the FBI). *ZD-Net*, December 5. http://reviews-zdnet.com/4520-6033_16-4206694.html.

Yahoo ups e-mail storage space to 2GB. 2004. *TechWeb.com* accessed June 15.

10

Lip-Synching at the Lectern: Information Technology and the Interpersonal Divide

Michael Bugeja

In the summer of 2005, I taught a two-day seminar for incoming communication students at Iowa State University to acquaint them with their new academic world. The goal of the ISU seminar was to help first-year students acclimate to the rigor of a residential Research I campus. As several prejournalism majors were in attendance, I opted to conduct a current events session. At the time, the debate about the nomination of John Roberts to the US Supreme Court was becoming intense in the aftermath of Harriet Meiers's withdrawal from the confirmation process. I asked students for their opinions. No one knew who John Roberts was. I informed them that President Bush had nominated him to replace Sandra Day O'Connor. An awkward silence ensued. "You heard that she has decided to retire, right?"

Then I noticed a cell phone in the pocket of a backpack. Inspiration struck. "Okay," I said, "put your weapons on your desks—every piece of portable technology that you have on your person or that you brought with you to class." Out came cell phones and other portable devices, such as iPods with myriad accessories. I asked them what other technology they owned or had access to at home or in residence halls and compiled a list that included desktops; laptops; handheld gadgets; MP3, CD and DVD players; video gaming devices; and much more. "You or your parents or guardians purchased these items for you so you would not fall into the digital divide," I said at the lectern. "You have access to these devices, but don't know the news." But those students knew *some* things. "Who remembers who lip-synched the wrong song on *Saturday Night Live*?" I asked. Nearly every hand in the class went up.

Students may not have heard about Sandra Day O'Connor's retirement, or even have known who she was; but they remembered Ashlee Simpson's live faux pas a year earlier. They also were acquainted with Jennifer Wilbanks, the runaway Georgia bride who fabricated a story about being kidnapped and sexually assaulted by a Hispanic man and white woman.

My research since 1999 has investigated the promises of communication technology, without which we were supposed to tumble into the so-called Digital Divide of social disenfranchisement. My theory is that we have tumbled into an interpersonal divide so isolatingly divisive that we are losing core values pertaining to community (how we interact with others at home, school, and work) and to the community of ideas (how we contribute to society, especially science). This chapter explores the promise and outcome of a society that believed technology would provide leisure and enlightenment rather than distraction and entertainment. We explored those outcomes in the orientation class that had never heard of our current chief justice of the Supreme Court, but who knew about lip-synching songstresses.

Instead of leading a current events discussion, I decided to introduce the class to Marshall McLuhan. "The medium is the moral," I quipped, alluding to the famous opening paragraph of *Understanding Media: the Extensions of Man*:

> In a culture like ours, long accustomed to splitting and dividing all things as a means of control, it is sometimes a bit of a shock to be reminded that, in operational and practical fact, the medium is the message. This is merely to say that the personal and social consequences of any medium—that is, of any extension of ourselves—result from the new scale that is introduced into our affairs by each extension of ourselves, or by any new technology. (McLuhan 2002, 7)

McLuhan was not only addressing how each medium distorts interpersonal messages by filtering the sense of sight through a television screen or by eliminating it entirely through the radio; he also was calling attention to how culture adapts to technology because human senses do—a powerful concept that is worth reconsidering in the digital age. In other words, if the iPod was produced to vend digital music, that is what it will do, despite its potential to perform a wider range of tasks. Moreover, consumers will adapt to that influence by purchasing tunes at $1 each, donning ear buds and ignoring peers on the campus green and professors in the wireless lecture hall. In sum, the iPod displaces students in their learning environment, changing the culture of the campus.

Duke University had to relearn the McLuhan lesson the hard way during its iPod giveaway to 1,650 first-year students in 2004. Since then, the university has had to defend its "iPod First-Year Experience," emphasizing that the Apple music player was the device of choice for a variety of educational tasks meant to keep pace with a mobile generation of learners. The institution could have taken a lesson from those learners rather than trying to send them lessons through a digital musical device. In a stinging editorial titled, "iPod program did not deliver," Duke's student newspaper proclaimed: "The much-hyped iPod program—for which the University spent $500,000 on iPods for the entire freshman class—was far from the overwhelming academic success the University hoped for, and the experiment should not continue next year" (*The Duke Chronicle online* 2005). The editorial criticized "the product itself," noting that iPods are great portable digital music players that "do not seem to translate well into academic use and benefit few students." That assertion was correct from a theoretical standpoint but inherently at odds with a statement by Tracy Futhey, Duke vice president for information technology, who touted the iPod program in a story carried on MSNBC—appropriately subtitled, "Experiment hopes to turn music players into study tools"—as being in line with Duke's plans to use more technology in teaching (Associated Press 2004). Student journalists saw the gadget for what it was, but the administration saw it for what it could be—perhaps more evidence of a generational divide.

> Children of the "Nintendo generation" (those born since 1990) view technology primarily as portable modes of entertainment. In fact, many do not distinguish between entertainment and communication—they are one and the same. Why not? They have been reared with speaking toys, interactive keyboards, ca-

> ble TV, arcade consoles, Internet play, game handsets, cellular phones, virtual reality goggles, and chat rooms. . . . Conversely, their parents and grandparents . . . view technology as they do books: repositories of fact. Those born between 1950 and 1980 generally do not distinguish between information and communication. (Bugeja 2005)

Initially, higher education invested in technology to keep students from falling into the digital divide. In *Digital Divide*, Pippa Norris asked: "Will digital inequalities prove a temporary problem that will gradually fade over time, as Internet connectivity spreads and 'normalizes,' or will this prove an enduring pattern generating a persistent division between info-haves and have-nots?" (Norris 2001, 11) The weakness in Norris's otherwise agenda-setting work was her inability to assess fully how media conglomerates would program technology for profit, disseminating infotainment rather than information. Even brilliant technology scholars such as Nicholas Negroponte, professor of media technology at the Massachusetts Institute of Technology, overlooked how commerce afflicts invention. Thus, he unwittingly revealed his own generational bias in waxing exuberant about the potential of the Nintendo generation in his 1995 book, *Being Digital*:

> The access, the mobility, and the ability to effect change are what will make the future so different from the present. The information superhighway may be mostly hype today, but it is an understatement about tomorrow. It will exist beyond people's wildest predictions. As children appropriate a global information resource, and as they discover that only adults need learner's permits, we are bound to find new hope and dignity in places where very little existed before. . . . The control bits of that digital future are more than ever before in the hands of the young. Nothing could make me happier. (Negroponte 1995)

Nothing could make me more apprehensive, especially since Negroponte's future has already arrived. Indeed, the Internet and assorted "killer applications" have fostered more access to information and more awareness about any number of topics. However, all that access and mobility have done little to effect enduring social change. Increasingly the current youth generation plays video games and socializes on the information superhighway where pop-ups, spams, and scams elevate hype to new levels of disenfranchisement. Child pornography, online gambling, and sexual predation exist beyond people's most dire predictions. As more students appropriate a global information resource and use it for entertainment, the time has come for the academy to reevaluate its investment in technology in light of reports about declining academic achievement. As the Associated Press reports, citing data from the Organization for Cooperation and Development, "The United States is losing ground in education, as peers across the globe zoom by with bigger gains in student achievement and school graduations" (Associated Press, Sept. 13, 2005). The AP dispatch also noted that American 15-year-olds are falling behind peers in Europe, Asia, and elsewhere in applying math skills to basic tasks, and added that we spend more per student on all levels of education than any other country except Switzerland.

Much of that investment is in technology at the college level. As the National Survey of Student Engagement documents, technology is being used more often in classrooms and course assignments (Figures 10.1 and 10.2). In the five years of its surveys, the number of students who have "used an electronic medium (listserv, chat group, Internet, instant messaging, etc.) to discuss or complete an assignment," has increased steadily, with 45% of first-year students nationally responding "quite a bit" or "very much" in 2001 as opposed to 52% in 2005, with senior percentages rising from 49% in 2001 to 60% in 2005 (NSSE reports 2001-05).

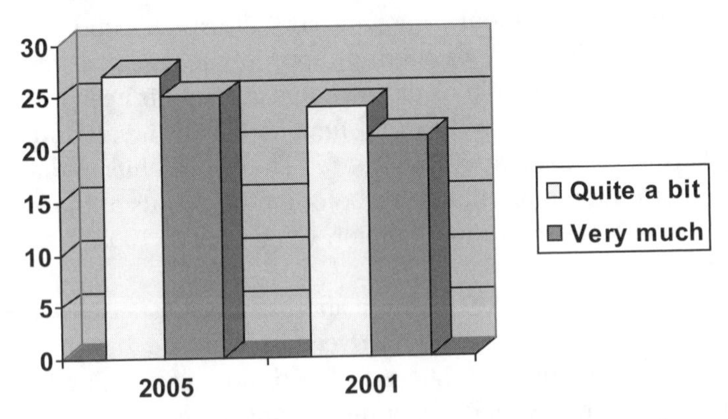

Figure 10.1 National Survey of Student Engagement: Used on electronic medium to discuss or complete an assignment. First Year. As early as 2001, a significant portion of first-year students were regularly using electronic media for assignments. Those rates rose in 2005 for first-year students.

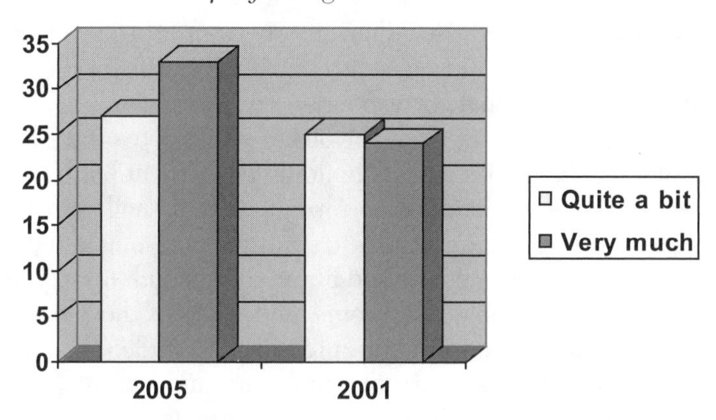

Figure 10.2 National Survey of Student Engagement: Used on electronic medium to discuss or complete an assignment. Senior Year. Seniors who used electronic media for assignments were relatively modest in 2001, assuming they had entered college in 1997–8 when the tech boom was starting to impact academe. However, 2005 seniors were accustomed to technology as first-year students in 2000–01 (see Figure 10.1), with a steep rise in the percentage who used electronic media for assignments "very much."

Information technology is designed by megacorporations such as Microsoft, Apple, Sony, Dell, and others whose goal is to raise profits, not intelligence quotients. McLuhan had to learn this lesson, when, like Negroponte, he pontificated about youth and the future that technology promised. McLuhan (1967) coined the phrase "global village," believing that electronic media "retribalized" people from various cultures and countries so that they became citizens of the planet rather than of a place, and thus would know more about world events than happenings in their hometowns. He wrote, "The new electronic interdependence recreates the world in the image of a global village" (McLuhan and Fiore 1996 [1967]).

Instead of a global village, we have inherited a global mall. Access to entertainment rather than information is built into applications of cell phones, handhelds, iPods, laptops, desktops, and the Internet. When used for communication, those devices often vend services or disseminate images of others seeking social contact, which substitutes for genuine interpersonal relationships. Consider the Facebook fad, which, unlike MySpace and other social networks, is restricted to educational settings. At Iowa State University, which has a total enrollment of 25,741, Facebook logs 20,247 registered users, with 85% of students at more than 2,000 institutions nationally—a McLuhanesque example of the culture that "information technology" has effected in academe. Facebook, an interactive, image-laden directory featuring groups that share lifestyles or attitudes, tallies 250 million hits every day and ranks 9th in overall traffic on the Internet (Chris Hughes, personal correspondence).

In 2006, I noted that issues involving social networking affect all levels of a university. Presidents building enrollment learn that technology rates higher than rigor or reputation in high school focus groups, pressuring administrators to continue investing in technology rather than faculty. Professors and librarians encounter improper use of technology in wireless classrooms, and some of those cases go to judiciary officials responsible for enforcing the student code. Career and academic advisers confront employers and parents who have screened Facebook and discovered what students have been up to in residence halls. I previously described Facebook as "a Janus-faced symbol of the online habits of students and the traditional objectives of higher education," predicting that student engagement will become more of a challenge as "first-year students enter our institutions weaned on high school versions of the Facebook and equipped with gaming devices, cell phones, iPods, and other portable technologies" (Bugeja 2006, C1).

Christine Rosen, a fellow at the Ethics and Public Policy Center in Washington DC, characterized the marketing culture that is saturating social networks. "Unlike real communities," she said, "most interactions in online groups do not take place face-to-face," so it is no surprise that "people who use networks like Facebook have a tendency to describe themselves like products" (Bugeja 2006, C1). To test that, I registered on the Iowa State Facebook and copied names of discussion groups that resembled direct mailing lists in *The Lifestyle Market Analyst,* a reference book that identifies potential audiences for advertisers. There are Baseball Fanatics (mailing list), Baseball Addicts (Facebook group), Outdoor Enthusiasts (both), Teddy Bear & Friends (list), Teddy Bear Club (group), Iowa Conservatives (list), Kickass Conservatives (group), Avid Sweatpants User (group), PC Gamers (both), Sail with Me (list), Sailing is Fun (group), Hot Ticket Teens (list), Brunettes Having More Fun (group). As one can discern, the medium again is the moral, with marketing culture saturating academic computing.

Those who teach understand how acute the problem has become in the wireless classroom. Michael Tracey, a journalism professor at the University of Colorado, recounts the time that he used the word "nostalgia" on a preexam review. His teaching assistant received e-mails asking what the term meant. "Students didn't even think to look it up," Tracey recalled. "I was tempted to reply, 'Nostalgia is yearning for a time when students at a major university would not have to ask that question'" (Michael Tracey, personal correspondence). Increasingly, in such a culture, professors find it difficult to engage students in discussions of current events. Tracey also recounted a class discussion during which he asked how many people had seen the previous night's PBS News Hour or read that day's *New York Times*. "A couple of hands went up out of about 140 students who were present," he recalls. "One student chirped, 'Ask them how many use Facebook.' I did. Every hand in the room went up. She then said, 'Ask them how many used it today.' I did. Every hand in the room went up. I was amazed."

The divide in perception between students and professors has not been assessed empirically. Nevertheless, technology that is supposed to enhance the learning process often impedes it because the devices displace teacher and student even when they are in the same physical location. The distractions have become so acute in wireless environs that about 20% of the professors in my school have inserted technology clauses in their syllabi. Excerpts:

- Anyone who engages in rude, thoughtless, selfish behavior, such as use of a cell phone for instant messages, games, etc., will have his or her cell phone confiscated until the next class session and will be excused from the class. The cell phone will be returned after the student apologizes to the class at the next class session.
- If your cellular phone is heard by the class . . . you are responsible for completing one of two options: 1. before the end of the class period you will sing a verse and chorus of any song of your choice or, 2. you will lead the next class period through a 10-minute discussion on a topic to be determined by the end of the class. (To the extent that there are multiple individuals in violation, duets will be accepted.)

Such warnings soon will be as standard as ones about cheating, which laptops also facilitate in wireless classes, with students filing answers to servers instead of inking them on palms. Small wonder that a gap is developing between professor and student. Technology was supposed to help us build a sense of community rather than displace us from our learning environment.

One of the earliest commentators about the value of digital community building, Howard Rheingold, wrote in *The virtual community: Homesteading on the electronic frontier* that he cared "about what happens in cyberspace, and to our freedoms in cyberspace, because I dwell there part of the time. The author's voice as a citizen and veteran of virtual community-building is one of the points of view presented in this book: I'm part of the story I'm describing, speaking as both native informant and as uncredentialed social scientist" (Rheingold 2000, xxxi). Two years later, Rheingold returned with a second book, *Smart mobs: The next social revolution*, in which he wrote more experientially and interpersonally about cybercitizenship: "Human discourse without eye contact has its dangers. Anyone who has experienced a misunderstanding via e-mail or witnessed a flame war in an online discussion knows that mediated communications, lacking the nuances carried by eye contact, facial expression, or tone of voice, increase the possibility of conflicts erupting from misunderstandings" (Rheingold 2002, 192).

Studies only now are documenting the so-called community-building experience. In "It's all about me: Why e-mails are so easily misunderstood," an aptly-titled article published in *The Christian Science Monitor*, Daniel Enemark noted that e-mail lacks body language and other interpersonal cues while creating a false sense of urgency that pressures people to send messages whose intent is often misconstrued by the receiver. Enemark cites a study by Profs. Justin Kruger of New York University and Nicholas Epley of the University of Chicago, who researched how sarcasm is detected in electronic messages, concluding that "[n]ot only do e-mail senders overestimate their ability to communicate feelings, but e-mail recipients also overestimate their ability to correctly decode those feelings" (Enemark 2006). As Epley states, "A typical e-mail has this feature of seeming like face-to-face communication. . . . It's informal and it's rapid, so you assume you're getting the same paralinguistic cues you get from spoken communication" (Enemark 2006).

Kruger and Epley's work indicates that e-mail not only filters interpersonal cues but also contributes to misunderstandings between correspondents. This is especially true in academe. Before the proliferation of e-mail in the mid- to late-1990s, cases in ombuds offices happened in real place and time; now, however, they are apt to contain an e-mail component such as an offending reference or inappropriate image. At its worst, the medium gives rise to cyberbullying, a new conflict variant at many institutions. In "How to Foil Cyberbullies," Jennifer Summerville and John C. Fischetti encourage professors to create "course agreements" that address the consequences of students' using inappropriate language or making threats or negative personal comments in e-mail and other digital forms of communication. "Instant messaging, blogs, and online chats often appear to be anonymous," they observe, "and a participant's sense of what is appropriate in those cyberspace interactions may differ from his or her view of suitable face-to-face encounters" (Summerville and Fischetti 2005, B36). Part of the problem with e-mail is the sheer number of messages that the typical user receives, the time it takes to sort through those messages, and the sense of urgency that accompanies these tasks. A recent survey conducted by the Iowa Association of Business and Industry noted that 66% of respondents indicated that they receive between 30 and 50 e-mails each day. A third indicated they receive more than 50 e-mails each business day. Seventy-seven percent reported that they spend between one and three hours each day sorting, dumping, answering, and otherwise processing e-mail (The *Ames* [Iowa] *Tribune* September 25, 2005). The survey also found that respondents anticipated quick replies to their electronic messages, with 56% expecting a response within a day and almost one-third of respondents expecting a reply within two or four hours. For these reasons, e-mail often intensifies incivility, a leading conflict issue at many institutions, because allegations come easily in virtual habitats in which one party is isolated at odd hours in front of a computer screen rather than present in front of a person during regular business hours. Yet in almost every case, resolution is achieved only through interpersonal intervention.

Robert B. Holmes, ombudsman at the University of Michigan, notes that his office has handled the full range of e-mail complaints, originating from prospective students as well as staff, faculty, administration, and alumni. "Some of our cases are based solely on e-mail, and some of those came to my attention via e-mail," he explains (Holmes, personal communication). It is easy to draw a line in the digital sand, he adds. "You're isolated alone at your computer and you may just say the first thing that comes to your mind." Holmes advises not to answer challenging e-mails, especially when they anger you, and to avoid sending electronically any semblance of bad news. He views these as occasions to seek resolution through a face-to-face conversation. Increasingly, however, my own research indicates that people not only are missing opportunities for face-to-face interaction, they also are losing their innate capacity to know when, where, and in what venue they should initiate communication.

This is due, in part, to technology's blurring of lines between home, school, and work. E-mail sends us assignments when we are at home and personal messages when we are at school or work. Cell phones do the same. If we don't know where we are, in real or virtual space, or if we are multitasking in both domains, then we cannot engage in focused dialogue. To understand the scope of this phenomenon, consider some statistics (Media Use in America 2003): Televisions are on seven and a half hours per day in the average household. We watch on average four hours of television per day and up to almost 50 per week or two months per year or nine years in a typical life-span. Some people use computers their entire eight-hour workday, but on average, the typical person spends 19 hours per month online. Add to these statistics iPod use, home video use, and video gaming. Now add eight hours of sleep. There are not enough hours in the day to practice interpersonal communication naturally.

The digital revolution happened quickly and seemingly without warning, changing academic culture overnight. In a 1988 study at the University of Massachusetts–Amherst, some 37.5% of students did not use a personal computer at all and only 22% owned a personal computer (Malaney 2004). By 1996, 94.3% of the undergraduates there used a personal computer and 45.1% of the students owned their own computer. Nevertheless, 33% of the students never used e-mail and 55% never used the Internet. By 2003, the latest assessment study at Amherst, 89.6% used e-mail every day or browsed the Web, spending an additional 3.46 hours per day on the Internet. Add this to the time that students spend watching television, playing video games, downloading music, and using mobile technologies like cell phones or personal digital assistants. Is it any wonder, then, that employers who once asked higher education to prepare technologically savvy graduates are calling on professors now to emphasize critical thinking? This is especially true in my discipline, journalism. The Poynter Institute notes that newsroom managers want schools to develop reporters with better critical thinking skills (Geisler 2005), including the ability to: think independently, refine oversimplifications, exercise fair-mindedness, evaluate assumptions, listen and read critically, and develop intellectual perseverance. These abilities seem at odds with the level of multitasking that is happening not only in the classroom but also in society. The Middletown Media Studies report states that "people spend more than double (129.7%) the time with media than they think they do—11.7 hours a day in total. . . . Because of media multitasking, total time in media usage is less than the sum of its parts," with tallies of "all media use by medium resulting in a staggering 15.4 hours per day" (Papper, et al. 2004, 4).

Critical thinking is essential in the sciences in particular. In explaining this at my own institution of science and technology, I often reference AIDS researcher David Ho, chief executive officer of the Aaron Diamond AIDS Research Center and Irene Diamond Professor at the Rockefeller University. Ho had access to cutting-edge medical technology and sophisticated computer systems, but did not rely on them in conceiving how to suppress viral replication through the use of multiple drug (antiretroviral) therapies. A bit of background: Ho was educated in the physical sciences as well as biology and had a passion for mathematics that would play a key role in turning a fatal disease into a treatable one. He would analyze treatments that succeeded in the lab but failed in patients. He came to understand that the HIV virus mutates rapidly, resisting each individual drug. That's when he and his team turned to mathematics, calculating the probabilities of the virus mutating simultaneously around multiple therapies. Finally, the odds were in the patient's favor.

Computers can calculate odds in split seconds. But when we rely on them, as we rely on laptops in wireless classrooms, we soon forget the role that critical thinking has played in the history of science. Machines simply cannot formulate key research questions or conceive the process by which to do so. In a recent interview, Ho expressed concern to me about critical thinking in the wireless classroom, stating, "We should be teaching our students to think creatively or to become innovators, not just test-takers."

That goal may be more difficult to attain than we realize, given the expenditures universities make in the name of universal access. For starters, you cannot turn off the wireless because there are too many overlapping signals in the typical university. Moreover, we encounter students there whose K-12 districts have been held "accountable" through multiple-choice testing, as legislatures cut budgets to higher education, resulting in ever larger class sizes where digital distractions are the greatest and where we rely again on computer-graded bubble tests emphasizing right answers rather than critical thinking.

David J. Skorton, M.D., president of Cornell University, believes wireless access properly used can enhance critical thinking. In a recent interview with me, he qualified that remark, noting that the volume of information accessible on the Web also makes "it difficult for the naive user to separate the wheat from the chaff." As for distraction in the wireless classroom, Skorton remarks that students have doodled since days of chalk and slate, "but the ability to check the weather or game scores or the headline news from their laptops during class puts an unprecedented barrier between the student and the instructor," impeding classroom discussion and the development of critical thinking.

The absence of thinking and focus is apparent outside the classroom and in the community, where we continue to consume media in record amounts and overuse technology to an unprecedented extent. A Stanford University School of Medicine study reports that one in eight Americans is addicted to the Internet (UPI October 17, 2006). A Ball State University study reports that consumers typically use more than one medium at a time, "preferring to work on the computer while the radio plays or surf the Web as the television blares in the background" (Center for Media Design January 1, 2006).

We require new theories to grasp this level of multitasking consumption. Distraction theory posits that externally generated and unanticipated random events can interrupt the cognitive process, causing a person to attend to the disruption and interrupting one's focus (Corragio 1990). Who can predict the distractions that may occur when two mobile and disembodied multitaskers use cell phones in different geographic locations? First, the number of real-time distractions doubles when two people communicate from two different locales. That doubling effect intensifies when one is in a car with children driving in traffic and another in a public setting, like a park. It is easy to conceive of other high-tech scenarios that place communicators in environments at odds with interpersonal dialogue, merely because of the mobility of technology. But are users responding the way that distraction theory postulated before the digital revolution?

Only a few years ago, if a cell phone call in public was disrupted, often that cell phone user would quickly end the call and then apologize to the person standing in his or her interpersonal space. That etiquette changed. When interrupted, cell phone users usually do not apologize to the person in the physical place but to the one in virtual space, on the other end of the call. More recently, even that etiquette has changed—so much so that interpersonal interaction has been obliterated in the process. A cell phone user now typically ignores any person in his or her physical presence. This is what I call "the 'it' factor." The cell phone user turns anyone within earshot into an inanimate object, an "it," as if that person does not exist—a practice, among others, that 60% of cell phone users find disturbing (Swanbrow 2005). Alex Halavais, assistant professor of informatics at the State University of New York at Buffalo, stated, "It's obviously a widespread frustration. The idea of a public and private space and time is getting confusing" (Melendez 2005).

Every day in the company of cell phone users, we are witness to the obliteration of such interpersonal values as privacy and discretion. Initially we bought cell phones for safety reasons, to carry with us if we rode horses in case an accident happened on the trail or in case our cars broke down on the dangerous interstate. Soon we were buying cell phones for our kindergarteners and preteens because we wanted them to have access to us in case they were abducted, as the news media tell us often occurs and which, in reality, rarely occurs. One "Amber Alert" is one too many, of course, but that is not the point: our concerns about safety *are*. The same parents who buy cell phones for their children's safety use those cell phones to order pizza while driving in a rainstorm, endangering everyone. That danger has risen dramatically. Researchers from the National Highway Traffic Safety Administration and the Virginia Tech Transportation Institute have documented that 8 of 10 collisions or near-crashes involve a lack of attention while driving. They noted that drivers "talk on their cell phones. Check their e-mail or send text messages. Help their kids with the backseat DVD player . . . " (Thomas 2006).

The latest parental practice involves the installation of DVD players in family vans. Car dealers install these devices so that parents on long trips do not have to hear children asking, "Are we there yet? Are we there yet?" There are safety concerns associated with the distraction that DVD players cause in moving vehicles, especially when drivers multitask. In addition, educators are only now beginning to realize that these media-saturated children will become our students in a few short years. How will we engage them interpersonally when society has moved within our lifetimes from family time (parents and children emphasizing togetherness), to quality time (parents advocating for fewer but more focused interpersonal interactions with children), to media time (amusing or entertaining children in what spare time we have in our digital day)?

To prepare for these future students, we must pay as much attention to interpersonal skills as to digital ones. We can require first-year students to take courses that enhance "interpersonal intelligence" and teach when, where, and for what purpose technology use is appropriate and inappropriate. Such a class might investigate how students can balance their use of technology in a world saturated with it. They might take an inventory of their digital devices and ascertain whether they are advancing or subverting their own priorities. They might research the short- and long-term effects of overexposure to media or overreliance on technology. They might analyze the consequences of mobile technology use, and question whether such use is proper for the occasion and safe for the surroundings. They could also observe how technology is used in those surroundings, and monitor their own use throughout the term, noting any changes in attitude or behavior. A final exam might entail spending a week without electronic media doing service work for others, contributing to the communities in which their universities are located.

Above all, universities must stop emphasizing programming to incoming students and start deprogramming them from technology overuse so that they fully appreciate the benefits of a residential campus. Administrators must reassess the investment in technology, too, and ascertain whether such funding is at the expense of faculty appointments, and decide whether more professors or greater electronic access is essential to engagement. Otherwise, those of us left will lip-synch to distracted learners, displaced by the yet-unfulfilled promise of information technology. Information technology can reenfranchise society, but only if we grasp its influence on interpersonal relationships and adjust our priorities accordingly.

References

Ames (IA) Tribune. 2005. Managing the E-mail crush. Sept. 25. http://www.midiowanews.com/site/tab1.cfm?newsid=15274157&BRD=2700&PAG=461&dept_id=554 335&rfi=6 (accessed May 22, 2006).

Associated Press. 2004. Duke to give iPods to all freshmen: Experiment hopes to turn music players into study tools. July 20. http://www.msnbc.msn.com/id/5469970/ (accessed May 19, 2006).

Associated Press. 2005. U.S. World Position in Education Slipping. Sept. 13. http://www.ous.edu/workinggroups/EDP/work/CNNworldposition.pdf (accessed May 23, 2006).

Bugeja, Michael. 2005. *Interpersonal divide: The search for community in a technological age*. New York: Oxford University Press.

Bugeja, Michael. 2006. Facing the Facebook. *Chronicle of Higher Education,* Jan. 27.

Center for Media Design. Engaging the Ad-Supported Media: A Middletown Media Studies Whitepaper. http://www.bsu.edu/webapps2/cmdreports/product_select.asp?product_id=7 (accessed December 6, 2006).

Corragio, L. 1990. *Deleterious effect of intermittent interruptions on the task performance of knowledge workers: A laboratory investigation.* Unpublished doctoral dissertation. Tucson, Arizona: University of Arizona.

Duke Chronicle. 2005. Ipod program did not deliver. *Duke Chronicle online,* Feb. 28. http://www.dukechronicle.com/media/storage/paper884/news/2005/02/28/EditorialstaffEditorials/Ipod-Program.Did.Not.Deliver-1472460.shtml?norewrite200605191139& sourcedomain=www.dukechronicle.com (accessed May 19, 2006).

Enemark, Daniel. 2006. It's all about me: Why e-mails are so easily misunderstood. *Christian Science Monitor.* May 15. http://www.csmonitor.com/2006/0515/p13s01-stct.html (accessed May 19, 2006).

Exploring Different Dimensions of Student Engagement, 2005 Annual Survey Results. http://nsse.iub.edu/pdf/NSSE2005_annual_report.pdf (accessed May 18, 2006).

Geisler, Jill. 2005. Critical thinking: what do you mean by that? Poynter online, April 26. http://www.poynter.org/column.asp?id=34&aid=81581 (accessed May 19, 2006).

Iowa State University. 2006. Student engagement PowerPoint. http://www.iastate.edu/~inst_res_info/PDFfiles/PCR/0106.pdf (accessed May 19, 2006).

Malaney, Gary D. 2004. Student internet use at UMass Amherst. http://www.studentaffairs.com/ejournal/Winter_2004/StudentInternetUse.html (accessed Sept. 21, 2005).

McLuhan, Marshall. 2002. *Understanding media: The extensions of man.* Cambridge, Mass: MIT Press.

McLuhan, Marshall, and Quentin Fiore. 1996 [1967]. *The medium is the massage: An inventory of effects*. San Francisco: HardWired.

Media Use in America. 2003. *Issue Briefs.* Universal City, Calif.: Mediascope Press. http://www.mediascope.org/pubs/ibriefs/mua.htm (accessed Sept. 21, 2005).

Melendez, Michelle M. 2005. The gall—they think my world is their phone booth! *Newhouse News Service*, July 28. http://www.newhousenews.com/archive/melendez072805.html (accessed May 22, 2005).

National Survey of Student Engagement project. 2001-05. http://nsse.iub.edu/html/current_users.cfm (accessed May 22, 2006).

Norris, Pippa. 2001. *Digital divide.* Cambridge, England: Cambridge University Press.

Negroponte, Nicholas. 1995. *Being digital.* New York: Knopf.

Papper, Robert A. et al. 2004. Middletown Media Studies, Spring. http://www.bsu.edu/icommunication/news/iDMAaJournal.pdf (accessed May 22, 2005).

Rheingold, Howard. 2000. *The virtual community: Homesteading on the electronic frontier*. Cambridge, Massachusetts: MIT Press.

Rheingold, Howard. 2002. *Smart mobs: The next social revolution*. Cambridge, Massachusetts: Perseus.

Summerville, Jennifer, and John C. Fischetti. 2005. How to foil cyberbullies. *Chronicle of Higher Education* 51:B36.

Swanbrow, Diane. 2005. Cell phone survey shows love-hate relationship. *University of Michigan News Service*, March 21. http://www.umich.edu/~urecord/0405/Mar21_05/06.shtml (accessed May 22, 2006).

Thomas, Ken. 2006. Distracted driving main crash course. *Associated Press*, April 21. http://www.post-gazette.com/pg/06111/683964-147.stm (accessed April 28, 2006).

United Press International. 2006. Scientists study Internet addiction. *Science Daily*, October 17. http://www.sciencedaily.com/upi/index.php?feed=Science&article=UPI-1-20061017-14405700-bc-us-webaddicts.xml (accessed December 6, 2006).

PART 3

Space Exploration: Reasons and Risks

11

An Historical Overview of US Manned Space Exploration

Roger D. Launius

Introduction

The US space program emerged in large part because of the pressures of national security during the Cold War with the Soviet Union (reviewed in Burrows 1998; McCurdy 1997; Launius 1998). From the latter 1940s, the Department of Defense had pursued research in rocketry and upper atmospheric sciences as a means of assuring American leadership in technology. The civilian side of the space effort began in 1952 when the International Council of Scientific Unions established a committee to arrange an International Geophysical Year (IGY) for the period July 1, 1957, to December 31, 1958. After years of preparation, on July 29, 1955, the US scientific community persuaded President Dwight D. Eisenhower to approve a plan to orbit a scientific satellite as part of the IGY effort. With the launch of *Sputnik I* and *II* by the Soviet Union in the fall of 1957, and the American orbiting of Explorer 1 in January 1958, the space race commenced and did not abate until near the end of the Cold War—although there were lulls in the competition (reviewed in Bulkeley 1991; Divine 1993; Dickson 2001). The most visible part of this competition was the human spaceflight program—with the Moon landings by Apollo astronauts as de rigueur—but the effort also entailed robotic missions to several planets of the solar system, military and commercial satellite activities, and other scientific and technological labors (reviewed in Logsdon 1970; McDougall 1985; Murray and Cox 1989; Chaikin 1994; Schorn 1998). In the post–Cold War era, the space exploration agenda underwent significant restructuring and led to such cooperative ventures as the International Space Station and the development of launchers, science missions, and applications satellites through international consortia (reviewed in Launius 2003a; Jenkins 2001; Heppenheimer 1999; Heppenheimer 2002; Harland 2004). This overview will examine the historical background of human space exploration, focusing on the history of NASA and the evolution of its activities in the last fifty years.

Rationales for Spaceflight

From the defining event of Sputnik in 1957, five major themes—and only these five—have been of use in justifying a large-scale spaceflight agenda: 1. scientific discovery and understanding; 2. national security and military applications; 3. economic competitiveness and commercial applications; 4. human destiny/survival of the species; 5. national prestige/geopolitics. Specific aspects of these five rationales have fluctuated over time, but they remain the only reasons for the endeavor that have any saliency whatsoever (Launius 2006).

The first and most common rationale for spaceflight is that an integral part of human nature is a desire for discovery and understanding. At one level, there exists the ideal of the pursuit of abstract scientific knowledge—learning more about the universe to expand the human mind—and pure science and exploration of the unknown will remain an important aspect of spaceflight well into the foreseeable future. This goal clearly motivated the scientific probes sent to all of the planets of the solar system. It propels a wide range of efforts to explore Mars, Jupiter, and Saturn in the twenty-first century (Siddiqi 2002). It energized such efforts as the Hubble Space Telescope, which has revolutionized knowledge of the universe since its deployment in 1990 (Smith 1989; DeVorkin and Smith 2004).

From the beginning, science has been a critical goal in spaceflight. The National Aeronautics and Space Act of 1958 that created NASA stated that its mandate included “the expansion of human knowledge of phenomena in the atmosphere and space.” This idea has continually drawn verbal and fiscal support, but it has proven less important than the pursuit of knowledge that enables some practical social or economic payoff (Logsdon 1998a).

Even the Apollo missions to the Moon, certainly inaugurated as a Cold War effort to best the Soviet Union and demonstrate the power of the United States, succeeded in enhancing scientific understanding. The scientific experiments placed on the Moon and the lunar soil samples returned through Project Apollo have provided grist for scientists' investigations of the solar system ever since. In that case, and many others, a linkage between the spirit of and the need for scientific inquiry and the spirit of and need for exploration served as strong synergetic forces for human spaceflight (Compton 1989; Harland 1999; Wilhelms 1993; Spudis 1996; Beattie 2001). The performance of scientific experiments on the Space Shuttle and the science program envisioned for the ISS demonstrated the same linkages at the beginning of the 21st century.

A second rationale for national defense and military space activity has also proved useful for spaceflight advocates. From the beginning, national leaders sought to use space to ensure US security from nuclear holocaust. For instance, in 1952, a popular conception of the US-occupied space station was as a platform from which to observe the Soviet Union and the rest of the globe in the interest of national security. The US military also argued for a human capability to fly in space for rapid deployment of troops to hot spots anywhere around the Earth. The human spaceflight enterprise also gained energy from Cold War rivalries in the 1950s and 1960s, as international prestige, translated into American support from nonaligned nations, found an important place in the space policy agenda. Human spaceflight also had a strong military nature during the 1980s, when astronauts from the military services deployed reconnaissance satellites into Earth orbit from the Space Shuttle. A human military presence in space promises to remain a compelling aspect of spaceflight well into the 21st century (Launius 2003b, 26–35, 114–121).

The third rationale of economic competitiveness and commercial applications also represents a useful justification for spaceflight that the public accepts. Space technologies, especially the complex human spaceflight component, demand a skilled and well-trained workforce whose talents are disseminated to the larger technological and economic base of the nation. The Apollo program, for example, served as an economic engine fueling the Southern states' economic growth. In recent years, the economic rationale has become stronger and even more explicit as space applications, especially communications satellites, become increasingly central for maintaining United States global economic competitiveness (Slotten 2002; Whalen 2002; Butrica 1997. Ronald Reagan's administration particularly emphasized enlarging the role of the private sector, and its priorities have remained in place ever since. One of the key initiatives in this effort for human spaceflight is tourism, a major aspect that envisages hotels in Earth orbit and lunar vacation packages. While this has yet to find realization, it remains a tantalizing possibility for the 21st century (Collins 2000).

The fourth imperative for spaceflight has revolved around human destiny. With the Earth so well known, advocates argue, exploration and settlement of the Moon and Mars is the next logical step in human exploration. Humans must question and explore and discover or die, advocates for this position insist. The terrifying implication of this rationale is that humanity will not survive if it does not become multiplanetary. Carl Sagan wrote eloquently about the last perfect day on Earth, before the Sun would fundamentally change and end our ability to survive on this planet (1980, 231–232). While this will happen billions of years in the future, any number of catastrophes could end life on Earth beforehand. The most serious threat is from an asteroid or meteor impact. Throughout history, asteroids and comets have struck Earth, and a great galactic asteroid probably killed the dinosaurs when an object only six to nine miles wide left a crater 186 miles wide in Mexico's Yucatán Peninsula.

In time, a really big one will again hit Earth with disastrous consequences. Efforts to catalogue all Earth-crossing asteroids, track their trajectories, and develop countermeasures to destroy or deflect objects on a collision course with Earth are important, but to ensure the survival of the species, humanity must build outposts elsewhere. Astronaut John Young said it best, paraphrasing Pogo, "I have met an endangered species, and it is us" (Young 2003).

Finally, national prestige and concern for geopolitical relations has dominated so many of the spaceflight decisions that it sometimes seems trite to suggest that it has been an impressive rationale over the years. Yet, there is more to it than that, for while all recognize that prestige sparked and sustained the space race of the 1960s, we too often fail to recognize that it continues to motivate support for NASA's programs. The United States went to the Moon for prestige purposes, but it also built the Space Shuttle and embarked on the space station for the same reason (Kennedy 1962).

Prestige may well ensure that no matter how difficult the challenges and how overbearing the obstacles, the United States will continue to fly in space indefinitely. In the aftermath of the *Columbia* accident on February 1, 2003, when it appeared that all reasons for human spaceflight should be questioned, no one seriously considered ending the program. Instead, support for the effort came from all quarters. Even President George W. Bush, who had previously been silent on spaceflight, stepped forward at a news conference on the day of the accident to say that "The cause in which they died will continue. Mankind is led into the darkness beyond our world by the inspiration of discovery and the longing to understand. Our journey into space will go on."

Are these sufficient rationales to sustain spaceflight indefinitely? While Americans want the endeavor's fruits, too many are unwilling to invest in it. The rationales, as real as they might be, are not compelling enough to sustain an expansive program indefinitely, and the effort has stumbled for more than thirty years after its initial acceptance.

The Space Policy Debate in the 1950s

The Role of Adventure and Discovery

There seems to be little doubt that adventure and discovery, the promise of exploration and colonization, were the motivating forces behind the small cadre of early space-program advocates in the United States prior to the 1950s. Most supporters of aggressive space exploration invoked an extension of the popular notion of the American frontier with its then attendant positive images of territorial and scientific discovery, exploration, colonization, and use. This is an expression of Frederick Jackson Turner's "Frontier Thesis," which guided inquiry into much of American history for a generation (1920). Indeed, the image of the American frontier has been an especially evocative and somewhat romantic—as well as popular—argument to support the aggressive exploration of space. It plays to the public's conception of "westering" and the settlement of the American continent by Europeans from the east, which was a powerful metaphor of national identity until the 1970s.

The space promoters of the 1950s and 1960s intuited that this set of symbols provided a vigorous explanation and justification of their efforts. The frontier metaphor was probably appropriate for what they wanted to accomplish. It conjured up an image of self-reliant Americans moving westward in sweeping waves of discovery, exploration, and conquest, culminating in the settlement of an untamed wilderness. In the process of movement, the Europeans who settled North America became, in their own eyes, a unique people on Earth, imbued with virtue and a sense of justice. The frontier model has always carried with it the ideals of optimism, democracy, and right relationships. It has been almost utopian in its expression, and it should come as no surprise that those people seeking to create perfect societies in the seventeenth, eighteenth, and nineteenth centuries—the Puritans, the Mormons, the Shakers, the Moravians, the Fourians, the Icarians, the followers of Horace Greeley—often went to the frontier to achieve their goals.

The frontier ideal also summoned in the popular mind a wide range of vivid and memorable tales of heroism, each a morally justified step of progress toward the modern democratic state. While this ideal reduced the complexity of events to a relatively static morality play, avoided matters that challenged or contradicted the myth, viewed Americans moving westward as inherently good and their opponents as evil, and ignored the cultural context of westward migration, it served a critical unifying purpose for the nation. Those who were persuaded by this metaphor—and most white Americans in 1960 did not challenge it—embrace the vision of space exploration. This frontier imagery was overtly mythic. Myths, however, are important to the maintenance of any society, for they are stories that symbolize an overarching ideology and moral consciousness. As James Oliver Robertson observes, "Myths are the patterns of behavior, or belief, and/or perception—which people have in common. Myths are not deliberately, or necessarily consciously, fictitious" (Robertson 1980, xv). Myth, therefore, is not so much a fable or a falsehood, as it is a story, a kind of poetry, about events and situations that have great significance for the people involved. Myths are, in fact, essential truths for the members of a cultural group who hold them, enact them, or perceive them. They are sometimes expressed in narratives, but in literate societies like the United States, they are also apt to be embedded in ideologies.

The Role of Popular Conceptions of Space Travel

If the frontier metaphor of space exploration conjured up romantic images of an American nation progressing to something for the greater good, the space advocates of the Eisenhower era also sought to convince the public that space exploration was an immediate possibility. It was seen in science fiction books and film, but more importantly, it was fostered by serious and respected scientists, engineers, and politicians. Deliberate efforts on the part of space boosters during the late 1940s and early 1950s helped to reshape the popular culture of space and to influence governmental policy. In particular, these advocates worked hard to overcome the level of disbelief that had been generated by two decades of "Buck Rogers"-type fantasies and to convince the American public that space travel might actually, for the first time in human history, be possible (Bainbridge 1976; Ley and Bonestell 1949).

The decade following World War II brought a sea change in perception, as most Americans went from skepticism about the probabilities of space flight to an acceptance of it as a near-term reality. This can be seen in the public opinion polls of the era. For instance, in December 1949, Gallup pollsters found that only 15% of Americans believed humans would reach the Moon within 50 years, while a whopping 70% believed that it would not happen within that time. By 1957, 41% believed firmly that it would not take longer than 25 years for humanity to reach the Moon, while only 25% believed that it would (Gallup 1972, 875, 1152). An important shift in perceptions had taken place during that era, and it was largely the result of a public relations campaign based on the real possibility of spaceflight, coupled with well-known advances in rocket technology.

There were many ways in which the US public became aware that flight into space was a possibility, ranging from science fiction literature and film, to speculations by science fiction writers about possibilities already real, to serious discussions of the subject in respected popular magazines. Among the most important efforts was that of the handsome German émigré, Wernher von Braun, working for the Army at Huntsville, Alabama. Von Braun, in addition to being a superbly effective technological entrepreneur, managed to seize the powerful print and electronic communication media that the science fiction writers and film makers had been using in the early 1950s (1953), and no one was a more effective promoter of space exploration to the public.

In 1952, von Braun burst onto the public stage with a series of articles in *Collier's* magazine about the possibilities of spaceflight. The first issue of *Collier's* devoted to space appeared on March 22, 1952. An editorial suggested that spaceflight was possible, not just science fiction, and that it was inevitable that humanity would venture outward. Von Braun advocated the orbiting of humans, development of a reusable spacecraft for travel to and from Earth orbit, building a permanently inhabited space station, and finally, human exploration of the Moon and Mars by spacecraft departing from the space station. The series concluded with a special issue of the magazine devoted to Mars, in which von Braun and others described how to get there and predicted what might be found, based on recent scientific data (von Braun and Ryan 1954; Editorial 1952; Liebermann 1992, 141; Miller 1986).

Following close on the heels of the *Collier's* series, Walt Disney contacted von Braun and asked his assistance in the production of three shows for Disney's weekly television series. The first of these, "Man in Space," premiered on March 9, 1955, with an estimated audience of 42 million. The second show, "Man and the Moon," also aired in 1955 and sported the powerful image of a wheel-like space station as a launching point for a mission to the Moon. The final show, "Mars and Beyond," premiered on December 4, 1957, after the launching of *Sputnik I*. Von Braun appeared in all three films to explain his concepts for human spaceflight, while Disney's characteristic animation illustrated the basic principles and ideas with wit and humor (Liebermann 1992, 144-146; Smith 1978, 59-60; Ley 1961, 331). Both the *Collier's* articles and the Disney series helped to shape the public's perception of space exploration as something that was no longer fantasy.

The coming together of public perceptions of spaceflight as a near-term reality with the technological developments then being seen at White Sands and elsewhere, created an environment much more conducive to the establishment of an aggressive space program. The convincing of the American public that space flight was *possible* was one of the most critical components of the space policy debate of the 1950s. Without it, the aggressive exploration programs of the 1960s would never have happened. For such programs to be approved in the public policy arena, the public must have both an appropriate vision of the phenomenon with which society seeks to grapple, and confidence in the attainability of the goal. Indeed, space enthusiasts were so successful in promoting their image of human spaceflight as being just over the horizon that when other developments forced public policy makers to consider the space program seriously, alternative visions of space exploration remained ill-formed. Even advocates of different futures emphasizing robotic probes and applications satellites were obliged to discuss space exploration using the symbols of the human space travel vision that had been so well established in the minds of Americans by the promoters. The dichotomy of human versus robotic visions has been one of the central components of the US space program. Those who advocated a scientifically-oriented program using nonpiloted probes and satellites for weather, communications, and a host of other useful activities were never able to capture the imagination of the American public the way the human spaceflight advocates did (Rolland 1989; Bainbridge 1991).

The Role of Foreign Policy and National Security Issues

At the same time that space exploration advocates, both buffs and scientists, were generating an image of spaceflight as genuine possibility and no longer fantasy, and proposing how to accomplish a far-reaching program of lunar and planetary exploration, another critical element entered the picture: the role of spaceflight in national defense and international relations. Space partisans early began hitching their exploration vision to the political requirements of the Cold War, in particular to the belief that the nation that occupied the "high ground" of space would dominate the territories underneath it. In the first of the *Collier's* articles in 1952, the exploration of space was framed in the context of the cold war rivalry with the Soviet Union and concluded that "Collier's believes that the time has come for Washington to give priority of attention to the matter of space superiority. The rearmament gap between the East and West has been steadily closing. And nothing, in our opinion, should be left undone that might guarantee the peace of the world. It's as simple as that." The magazine's editors argued "that the U.S. must immediately embark on a long-range development program to secure for the West 'space superiority.' If we do not, somebody else will. That somebody else very probably would be the Soviet Union" (*Collier's* editorial 1952, 23).

Couple this sense of terror with the reality of the Soviet Union successfully testing an atomic bomb on August 29, 1949, in Semipalatinsk, Siberia, and the nightmare had become reality. This shock was still reverberating when the Soviets tested their first hydrogen bomb in the early 1950s. After an arms race that had a definite nuclear component and a series of hot and cold crises in the Eisenhower era, with the launching of *Sputnik* in 1957, the threat of holocaust for most Americans was now not just a possibility but a probability. One of President Lyndon Johnson's aides, George E. Reedy, summarized the feelings of many Americans at that time: "the simple fact is that we can no longer consider the Russians to be behind us in technology. It took them four years to catch up to our atomic bomb and nine months to catch up to our hydrogen bomb. Now we are trying to catch up to their satellite." Then Senator John F. Kennedy agreed during the 1960 presidential campaign that "if the Soviets control space they can control earth, as in past centuries the nation that controlled the seas dominated the continents" (Saegesser 1960, 12).

The linkage between the idea of progress manifested through the frontier, the selling of spaceflight as a reality in American popular culture, and the Cold War rivalries between the United States and the Soviet Union made possible the adoption of an aggressive space program by the early 1960s. The NASA effort through Project Apollo, with its emphasis upon human spaceflight and extraterrestrial exploration, emerged from these three major ingredients, with Cold War concerns the dominant driver behind monetary appropriations for space efforts.

The Space Race

The Cold War rivalry between the United States and the Soviet Union was the key that opened the door to aggressive space exploration, not as an end in itself, but as a means for the United States to achieve technological superiority in the eyes of the world. From the perspective of the twenty-first century, it is difficult to appreciate the near-hysterical concern about nuclear attack that preoccupied Americans in the 1950s. Far from being the "Happy Days" of the television sitcom, the United States was a dysfunctional nation preoccupied with death by nuclear war. Schools required children to practice civil defense techniques and shield themselves from nuclear blasts, in some cases by simply crawling under their desks. Communities practiced civil defense drills and families built personal bomb shelters in their backyards (May 1988, 93–94, 104–113). In the popular culture, nuclear attack was inexorably linked to the space above the United States, from which the attack would come.

With the launching of *Sputnik* in 1957, the sense of fear was palpable. It was the first time enemies could reach the United States with a radical new technology. In the contest over the ideologies and allegiances of the world's nonaligned nations, space exploration became contested ground (Launius, et al., 2000). Even while US officials congratulated the Soviet Union for this accomplishment, many Americans thought that the Soviet Union had staged a tremendous coup for the communist system at US expense. It was a shock, introducing the illusion of a technological gap and providing the impetus for the 1958 act creating NASA. Sputnik led directly to several critical efforts aimed at "catching up" to the Soviet Union's space achievements. Among these: a full-scale review of both the civil and military programs of the United States (scientific satellite efforts and ballistic missile development; establishment of a Presidential Science Advisor in the White House who had responsibility for overseeing the activities of the Federal government in science and technology; creation of the Advanced Research Projects Agency in the Department of Defense, and the consolidation of several space activities under centralized management; establishment of the National Aeronautics and Space Administration to manage civil space operations; passage of the National Defense Education Act to provide federal funding for education in the scientific and technical disciplines (Launius 1996a; Launius 1996b)).

More immediately, the United States launched its first Earth satellite on January 31, 1958, when *Explorer 1* documented the existence of radiation zones encircling the Earth. (Shaped by the Earth's magnetic field, what came to be called the Van Allen Radiation Belt partially dictates the electrical charges in the atmosphere and the solar radiation that reaches Earth.) It also began a series of scientific missions to the Moon and planets in the latter 1950s and early 1960s (Van Allen 1983; Von Benke 1997).

Because of this perception, Congress passed and President Dwight D. Eisenhower signed the National Aeronautics and Space Act of 1958 (Public Law 85-568, U.S. Statutes at Large 72 (1996): 426). This legislation established NASA with a broad mandate to explore and use space for "peaceful purposes for the benefit of all mankind" (Griffith 1962, 27–43). The core of NASA came from the earlier National Advisory Committee for Aeronautics with its 8,000 employees, an annual budget of $100 million, and its research laboratories. It quickly incorporated other organizations into the new agency, notably the space science group of the Naval Research Laboratory in Maryland, the Jet Propulsion Laboratory managed by the

California Institute of Technology for the Army, and the Army Ballistic Missile Agency in Huntsville, Alabama (Launius 1994, 29–41).

The Soviet Union, while not creating a separate organization dedicated to space exploration, infused money into its various rocket design bureaus and scientific research institutions. The chief beneficiaries of Soviet spaceflight enthusiasm were the design bureau of Sergei P. Korolev (the chief designer of the first Soviet rockets used for the Sputnik program) and the Soviet Academy of Sciences, which devised experiments and built the instruments that were launched into orbit. With huge investments in spaceflight technology urged by Soviet premier Nikita Khrushchev, the Soviet Union accomplished one public relations coup after another against the United States during the late 1950s and early 1960s (Siddiqi 2000; Harford 1997).

In the United States, within a short time after NASA's formal organization, the new agency also took over management of space exploration projects from other federal agencies and began to conduct space science missions—such as Project Ranger to send probes to the Moon, Project Echo to test the possibility of satellite communications, and Project Mercury to ascertain the possibilities of human spaceflight. Even so, these activities were constrained by a modest budget and a measured pace on the part of NASA leadership.

In an irony of the first magnitude, Eisenhower believed that the creation of NASA and the placing of so much power in its hands by the Kennedy administration during the Apollo program of the 1960s was a mistake. He remarked in a 1962 article: "Why the great hurry to get to the Moon and the planets? We have already demonstrated that in everything except the power of our booster rockets we are leading the world in scientific space exploration. From here on, I think we should proceed in an orderly, scientific way, building one accomplishment on another" (1962, 24). He later cautioned that the Moon race "has diverted a disproportionate share of our brainpower and research facilities from equally significant problems, including education and automation" (1964, 19). He believed that Americans had overreacted to the perceived threat.

The Heroic Years of US Spaceflight

During the first 15 years of the Space Age, the United States emphasized a civilian space exploration program consisting of several major components: human spaceflight initiatives—Mercury's single astronaut program (flights during 1961–1963) to ascertain if a human could survive in space; Project Gemini (flights during 1965–1966) with two astronauts to practice for space operations; and Project Apollo (flights during 1968-1972) to explore the Moon (Grimwood 1963; Link, 1965; Swenson, et al. 1966; Grimwood, et al. 1969; Ertel and Morse 1969; Morse and Bays 1973; Brooks and Ertel 1973; Hacker and Grimwood 1977; Benson and Faherty 1978; Ertel, et al. 1978; Brooks, et al 1966; Bilstein 1980; Levine 1982; Compton 1989; Fries 1992; Seamans 1996; Swanson 1999; Orloff 2000; Seamans 2005); robotic missions to the Moon (Ranger, Surveyor, and Lunar Orbiter), Venus (*Pioneer Venus*), Mars (*Mariner 4, Viking 1* and *2*), and the outer planets (*Pioneer 10* and *11*, *Voyager 1* and *2*) (Hall 1977; Newell 1980; Ezell and Ezell 1984; Naugle 1991); orbiting space observatories (*Orbiting Solar Observatory*, *Hubble Space Telescope*) to view the galaxy from space without the clutter of Earth's atmosphere (Chatterjee 1998); remote-sensing Earth-satellites for information gathering (Landsat satellites for environmental monitoring) (Mack 1990); applications satellites such as communications (*Echo 1*, *TIROS*, and *Telstar*) and weather monitoring instruments (Butrica 1997); an orbital workshop for astronauts, *Skylab* (Newkirk, et al. 1977; Compton and Benson 1983); a reusable spacecraft for traveling to and from Earth orbit, the Space Shuttle (Heppenheimer 1999; Morgan 2001).

The capstone of this effort was, of course, the human expedition to the Moon, Project Apollo. A unique confluence of political necessity, personal commitment and activism, scientific and technological ability, economic prosperity, and public mood made possible the May 25, 1961 announcement by President John F. Kennedy that the United States would carry out a lunar landing program before the end of the decade as a means of demonstrating its technological virtuosity (Cortright 1975; Lambright 1995; Reynolds 2002).

Project Apollo, backed by sufficient funding, was the tangible result of an early national commitment in response to a perceived threat to the United States by the Soviet Union. NASA leaders recognized that, while the size of the task was enormous, it was still technologically and financially within their grasp, but they had to move forward quickly. Accordingly, the space agency's annual budget increased from $500 million in 1960 to a high point of $5.2 billion in 1965. A comparable percentage of the $1.9 trillion federal budget in 2006 would have equaled more than $77 billion for NASA, whereas the agency's actual budget then stood at $16.6 billion. NASA's budget began to decline beginning in 1966 and continued a downward trend until 1975. NASA's fiscal year 1971 budget took a battering; forcing the cancellation of Apollo missions 18 through 20. With the exception of a few years during the Apollo era, the NASA budget has hovered at slightly less than 1% of all money expended by the US Treasury. Stability has been the norm as the annual NASA budget has incrementally gone up or down in relation to that 1% benchmark (see Figure 11.1).[1]

While there may be reason to accept that Apollo was transcendentally important at some sublime level, assuming a generally rosy public acceptance of it is at best a simplistic and ultimately unsatisfactory conclusion. Indeed, the public's support for space funding has remained remarkably stable at approximately 80% in favor of the status quo since 1965, with only one significant dip in support in the early 1970s. However, responses to funding questions on public opinion polls are extremely sensitive to the wording of questions and must be used cautiously (Roy, et al. 1999). Polls in the 1960s consistently ranked spaceflight near the top of those programs to be cut in the federal budget. Most Americans seemingly preferred doing something about air and water pollution, job training for unskilled workers, national beautification, and poverty, before spending federal funds on human spaceflight. As *Newsweek* stated: "The US space program is in decline. The Vietnam war and the desperate conditions of the nation's poor and its cities—which make spaceflight seem, in comparison, like an embarrassing national self-indulgence—have combined to drag down a program where the sky was no longer the limit."[2]

---

[1]These observations are based on calculations using the budget data included in the annual *Aeronautics and Space Report of the President, 2003 Activities* (Washington, DC: NASA Report, 2004), Appendix E, which contains this information for each year since 1959; "National Aeronautics and Space Administration President's FY 2007 Budget Request," FY 2007 Budget Request, February 6, 2006, p. 1, NASA Historical Reference Collection.

[2]*The Gallup Poll: Public Opinion. 19351971*, III: *19591971*, pp. 1952, 218384, 2209; *New York Times*, December 3, 1967; *Newsweek* is quoted in *Administrative History of NASA*, chap. 2, p. 48, NASA Historical Reference Collection.

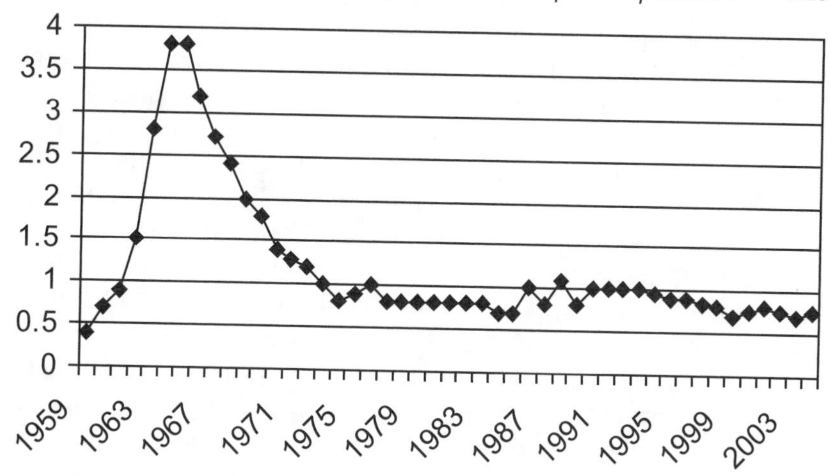

Figure 11.1 NASA budget as a percentage of federal budget.

Nor did lunar exploration in and of itself create a groundswell of popular support from the general public. The American public during the 1960s largely showed hesitancy to "race" the Soviets to the Moon, as shown in Figure 11.2. "Would you favor or oppose US government spending to send astronauts to the Moon?" these polls asked, and, in virtually all cases, a majority opposed doing so, even during

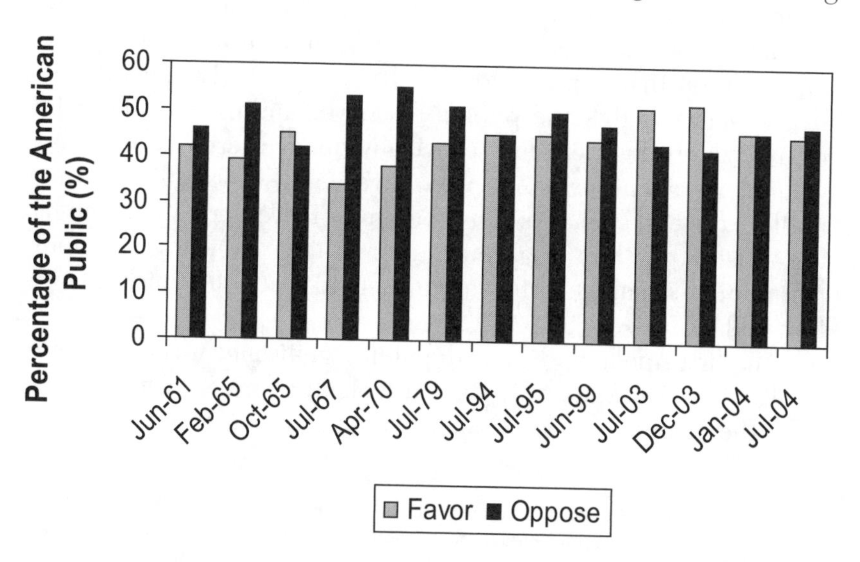

Figure 11.2 Should the government fund human trips to the moon?

the height of Apollo. At only one point, October 1965, did more than half of the public favor continuing human lunar exploration. In the post-Apollo era, the American public has continued to question the validity of undertaking human expeditions to the Moon.[3]

Some conclude from these opinion polls that even though the American public might have been generally unsupportive of human lunar exploration, Project Apollo—wrapped as it was in the bosom of American virtue, advocated by the most publicly wholesome of astronaut heroes, and hawked by everyone from journalists to Madison Avenue marketers—enjoyed consistent popularity. There is some evidence to suggest this, but it is, on the main, untrue. From the 1960s to near the present, using the polling data that exists, there is little evidence of public support for an expansive lunar exploration and colonization program. One must conclude from hard evidence that the United States undertook and carried out Apollo not because the public clamored for it during the 1960s, but because it served other purposes. Furthermore, the polling data suggests that, should the United States mount another human mission to the Moon in the future, it will also be because the mission serves a larger political, economic, or national defense agenda (Launius 2003b).

These statistics do not demonstrate unqualified support for NASA's effort to reach the Moon in the 1960s. They suggest, instead, that the political crisis that brought public support to the initial lunar landing decision was fleeting and within a short period the coalition that announced it had to retrench (Launius 2003a). It also suggests that the public was never enthusiastic about human lunar exploration, and especially about the costs associated with it. What enthusiasm it may have enjoyed waned over time, until by the end of the Apollo program in December 1972, one has the image of the program as something akin to a limping marathoner straining with every muscle to reach the finish line before collapsing.

The first Apollo mission with public significance was the flight of *Apollo 8*. On December 21, 1968, it took off atop a *Saturn V* booster from the Kennedy Space Center. Three astronauts were aboard—

[3]This analysis is based on a set of Gallup, Harris, NBC/Associated Press, CBS/New York Times, and ABC/USA Today polls conducted throughout the 1960s, copies available in the NASA Historical Reference Collection.

Frank Borman, James A. Lovell, Jr., and William A. Anders—for a historic mission to orbit the Moon. After *Apollo 8* made one and a half Earth orbits, its third stage began a burn to put the spacecraft on a lunar trajectory. It orbited the Moon on December 24–25. The crew undertook a Christmas Eve broadcast, sending an image of Earth from lunar orbit while reading the first part of the Bible—"God created the heavens and the Earth, and the Earth was without form and void"—before sending Christmas greetings to humanity. The next day they fired the boosters for a return flight and "splashed down" in the Pacific Ocean on December 27. That flight was such an enormously significant accomplishment because it came at a time when American society was in crisis over Vietnam, race relations, urban problems, and a host of other difficulties. And, if only for a few moments, the nation united as one to focus on this epochal event. Two more Apollo missions occurred before the climax of the program, confirming that the time had come for a lunar landing (Zimmerman 1998).

That landing came during the flight of *Apollo 11*, which lifted off on July 16, 1969, and, after confirmation that the hardware was working well, began the three-day trip to the Moon. Then, at 4:18 PM EST on July 20, 1969, the lunar module—with astronauts Neil Armstrong and Buzz Aldrin aboard—landed on the lunar surface, while Michael Collins orbited overhead in the Apollo command module. After checkout, Armstrong set foot on the surface, telling millions who saw and heard him on Earth that it was "one small step for [a] man—one giant leap for mankind." Aldrin soon followed him out and the two plodded around the landing site in the 1/6 lunar gravity, planted an American flag (but omitted claiming the land for the United States as had been routinely done during European exploration of the Americas), collected soil and rock samples, and set up scientific experiments. The next day they launched back to the Apollo capsule orbiting overhead and began the return trip to Earth, splashing down in the Pacific on July 24 (Armstrong, et al. 1970; Aldrin 1973; Collins 1974).

Five more landing missions followed at approximately six-month intervals through December 1972, each of them increasing the time spent on the Moon. The scientific experiments placed on the Moon and the lunar soil samples returned have provided grist for scientists' investigations ever since. The scientific return was significant, with the latter Apollo missions using a lunar rover to enhance substantially the ability to undertake scientific investigation. None of them, however, equalled the excitement of *Apollo 11*.

Project Apollo in general should be viewed as a watershed in the nation's history. It was an endeavour that demonstrated both the technological and economic virtuosity of the United States and established national preeminence over rival nations—the primary goal of the program when first envisioned by the Kennedy administration in 1961. At the same time, the Apollo program, while an enormous achievement, left a divided legacy for NASA and the aerospace community. The perceived "golden age" of Apollo created for the agency an expectation that the statement of any major space goal by the president would always bring NASA a broad consensus of support and provide it with the resources and license to dispense them as it saw fit. Something most NASA officials did not understand at the time of the Moon landing in 1969, however, was that Apollo had not been conducted under normal political circumstances and that the exceptional circumstances surrounding Apollo would not be repeated.[4]

Sustained Space Exploration?

After Apollo—and the interlude of Skylab—the space program went into a holding pattern as nearly a decade passed. During that time it moved from its earlier heroic age to one that was more characterized as a "routinization" of activities, perspectives, and processes; it was an institutionalizing of critical elements from a remarkably fertile, heroic time not at all unlike that analyzed by longshoreman philosopher Eric Hoffer in *The True Believer* (Hoffer 1951, 3–23, 137–155; Weber 1968, 46).

The Space Shuttle was intended to make spaceflight routine, safe, and relatively inexpensive. Although NASA considered a variety of configurations, some of them quite exotic, it settled on a stage-and-one-half partially reusable vehicle with an approved development price tag of $5.15 billion. On January 5, 1972, President Nixon announced the decision to build a space shuttle. He did so for both po-

[4]As an example, see the argument made by George M. Low, Team Leader, to Richard Fairbanks, Director, Transition Resources and Development Group, "Report of the NASA Transition Team," December 19, 1980, NASA Historical Reference Collection, advocating strong presidential leadership to make everything right with the US space program.

litical reasons and national prestige purposes. Politically, it would help a lagging aerospace industry in key states he wanted to carry in the next election, especially California, Texas, and Florida.[5] Supporters—especially Caspar W. Weinberger, who later became Ronald Reagan's defense secretary—argued that building the Shuttle would reaffirm America's superpower status and help restore confidence, at home and abroad, in America's technological genius and will to succeed. This was purely an issue of national prestige.[6]

The prestige factor belies a critical component. United States leaders supported the Shuttle not on its merits but on the image it projected. In so doing, the space shuttle that emerged in the early 1970s was essentially a creature of compromise that consisted of three primary elements: a delta-winged orbiter spacecraft with a large crew compartment, a cargo bay 15 by 60 feet in size, and three main engines (two solid rocket boosters (SRBs) and an external fuel tank housing the liquid hydrogen and oxidizer burned in the main engines). The orbiter and the two solid rocket boosters were reusable. The Shuttle was designed to transport approximately 45,000 tons of cargo into low-Earth orbit, 115 to 250 statute miles above the Earth. It could also accommodate a flight crew of up to 10 persons (although a crew of seven would be more common) for a basic space mission of seven days. The orbiter was designed so that, during a return to Earth, it had a cross-range maneuvering capability of 1,265 statute miles to meet requirements for liftoff and landing at the same location after only one orbit (Draper, et al. 1971; Mathews 1972; Logsdon 1986; Pace 1982; Scott 1979).

After a decade of development, on April 12, 1981, *Columbia* took off for the first orbital test mission. It was successful, and after only the fourth flight, in 1982, President Ronald Reagan declared the sys-

---

[5]George M. Low, NASA Deputy Administrator, Memorandum for the Record, "Meeting with the President on January 5, 1972," January 12, 1972, NASA Historical Reference Collection. The John Erlichman interview by John M. Logsdon, May 6, 1983, NASA Historical Reference Collection, emphasizes the political nature of the decision. This aspect of the issue was also brought home to Nixon by other factors such as letters and personal meetings. See Frank Kizis to Richard M. Nixon, March 12, 1971; Noble M. Melencamp, White House, to Frank Kizis, April 19, 1971, both in Record Group 51, Series 69.1, Box 51-78-31, National Archives and Records Administration, Washington, DC.

[6]Caspar W. Weinberger, Memorandum for the President, via George Shultz, "Future of NASA," August 12, 1971, White House, Richard M. Nixon, President, 1968–1971 File, NASA Historical Reference Collection.

tem "operational." It would henceforth carry all US government payloads; military, scientific, and even commercial satellites could all be deployed from its payload bay (Jenkins 2001).

The Shuttle soon proved disappointing. By January 1986, there had been only 24 Shuttle flights, although in the 1970s NASA had projected more flights than that every year. Critical analyses agreed that the Shuttle had proven to be neither cheap nor reliable, both primary selling points, and that NASA should never have used those arguments in building a political consensus for the program. In some respects, therefore, many agreed that the effort had been both a triumph and a tragedy. The program had been an engagingly ambitious program that had developed an exceptionally sophisticated vehicle, one that no other nation on Earth could have built at the time. As such it had been an enormously successful program. At the same time, the Shuttle was essentially a continuation of space spectaculars, *à la* Apollo, and its much-touted capabilities had not been realized. It made far fewer flights and conducted far fewer scientific experiments than NASA had publicly predicted (Launius 2006b).

All of these criticisms reached crescendo proportions following the loss of *Challenger* during launch on January 28, 1986. Although it was not the entire reason, the pressure to get the Shuttle schedule more in line with earlier projections throughout 1985 prompted NASA workers to accept operational procedures that fostered shortcuts and increased the opportunity for disaster. The accident, traumatic even under the best of situations, was made that much worse because *Challenger*'s crew members represented a cross section of the American population in terms of race, gender, geography, background, and religion. The explosion became one of the most significant events of the 1980s, as billions around the world saw the accident on television and empathized with any one or more of the crew members killed (Vaughan 1996).

With the *Challenger* accident, the Shuttle program went into a two-year hiatus while NASA worked to redesign the system. The Space Shuttle finally returned to flight without further incident on September 29, 1988 (Logsdon 1998). Through February 2003, NASA had launched a total of 114 Shuttle missions, with two tragic accidents. *Challenger* was lost during launch and *Columbia* during reentry on February 1, 2003. Each undertook scientific and technological experiments ranging from the deployment of important space probes like the *Magellan* Venus radar mapper in 1989 and the Hubble Space Telescope in 1990, through the flights of "Spacelab," to a dramatic three-person EVA in 1992 to retrieve a satellite and bring it back to Earth for repair, to the exciting missions visiting the Russian space station *Mir*, and the orbital construction of an International Space Station. Through all of these activities, a good deal of realism about what the Shuttle could and could not do began to emerge (Launius 2004).

Space Science Comes of Age

Throughout the 1960s and 1970s, robotic exploration of other planets took second stage to the human effort, but there were notable successes. Two of the most spectacular were the Viking mission to Mars and the Voyager "grand tour" of the outer planets. Launched in 1975 from the Kennedy Space Center, Florida, *Viking 1* spent nearly a year cruising to Mars, placed an orbiter in operation around the planet, and landed on July 20, 1976, on the Chryse Planitia (Golden Plains), with *Viking 2* following in September 1976. These were the first sustained landings on another planet in the solar system. While one of the most important scientific activities of this project involved an attempt to determine whether there was life on Mars, the scientific data returned mitigated against the possibility. The two landers continuously monitored weather at the landing sites and found both exciting cyclical variations and an exceptionally harsh climate that prohibited the possibility of life. The failure to find any evidence of life on Mars, past or present, devastated the optimism of scientists and led to a twenty-year hiatus in the exploration of Mars (Launius 1998b). But the Voyager missions to the outer solar system proved more exciting. Two Voyager probes were launched from Kennedy Space Center in 1977 to image Jupiter and Saturn. As the mission progressed, with the successful achievement of all its early objectives, additional flybys of the two outermost giant planets, Uranus and Neptune, proved possible—and irresistible—to mission scientists. Eventually, between them, *Voyager 1* and *Voyager 2* explored all the giant outer planets, 48 of their moons, and the unique systems of rings and magnetic fields those planets possess (Dethloff and Schorn 2003).

While space science did not make news in the 1980s, as the last decade of the twentieth century dawned, NASA moved forward with its "Great Observatories" program and astounded the science world with its findings. The $2 billion Hubble Space Telescope was the first of these "Great Observatories," launched from the Space Shuttle in April 1990. A key component of it was a precision-ground 94-inch primary mirror shaped to within microinches of perfection from ultralow expansion titanium silicate glass with an aluminum-magnesium fluoride coating. The first photos provided bright, crisp images against the black background of space, much clearer than pictures of the same target taken by ground-based telescopes. Controllers then began moving the telescope's mirrors to better focus images. Although the focus sharpened slightly, the best image still had a pinpoint of light encircled by a hazy ring or "halo" (Smith 1989; Gribbin and Goodwin 1998; Chaisson 1994).

At first many believed that the spherical aberration would cripple the 43-foot-long telescope, and NASA received considerable negative publicity, but soon scientists found a way to work around the abnormality with computer enhancement. Because of the difficulties with the mirror, in December 1993, NASA launched the Shuttle *Endeavour* on a repair mission to insert corrective lenses into the telescope and to service other instruments. During a weeklong mission, astronauts conducted a record five space walks to repair the spacecraft. The first reports from the Hubble spacecraft indicated that the images being returned afterward were more than an order of magnitude greater than those obtained before (Tatarewicz 1998).

In the late 1980s and throughout the 1990s, a new enthusiasm for planetary exploration transformed our knowledge of the solar system. Numerous projects came to fruition during the period. For example, the highly successful *Magellan* mission to Venus provided significant scientific data about that planet (Roth and Wall 1995; Grinspoon 1997). Another such project was the *Galileo* mission to Jupiter, which, even before reaching its destination, had become a source of great concern to both NASA and public officials because not all of its systems were working properly, but it returned enormously significant scientific data (Fischer 2001; Hanlon 2001; Harland 2000; Barbieri, et al. 1997).

Finally, Mars exploration received new impetus beginning on July 4, 1997, when *Mars Pathfinder* successfully landed on Mars, the first return to the red planet since Viking in 1976. Its small, 23-pound robotic rover, named *Sojourner*, departed the main lander and began to record weather patterns, atmospheric opacity, and the chemical composition of rocks washed down into the Ares Vallis flood plain, an ancient outflow channel in Mars's northern hemisphere. This vehicle completed its projected milestone 30-day mission on August 3, 1997, capturing far more data on the atmosphere, weather, and geology of Mars than scientists had expected. In all, the *Pathfinder* mission returned more than 1.2 gigabits (1.2 billion bits) of data and over 10,000 tantalizing pictures of the Martian landscape (Pritchett and Muirhead 1998; Reeves-Stevens, et al. 2002; Muirhead and Simon 1999; Mishkin 2003).

A new portrait of the Martian environment began to emerge in the years since *Pathfinder* because of a succession of spacecraft and their new data relating to weather patterns, atmospheric opacity, and the chemical composition of rocks. Indeed, *Mars Global Surveyor*, which began orbiting and mapping the Martian surface in March 1998, imaged gullies on Martian cliffs and crater walls and suggested that liquid water has seeped onto the surface in the geologically recent past. This was confirmed by *Mars Odyssey 2001*, another NASA orbiter, which found that hydrogen-rich regions are located in areas known to be very cold and where ice should be stable. This relationship between high hydrogen content and regions of predicted ice stability led scientists to conclude that the hydrogen is, in fact, in the form of ice. Only time and more research will tell if these findings will be borne out. More recently, two Mars rovers that landed in 2004 have greatly enhanced knowledge of the red planet with a mission more than two years in duration (Bergreen 2000; Caidin 1997; Collins 1990; Hartmann 2003; Morton 2002; Portree 2001; Sheehan 1996; Wilford 1990; Zubrin and Wagner 1996; Squyres 2005; Mishkin 2003).

Seeking a Permanent Presence in Space: The International Space Station

In 1984, as part of its interest in reinvigorating the space program, the Reagan administration called for the development of a permanently-occupied space station. At first projected to cost $8 billion, in part because of tough Washington politics, within five years the projected costs had more than tripled and the station had become too expensive. NASA pared away at the station budget, and in the end the project was satisfactory to almost no one. In 1993 the international situation allowed NASA to negotiate a landmark decision to include Russia in the building of an International Space Station (ISS).

By 1998, the first elements had been launched and in 2000 the first crew went aboard. At the beginning of the 21st century, the effort involved 16 nations, but the ISS was a shadow of what had been intended, caught in the backwash of another loss of a shuttle and the inability to complete construction and resupply. ISS has consistently proven a difficult issue as policy makers wrestled with competing political agendas without consensus (Launius 2003c).

The *Columbia* accident of 2003, which resulted in the deaths of seven astronauts, grounded the Space Shuttle fleet and thereby placed construction of the ISS on hold. Access to the station, thereafter, came only through the use of the Russian *Soyuz* capsule, a reliable but limited vehicle whose technology extended back to the 1960s. Because of this limitation, the ISS crew was cut to two members in May 2003, a skeleton workforce designed to keep the station operational as further deliberations took place, as the Shuttle program underwent organizational reform and technical modifications, and as a policy debate over the long-term viability of human spaceflight took place.

On January 14, 2004, President George W. Bush announced a vision of space exploration that called for humans to reach for the Moon and Mars during the next thirty years. As stated at the time, the fundamental goal of this vision is to advance US scientific, security, and economic interests through a robust space exploration program. In support of this goal, the United States will: implement a sustained and affordable human and robotic program to explore the solar system and beyond; extend human presence across the solar system, starting with a human return to the Moon by the year 2020, in preparation for human exploration of Mars and other destinations; develop the innovative technologies, knowledge, and infrastructures both to explore and to support decisions about the destinations for human exploration; and promote international and commercial participation in exploration to further US scientific, security, and economic interests. In so doing, the president called for completion of the ISS and retirement of the Space Shuttle fleet by 2010. Resources expended there would go toward creating the enabling technologies necessary to return to the Moon and eventually to Mars. He also proposed a small increase in the NASA budget to help make this a reality. By 2006, however, it had become highly uncertain that the initiative could be realized. It appeared increasingly that this proposal would follow the path of the aborted Space Exploration Initiative (SEI) announced with great fanfare in 1989 but derailed in the early 1990s (Sietzen and Cowing 2004; Cornelius 2005; Mendell, 2005).

Conclusion

The combination of technological and scientific advancement, political competition with the Soviet Union, and changes in popular opinion about spaceflight came together in a focused way in the 1950s to affect public policy in favor of an aggressive space program. This found tangible expression in the efforts of the 1950s and 1960s to move forward with an expansive space program and the budgets necessary to support it. After that initial rise of effort, space exploration reached an equilibrium in the 1970s that it has sustained to the present. The American public is committed to a measured program that includes a modest level of human and robotic missions, earth science activities, and technology development efforts. A longstanding fascination with discovery and investigation has nourished much of the interest by the peoples of the United States in spaceflight. At the dawn of the 21st century, however, support for human space exploration is soft and perhaps it will collapse in the coming years. George W. Bush's plan to return to the Moon between 2015 and 2020 is problematic, although with sufficient diligence and resources, virtually anything humans can imagine in spaceflight may be achieved. There is, however, reason to question whether sufficient diligence or resources will be available for this initiative.

References

Aldrin, Buzz. 1973. *Return to Earth.* New York: Bantam Books.

Armstrong, Neil A. et al. 1970. *First on the Moon: A voyage with Neil Armstrong, Michael Collins, and Edwin E. Aldrin, Jr.* Boston: Little, Brown.

Bainbridge, William Sims. 1976. *The spaceflight revolution: A sociological study.* New York: John Wiley & Sons.

Bainbridge, William Sims. 1991. *Goals in space: American values and the future of technology.* Albany: State University of New York Press.

Barbieri, Casare et al., eds. 1997. *The three Galileos: The man, the spacecraft, the telescope.* Dondrecht, The Netherlands: Kluwer Academic.

Beattie, Donald A. 2001. *Taking science to the Moon: Lunar experiments and the Apollo Program.* Baltimore: Johns Hopkins University Press.

Benson, Charles D., and William Barnaby Faherty. 1978. *Moonport: A history of Apollo launch facilities and operations.* Washington, DC: NASA SP-4204.

Bergreen, Laurence. 2000. *Voyage to Mars: NASA's search for life beyond Earth.* New York: Riverhead Books.

Bilstein, Roger E. 1980. *Stages to Saturn: A technological history of the Apollo/Saturn launch vehicles.* Washington, DC: NASA SP-4206.

Brooks, Courtney G., and Ivan D. Ertel. 1973. *October 1, 1964–January 20, 1966.* Vol. 3 of *The Apollo spacecraft: A chronology.* Washington, DC: NASA SP-4009.

Brooks, Courtney G. et al. 1966. *Chariots for Apollo: A history of manned lunar spacecraft.* Washington, DC: NASA SP-4201.

Bulkeley, Rip. 1991. *The Sputniks crisis and early United States space policy: A critique of the historiography of space.* Bloomington: Indiana University Press.

Burrows, William E. 1998. *This new ocean: The story of the first space age.* New York: Random House.

Bush, George W. 2003. The cabinet room, Feb. 1. *NASA Historical Reference Collection.*

Butrica, Andrew J. 1997. *Beyond the ionosphere: Fifty years of satellite communications.* Washington, DC: NASA SP-4217.

Caidin, Martin et al. 1997. *Destination Mars: In art, myth, science.* New York: Penguin Studio.

Chaikin, Andrew. 1994. *A man on the Moon: The voyages of the Apollo astronauts.* New York: Viking.

Chaisson, Eric J. 1994. *The Hubble wars: Astrophysics meets astropolitics in the two-billion-dollar struggle over the Hubble Space Telescope.* New York: Harper-Collins.

Chatterjee, Shami. 1998. *Exploring the Universe: The NASA Great Observatory Program.* Washington, DC: NASA EP-1998-12-384-HQ.

Collins, Michael. 1974. *Carrying the fire: An astronaut's journeys.* New York: Farrar, Straus and Giroux.

Collins, Michael. 1990. *Mission to Mars: An astronaut's vision of our future in space.* New York: Grove Weidenfeld.

Collins, Patrick. 2000. The space tourism industry in 2030. In *Space 2000: Proceedings of the Seventh International Conference and Exposition on Engineering, Construction, Operations, and Business in Space*, ed. Stewart W. Johnson et al., 594-603. Reston, VA: American Society of Civil Engineers.

Compton, W. David. 1989. *Where no man has gone before: A history of Apollo lunar exploration missions.* Washington, DC: NASA SP-4214.

Compton, W. David, and Charles D. Benson. 1983. *Living and working in space: A history of Skylab.* Washington, DC: NASA SP-4208.

Cornelius, Craig. 2005. Science in the national vision for space exploration: Objectives and constituencies of the "discovery-driven" paradigm. *Space Policy* 21:41–48.

Cortright, Edgar M., ed. 1975. *Apollo expeditions to the Moon.* Washington, DC: NASA SP-350.

Dethloff, Henry C., and Ronald A. Schorn. 2003. *Voyager's grand tour: To the outer planets and beyond.* Washington, DC: Smithsonian Institution Press.

DeVorkin, David H., and Robert W. Smith. 2004. *The Hubble Space Telescope: Imaging the Universe.* Washington, DC: National Geographic.

Dickson, Paul. 2001. *Sputnik: The shock of the century.* New York: Walker and Co.

Divine, Robert A. 1993. *The Sputnik challenge: Eisenhower's response to the Soviet satellite.* New York: Oxford University Press.

Draper, Alfred C. et al. 1971. A delta shuttle orbiter. *Astronautics and Aeronautics* 9:26–35.

Editorial. 1952. What are we waiting for? *Collier's* (March 29):23.

Eisenhower, Dwight D. 1962. Are we headed in the wrong direction? *Saturday Evening Post,* Aug. 11.

Eisenhower, Dwight D. 1964. Why I am a Republican. *Saturday Evening Post,* April 11.

Ertel, Ivan D., and Mary Louise Morse. 1969. *The Apollo spacecraft: A chronology.* Vol. 1. Washington, DC: NASA SP-4009.

Ertel, Ivan D. et al. 1978. *January 21 – July 13, 1974.* Vol. 4 of *The Apollo spacecraft: A chronology.* Washington, DC: NASA SP-4009.

Ezell, Edward Clinton, and Linda Neuman Ezell. 1984. *On Mars: Exploration of the red planet, 1958–1978.* Washington, DC: NASA SP-4212.

Fischer, Daniel. 2001. *Mission Jupiter: The spectacular journey of the Galileo spacecraft.* New York: Copernicus Books.

Fries, Sylvia D. 1992. *NASA engineers and the age of Apollo.* Washington, DC: NASA SP-4104.

Gallup, George H. 1972. *The Gallup Poll: Public opinion, 1935–1971.* New York: Random House.

Gribbin, John, and Simon Goodwin. 1998. *Origins: Our place in Hubble's universe.* Woodstock, NY: The Overlook Press.

Griffith, Alison. 1962. *The National Aeronautics and Space Act: A study of the development of policy.* Washington, DC: Public Affairs Press.

Grimwood, James M. 1963. *Project Mercury: A chronology.* Washington, DC: NASA SP-4001.

Grimwood, James M. et al. 1969. *Project Gemini technology and operations: A chronology.* Washington, DC: NASA SP-4002.

Grinspoon, David Harry. 1997. *Venus revealed: A new look below the clouds of our mysterious twin planet.* Reading, MA: Addison-Wesley.

Hacker, Barton C., and James M. Grimwood. 1977. *On the shoulders of titans: A history of Project Gemini.* Washington, DC: NASA SP-4203.

Hall, R. Cargil. 1977. *Lunar impact: A history of Project Ranger*. Washington, DC: NASA SP-4210.

Hanlon, Michael. 2001. *The worlds of Galileo: The inside story of NASA's mission to Jupiter*. New York: St. Martin's Press.

Harford, James J. 1997. *Korolev: How one man masterminded the Soviet drive to beat America to the Moon*. New York: John Wiley and Sons.

Harland, David M. 1999. *Exploring the Moon: The Apollo expeditions*. Chichester, UK: Springer-Praxis.

Harland, David M. 2000. *Jupiter odyssey: The story of NASA's Galileo mission*. Chichester, England: Springer-Praxis.

Harland, David M. 2004. *The story of the space shuttle*. Chichester, UK: Springer-Praxis.

Hartmann, William K. 2003. *A traveler's guide to Mars*. San Francisco: Workman Publishing Company.

Heppenheimer, T. A. 1999. *The space shuttle decision: NASA's search for a reusable space vehicle*. Washington, DC: NASA SP-4221.

Heppenheimer, T. A. 2002. *History of the space shuttle*. Vol. 2 of *Development of the space shuttle, 1972-1981*. Washington, DC: Smithsonian Institution Press.

Hoffer, Eric. 1951. *The true believer: Thoughts on the nature of mass movements*. New York: Harper & Row.

Hudson, H.E. 1990. *Communication satellites: Their development and impact*. New York: Free Press.

Jenkins, Dennis R. 2001. *Space shuttle: The history of the national space transportation system, the first 100 missions*, 3rd ed. Cape Canaveral, FL: Dennis R. Jenkins.

Kennedy, John F. 1960. If the Soviets control space. *Missiles and Rockets* 7:12–13.

Kennedy, John F. 1962. Address at Rice University on the nation's space effort, Sept. 12. http://www.cs.umb.edu/jfklibrary/j091262.htm (accessed Oct. 27, 2002).

Lambright, W. Henry. 1995. *Powering Apollo: James E. Webb of NASA*. Baltimore: Johns Hopkins University Press.

Launius, Roger D. 1994. *NASA: A history of the U.S. civil space program*. Malabar, FL: Krieger Publishing Co.

Launius, Roger D. 1996a. Eisenhower, Sputnik, and the creation of NASA: Technological elites and the public policy. *Prologue: Quarterly of the National Archives and Records Administration* 28:127–143.

Launius, Roger D. 1996b. Space program. In *Dictionary of American history: Supplement*, vol. 2, ed. Robert H. Ferrell and Joan Hoff, 221–223. New York: Charles Scribner's Sons Reference Books.

Launius, Roger D. 1998a. *Frontiers of space exploration*. Westport, CT: Greenwood Press.

Launius, Roger D. 1998b. "Not too wild a dream": NASA and the quest for life in the solar system. *Quest: The History of Spaceflight Quarterly* 6:17–27.

Launius, Roger D. 2003a. Kennedy's space policy reconsidered: A Post–Cold War perspective. *Air Power History* 50:16–29.

Launius, Roger D. 2003b. Public opinion polls and perceptions of U.S. human spaceflight. *Space Policy* 19:163–175.

Launius, Roger D. 2003c. *Space stations: Base camps to the stars*. Washington, DC: Smithsonian Books.

Launius, Roger D. 2004. After Columbia: The space shuttle program and the crisis in space access. *Astropolitics* 2:277–322.

Launius, Roger D. 2006a. Compelling rationales for spaceflight? History and the search for relevance. In *Critical issues in the history of spaceflight*, ed. Steven J. Dick and Roger D. Launius, 38–70. Washington, DC: NASA SP-2006-4702.

Launius, Roger D. 2006b. The space shuttle–twenty-five years on: What does it mean to have reusable access to space? *Quest: The History of Spaceflight Quarterly* 13:4–20.

Launius, Roger D. et al., eds. 2000. *Reconsidering Sputnik: Forty years since the Soviet satellite.* Amsterdam: Harwood Academic Publishers.

Levine, Arnold S. 1982. *Managing NASA in the Apollo era*. Washington, DC: NASA SP-4102.

Ley, Willy. 1961. *Rockets, missiles, and space travel.* New York: The Viking Press.

Ley, Willy, and Chesley Bonestell. 1949. *The conquest of space.* New York: Viking.

Liebermann, Randy L. 1992. The Collier's and Disney series. In *Blueprint for space*, ed. Frederick I. Ordway III and Randy L. Liebermann, 135–146. Washington, DC: Smithsonian Institution Press.

Link, Mae Mills. 1965. *Space medicine in Project Mercury*. Washington, DC: NASA SP-4003.

Logsdon, John M. 1970. *The decision to go to the Moon: Project Apollo and the national interest.* Cambridge, MA: MIT Press.

Logsdon, John M. 1986. The space shuttle program: A policy failure. *Science* 232:1099–1105.

Logsdon, John M. 1998a. Return to flight: Richard H. Truly and the recovery from the Challenger accident. In *From engineering science to big science: The NACA and NASA Collier Trophy research project winners*, ed. Pamela E. Mack, 345–364. Washington, DC: NASA SP-4219.

Logsdon, John. 1998b. The legislative origins of the National Aeronautics and Space Act of 1958: Proceedings of an oral history workshop. Washington, DC: Monographs in Aerospace History, No. 8.

Mack, Pamela E. 1990. *Viewing the Earth: The social construction of Landsat*. Cambridge, MA: MIT Press.

Mathews, Charles W. 1972. The space shuttle and its uses. *Aeronautical Journal* 76:19–25.

May, Elaine Tyler. 1988. *Homeward bound: American families in the Cold War Era*. New York: Basic Books.

McCurdy, Howard E. 1997. *Space and the American imagination*. Washington, DC: Smithsonian Institution Press.

McDougall, Walter A. 1985. *The heavens and the Earth: A political history of the Space Age*. New York: Basic Books.

Mendell, Wendell. 2005. The vision for human spaceflight. *Space Policy* 21:7–10.

Miller, Ron. 1986. Days of future past. *Omni* 9:76–81.

Mishkin, Andrew. 2003. *Sojourner: An insider's view of the Mars Pathfinder mission*. New York: Berkeley Publishing Group.

Morgan, Clay. 2001. *Shuttle-Mir: The U.S. and Russia share history's highest stage*. Washington, DC: NASA SP-2001-4225.

Morse, Mary Louise, and Jean Kernahan Bays. 1973. *The Apollo spacecraft: A chronology*. Vol. 2. Washington, DC: NASA SP-4009.

Morton, Oliver. 2002. *Mapping Mars: Science, imagination, and a birth of a world.* New York: Picador Press.

Muirhead, Brian K., and William L. Simon. 1999. *High velocity leadership: The Mars Pathfinder approach to faster, better, cheaper*. New York: Harper Business.

Murray, Charles A., and Catherine Bly Cox. 1989. *Apollo, the race to the Moon*. New York: Simon and Schuster.

Naugle, John E. 1991. *First among equals: The selection of NASA space science experiments*. Washington, DC: NASA SP-4215.

Newell, Homer E. 1980. *Beyond the atmosphere: Early years of space science*. Washington, DC: NASA SP-4211.

Newkirk, Roland W. et al. 1977. *Skylab: A chronology*. Washington, DC: NASA SP-4011.

Orloff, Richard W. 2000. *Apollo by the numbers: A statistical reference*. Washington, DC: NASA SP-2000-4029.

Pace, Scott. 1982. Engineering design and political choice: The space shuttle, 1969-1972. Master's Thesis, MIT.

Portree, David S. F. 2001. *Humans to Mars: Fifty years of mission planning, 1950-2000*. Washington, DC: NASA SP-2001-4521.

Pritchett, Price, and Brian Muirhead. 1998. *The Mars Pathfinder approach to "faster-better-cheaper."* Dallas, TX: Pritchett and Associates.

Reeves-Stevens, Judith et al. 2002. *Going to Mars: The untold story of Mars Pathfinder and NASA's bold new mission for the 21st century*. New York: Harper-Collins.

Reynolds, David West. 2002. *Apollo: The epic journey to the Moon*. New York: Harcourt.

Robertson, James Oliver. 1980. *American myth, American reality.* New York: Hill and Wang.

Rolland, Alex. 1989. Barnstorming in space: The rise and fall of the romantic era of spaceflight, 1957–1986. In *Space policy reconsidered*, ed. Radford Byerly Jr., 33–52. Boulder, CO: Westview Press.

Roth, Ladislav E., and Stephen D. Wall. 1995. *The face of Venus: The Magellan radar-mapping mission.* Washington, DC: NASA SP-520.

Roy, Stephanie A. et al. 1999. The complex fabric of public opinion on space. IAF-99-P.3.05, International Astronautical Federation annual meeting, Amsterdam, The Netherlands, Oct. 5.

Saegesser, Lee D. High ground advantage. Unpublished statement, *NASA Historical Reference Collection.*

Sagan, Carl. 1980. *Cosmos.* New York: Random House.

Schorn, Ronald A. 1998. *Planetary astronomy: From ancient times to the third millennium.* College Station: Texas A&M University Press.

Scott, Harry A. 1979. Space shuttle: A case study in design. *Astronautics & Aeronautics* 17:54–58.

Seamans, Robert C., Jr. 1996. *Aiming at targets: The autobiography of Robert C. Seamans, Jr.* Washington, DC: NASA SP-4104.

Seamans, Robert C., Jr. 2005. *Project Apollo: The tough decisions.* Washington, DC: NASA SP-2005-4537, Monographs in Aerospace History, No. 37.

Sheehan, William. 1996. *The planet wars: A history of observation & discovery.* Tucson: University of Arizona Press.

Siddiqi, Asif A. 2000. *Challenge to Apollo: The Soviet Union and the space race, 1945–1974.* Washington, DC: NASA SP-2000-4408.

Siddiqi, Asif A. 2002. *Deep space chronicle: Robotic exploration missions to the planets.* Washington, DC: NASA SP-2002-4524.

Sietzen, Frank, Jr., and Keith L. Cowing. 2004. *New moon rising: The making of the Bush space vision.* Burlington, Ontario: Apogee Books.

Slotten, Hugh R. 2002. Satellite communications, globalization, and the Cold War. *Technology and Culture* 43:315–360.

Smith, David R. 1978. They're following our script: Walt Disney's trip to tomorrowland. *Future* (May):59–60.

Smith, Robert W. 1989. *The space telescope: A study of NASA, science, technology, and politics.* New York: Cambridge University Press.

Spudis, Paul D. 1996. *The once and future Moon.* Washington, DC: Smithsonian Institution Press.

Squyres, Steve. 2005. *Roving Mars: Spirit, opportunity, and the exploration of the red planet.* New York: Berkeley Books.

Swanson, Glen E., ed. 1999. *Before this decade is out . . . : Personal reflections on the Apollo Program.* Washington, DC: NASA SP-4223.

Swenson, Loyd S., Jr. et al. 1966. *This new ocean: A history of Project Mercury.* Washington, DC: NASA SP-4201.

Tatarewicz, Joseph N. 1998. The Hubble Space Telescope servicing mission. In *From engineering science to big science: The NACA and NASA Collier Trophy research project winners*, ed. Pamela E. Mack, 365–396. Washington, DC: NASA SP-4219.

Turner, Fredrick Jackson. 1920. *The frontier in American history*. New York: Holt, Rinehart, and Winston.

Van Allen, James A. 1983. *Origins of magnetospheric physics*. Washington, DC: Smithsonian Institution Press.

Vaughan, Diane. 1996. *The Challenger launch decision: Risky technology, culture, and deviance at NASA*. Chicago: University of Chicago Press.

Von Benke, Matthew J. 1997. *The politics of space: A history of U.S.–Soviet/Russian competition and cooperation in space*. Boulder, CO: Westview Press.

von Braun, Wernher. 1953. *The Mars project*. Urbana: University of Illinois Press.

von Braun, Wernher, and Cornelius Ryan. 1954. Can we get to Mars? *Collier's* 133:22–28.

Weber, Max. 1968. The pure types of legitimate authority. In *Max Weber on charisma and institution building: Selected papers*, ed. S. N. Eisenstadt, 46–47. Chicago: University of Chicago Press.

Whalen, David J. 2002. *The origins of satellite communications, 1945–1965*. Washington, DC: Smithsonian Institution Press.

Wilford, John Noble. 1990. *Mars beckons: The mysteries, the challenges, the expectations of our next great adventure in space*. New York: Alfred A. Knopf.

Wilhelms, Don E. 1993. *To a rocky Moon: A geologist's history of lunar exploration*. Tucson: University of Arizona Press.

Young, John W. 2003. The big picture: Ways to mitigate or prevent very bad planet Earth events. *Space Times* 42:22–23.

Zimmerman, Robert. 1998. *Genesis: The story of Apollo 8*. New York: Four Walls Eight Windows.

Zubrin, Robert, and Richard Wagner. 1996. *The case for Mars: The plan to settle the red planet and why*. New York: The Free Press.

12

Technical and Managerial Factors in the NASA *Challenger* and *Columbia* Losses: Looking Forward to the Future

Nancy G. Leveson

The well-known George Santayana quote, "Those who cannot learn from history are doomed to repeat it" (Santayana 1905) seems particularly apropos when considering NASA and the manned space program. The Rogers Commission study of the Space Shuttle *Challenger* accident concluded that the root cause of the accident was an accumulation of organizational problems (Rogers 1986). The commission was critical of management complacency, bureaucratic interactions, disregard for safety, and flaws in the decision-making process. It cited various communication and management errors that affected the critical launch decision on January 28, 1986, including a lack of problem-reporting requirements; inadequate trend analysis; misrepresentation of criticality; lack of adequate resources devoted to safety; lack of safety personnel involvement in important discussions and decisions; and inadequate authority, responsibility, and independence of the safety organization.

Despite a sincere effort to fix these problems after the *Challenger* loss, 17 years later almost identical management and organizational factors were cited in the *Columbia* Accident Investigation Board (CAIB) report. These are not two isolated cases. In most of the major accidents in the past 25 years (in all industries, not just aerospace), technical information on how to prevent the accident was known and often even implemented. But in each case, the potential engineering and technical solutions were negated by organizational or managerial flaws.

Large-scale engineered systems are more than just a collection of technological artifacts (Leveson 1995). They are a reflection of the structure, management, procedures, and culture of the engineering organization that created them. They are also, usually, a reflection of the society in which they were created. The causes of accidents are frequently, if not always, rooted in the organization—its culture, management, and structure. Blame for accidents is often placed on equipment failure or operator error without recognition of the social, organizational, and managerial factors that made such errors and defects inevitable. To truly understand why an accident occurred, it is necessary to examine these factors. In doing so, common causal factors may be seen that were not visible by looking only at the direct, proximal causes. In the case of the *Challenger* loss, the proximal cause[1] was the failure of an O-ring to control the release of propellant gas (the O-ring was designed to seal a tiny gap in the field joints of the solid rocket motor that is created by pressure at ignition). In the case of *Columbia*, the proximal cause was very different—insulation foam coming off the external fuel tank and hitting and damaging the heat-resistant surface of the orbiter. These proximal causes, however, resulted from the same engineering, organizational, and cultural deficiencies, and they will need to be fixed before the potential for future accidents can be reduced.

---

[1]Although somewhat simplified, the *proximal* accident factors can be thought of as those events in a chain of linearly related events leading up to an accident. Each event is directly related in the sense that if the first event had not occurred, then the second one would not have—e.g., if a source of ignition had not been present, then the flammable mixture would not have exploded. The presence of the ignition source is a proximal cause of the explosion as well as the events leading up to the ignition source becoming present. The *systemic* factors are those that explain why the chain of proximal events occurred. For example, the proximal cause might be a human error, but that does not provide enough information to avoid future accidents (although it does provide someone to blame). The systemic causes of that human error might point to poor system design, complacency, inadequate training, the culture values around work, etc.

This essay examines the technical and organizational factors leading to the *Challenger* and *Columbia* accidents and what we can learn from them. While accidents are often described in terms of a chain of directly related events leading to the loss, examining this chain does not explain why the events themselves occurred. In fact, accidents are better conceived as complex processes involving indirect and nonlinear interactions among people, societal and organizational structures, engineering activities, and physical system components (Leveson 2006). They are rarely the result of a chance occurrence of random events, but usually result from the migration of a system (organization) toward a state of high risk where almost any deviation will result in a loss. Understanding enough about the *Challenger* and *Columbia* accidents to prevent future ones, therefore, requires not only determining what was wrong at the time of the losses, but also why the high standards of the Apollo program deteriorated over time and allowed the conditions cited by the Rogers Commission as the root causes of the *Challenger* loss and why the fixes instituted after *Challenger* became ineffective over time—i.e., why the manned space program has a tendency to migrate to states of such high-risk and poor decision-making processes that an accident becomes almost inevitable.

One way of describing and analyzing these dynamics is to use a modeling technique, developed by Jay Forrester in the 1950s, called System Dynamics. System dynamics is designed to help decision makers learn about the structure and dynamics of complex systems, to design high-leverage policies for sustained improvement, and to catalyze successful implementation and change. Drawing on engineering control theory, system dynamics involves the development of formal models and simulators to capture complex dynamics and to create an environment for organizational learning and policy design (Sterman 2000).

Figure 12.1 shows a simplified system dynamics model of the NASA manned space program. Although a simplified model is used for illustration in this paper, we have a much more complex model with several hundred variables that we are using to analyze the dynamics of the NASA manned space program (Leveson, et al. 2006). The loops in Figure 12.1 represent feedback control loops where the "+" and "−" on the loops represent the relationship (positive or negative) between state variables: a "+" means the variables change in the same direction, while a "−" means they move in opposite direc-

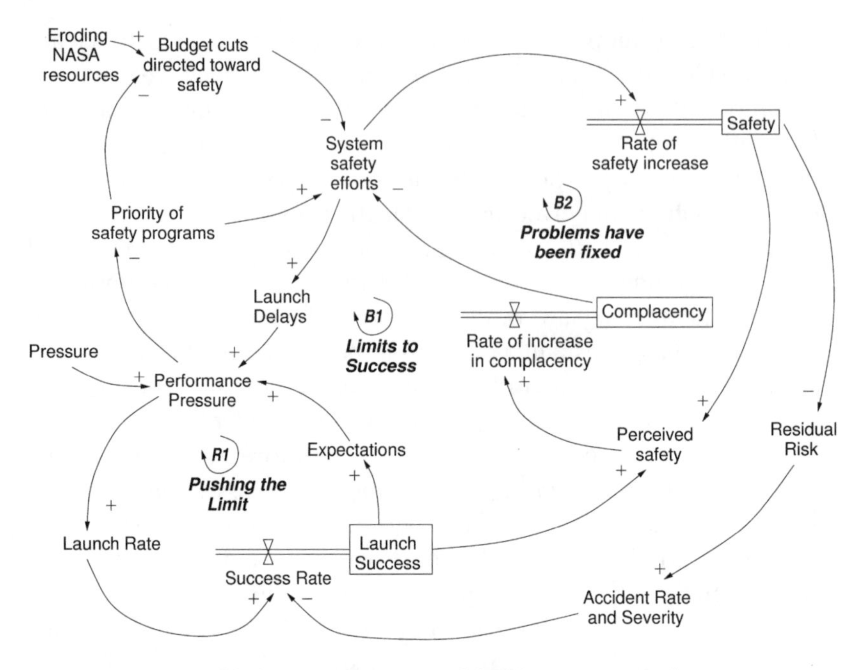

Figure 12.1 A simplified systems dynamics model of the NASA manned space program.

tions. There are three main variables in the model: safety, complacency, and success in meeting launch rate expectations. The model will be explained in the rest of the paper, which examines four general factors that played an important role in the accidents: the political and social environment in which decisions were made, the NASA safety culture, the NASA organizational structure, and the safety engineering practices in the manned space program.

Political and Social Factors

All engineering efforts take place within a political, social, and historical context that has a major impact on the technical and operational decision-making. Understanding the context in which decisions were made in the manned space program helps in explaining why bright and experienced engineers made what turned out to be poor decisions and what might be changed to prevent similar accidents in the future.

In the case of the Space Shuttle, political and other factors contributed to the adoption of a vulnerable design during the original approval process. Unachievable promises were made with respect to performance in order to keep the manned space flight program alive after Apollo and the demise of the Cold War. While these performance goals even then seemed unrealistic, the success of the Apollo program and the *can-do* culture that arose during it—marked by tenacity in the face of seemingly impossible challenges—contributed to the belief that these unrealistic goals could be achieved if only enough effort were expended. Performance pressures and program survival fears gradually led to an erosion of the rigorous processes and procedures of the Apollo program as well as the replacement of dedicated NASA staff with contractors who had dual loyalties (Leveson, et al. 2004). The Rogers Commission report on the *Challenger* accident concluded:

> The unrelenting pressure to meet the demands of an accelerating flight schedule might have been adequately handled by NASA if it had insisted upon the exactingly thorough procedures that were its hallmark during the Apollo program. An extensive and redundant safety program comprising interdependent safety, reliability, and quality assurance functions existed during and after the lunar program to discover any potential safety problems. Between that period and 1986, however, the program became ineffective. This loss of effectiveness seriously degraded the checks and balances essential for maintaining flight safety. (Rogers, et al. 1986, 152)

The goal of this essay is to provide an explanation for why the loss of effectiveness occurred, so that the pattern can be prevented in the future.

The Space Shuttle was part of a larger Space Transportation System concept that arose in the 1960s when Apollo was in development. The concept originally included a manned Mars expedition, a space station in lunar orbit, and an Earth-orbiting station serviced by a reusable ferry, or space shuttle. The funding required for this large an effort, on the order of that provided for Apollo, never materialized, and the concept was scaled back until the reusable Space Shuttle, earlier only the transport element of a broad transportation system, became the focus of NASA's efforts. In addition, to maintain its funding, the Shuttle had to be sold as performing a large number of tasks, including launching and servicing satellites, which required compromises in the design. The compromises contributed to a design that was more inherently risky than was necessary. NASA also had to make promises about performance (number of launches per year) and cost per launch that were unrealistic. An important factor in both accidents was the pressures exerted on NASA by an unrealistic flight schedule with inadequate resources and by commitments to customers. The nation's reliance on the Shuttle as its principal space launch capability, which NASA sold in order to get the money to build the Shuttle, created a relentless pressure on NASA to increase the flight rate to the originally promised 24 missions a year.

Budget pressures added to the performance pressures. Budget cuts occurred during the life of the Shuttle—for example, amounting to a 40% reduction in purchasing power over the decade before the *Columbia* loss. At the same time, the budget was occasionally raided by NASA itself to make up for overruns in the International Space Station program. The later budget cuts came at a time when the Shuttle was aging and costs were actually increasing. The infrastructure, much of which dated back to the Apollo era, was falling apart before the *Columbia* accident. In the past 15 years of the Shuttle program, uncertainty about how long the Shuttle would fly added to the pressures to delay safety upgrades and improvements to the Shuttle program infrastructure.

Budget cuts without concomitant cuts in goals led to trying to do too much with too little. NASA's response to its budget cuts was to defer upgrades and to attempt to increase productivity and efficiency rather than eliminate any major programs. By 2001, an experienced observer of the space program described the Shuttle workforce as "The Few, the Tired" (Gehman, et al. 2003).

NASA Shuttle management also had a belief that less safety, reliability, and quality assurance activity would be required during routine Shuttle operations. Therefore, after the successful completion of the orbital test phase and the declaration of the Shuttle as "operational," several safety, reliability, and quality assurance groups were reorganized and reduced in size. Some safety panels, which were providing safety review, went out of existence entirely or were merged.

One of the ways to understand the differences between the Apollo and Shuttle programs that led to the loss of effectiveness of the safety program is to use the system dynamics model in Figure 12.1.[2] The control loop in the lower left corner of the model, labeled R1 or *Pushing the Limit*, shows how, as external pressures increased, performance pressure increased, which led to increased launch rates and success, which in turn led to increased expectations and increasing performance pressures. The larger loop B1 is labeled *Limits to Success* and explains how the performance pressures led to failure. The upper left loop represents part of the safety program dynamics. The external influences of budget cuts and increasing performance pressures reduced the priority of system safety practices and led to a decrease in system safety efforts.

The safety efforts also led to launch delays, which produced increased performance pressures and more incentive to reduce the safety efforts. At the same time, problems were being detected and fixed, which led to a belief that *all* the problems would be detected and fixed (and that the most important ones had been) as depicted in loop B2, labeled *Problems have been Fixed*. The combination of the decrease in system safety program priority leading to budget cuts in the safety activities, along with the complacency denoted in loop B2, which also contributed to the reduction of system safety efforts, eventually led to a situation of (unrecognized) high risk where, despite effort by the operations workforce, an accident became almost inevitable.

One thing not shown in the simplified model is that delays can occur along the arrows in the loops. While reduction in safety efforts and lower prioritization of safety concerns may eventually lead to accidents, accidents do not occur for a while, so false confidence is created that the reductions are having no impact on safety. Pressures increase to reduce the safety program activities and priority even further as the external performance and budget pressures mount, leading almost inevitably to a major accident.

Figure 12.2 shows the outputs from the simulation of our complete NASA system dynamics model. The upward-pointing arrows on

[2]Many other factors are contained in the complete model, including system safety status and prestige; shuttle aging and maintenance; system safety resource allocation; learning from incidents; system safety knowledge, skills, and staffing; and management perceptions. More information about this model can be found in Nancy Leveson, et al., "Engineering Resilience into Safety Critical Systems" pages 95–124 in *Resilience Engineering*, ed. Erik Hollnagel, David Woods, and Nancy Leveson, (London: Ashgate Publishing, 2006).

Accidents lead to a reevaluation of NASA safety and performance priorities but only for a short time:

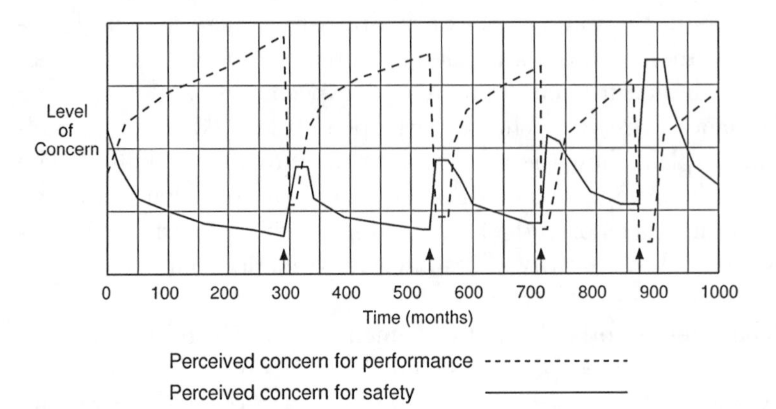

Attention to fixing systemic problems lasts only a short time after an accident

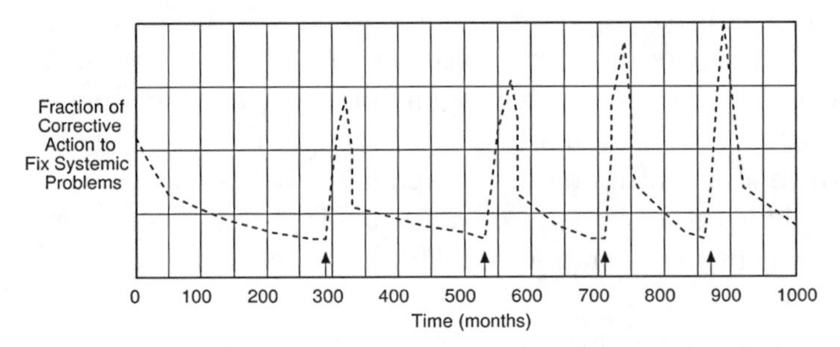

Attempts to fix systemic problems

Responses to accidents have little lasting impact on risk

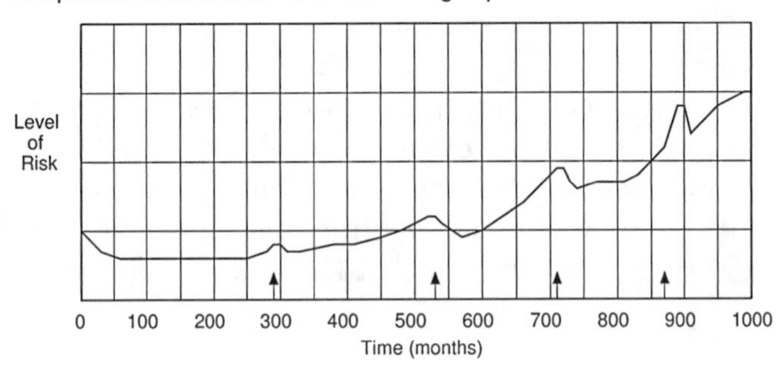

Figure 12.2 Results from running the NASA system dynamics model.

the x-axis represent the points in time when accidents or serious incidents occur during the simulation. Despite sincere attempts to fix the problems, the dysfunctional dynamics return very quickly after an accident using our model (and appear also to return quickly in the actual Shuttle program) if the factors underlying the drift toward high risk are not countered. In the top graph, the arrows represent the occurrence of accidents over time. Note that, while safety becomes of higher priority than performance for a very short time after an accident, performance quickly resumes its position of greater importance. The middle graph shows that concern about fixing the systemic problems that led to an accident also lasts only a short time after the accident. Finally, the bottom graph shows that the responses to accidents do not reduce risk significantly due to the first two patterns illustrated, plus others in the model.

One of the uses for such a model is to hypothesize changes in the organizational dynamics that might prevent this type of cyclical behavior. For example, preventing the safety engineering priorities and activities from being subject to the performance pressures might be achieved by anchoring the safety efforts outside the Shuttle program—e.g., by establishing and enforcing NASA-wide safety standards and not providing the Shuttle program management with the power to reduce the safety activities. This independence did not and still does not exist in the Shuttle program, although the problem is recognized and attempts are being made to fix it.

Safety Culture

A significant change in NASA after the Apollo era has been in the safety culture. Much of both the Rogers's Commission Report and the CAIB report are devoted to flaws in the NASA safety culture. The cultural flaws at the time of the *Challenger* accident either were not fixed or reoccurred before the *Columbia* loss. Many still exist in NASA today.

A *culture* can be defined as a shared set of norms and values. It includes the way we look at and interpret the world and events around us (our mental model) and the way we take action in a social context. *Safety culture* is that subset of an organizational or industry culture that reflects the general attitude and approaches to safety and risk management. It is important to note that trying to change culture and the behavior resulting from it without changing the environment in which it is embedded is doomed to failure. Superficial fixes that do not address the set of shared values and social norms, as well as deeper underlying assumptions, are likely to be undone over time (Shein 1986). Perhaps this partially explains why the changes at NASA after the *Challenger* accident intended to fix the safety culture, like the safety activities themselves, were slowly dismantled or became ineffective. Both the *Challenger* accident report and the CAIB report, for example, note that system safety was "silent" and ineffective at NASA despite attempts to fix this problem after the *Challenger* accident. Understanding the pressures and other influences that have twice contributed to a drift toward an ineffective NASA safety culture is important in creating an organizational infrastructure and environment that will resist pressures against applying good safety engineering practices and procedures in the future.

Risk and occasional failure has always been recognized as an inherent part of space exploration, but the way the inherent risk is handled at NASA has changed over time. In the early days of NASA and during the Apollo era, the belief was prevalent that risk and failure were normal aspects of spaceflight. At the same time, the engineers did everything they could to reduce risk (McCurdy 1994). People were expected to speak up if they had concerns, and risks were debated vigorously. "What if" analysis was a critical part of any design and review procedure. Sometime between those early days and the *Challenger* accident, the culture changed drastically. The Rogers Commission Report includes a chapter titled the "Silent Safety Program." Those on the Thiokol task force appointed to investigate problems that had been occurring with the O-rings on flights prior to the catastrophic *Challenger* flight complained about lack of management support and cooperation for the O-ring team's efforts. One memo started with the word "HELP!" and complained about the O-ring task force being "constantly delayed by every possible means"; the memo ended with the words "This is a red flag" (Rogers, et al., 1986).

The CAIB report notes that at the time of the *Columbia* loss, "Managers created huge barriers against dissenting opinions by stating preconceived conclusions based on subjective knowledge and experience, rather than solid data." An indication of the prevailing culture at the time of the *Columbia* accident can be found in the reluctance of the debris assessment team—created after the launch of *Columbia* to assess the damage caused by the foam hitting the wing—to adequately express their concerns. Members told the CAIB that "by raising contrary points of view about Shuttle mission safety, they would be singled out for possible ridicule by their peers and managers" (Gehman, et al. 2003, 169).

In an interview shortly after he became Center Director at the NASA Kennedy Space Center after the *Columbia* loss, Jim Kennedy suggested that the most important cultural issue the Shuttle program faces is establishing a feeling of openness and honesty with all employees where everybody's voice is valued (Leveson, et al. 2004). Statements during the *Columbia* accident investigation and anonymous messages posted on the NASA Watch Web site document a lack of trust leading to a reluctance of NASA employees to speak up. At the same time, a critical observation in the CAIB report focused on the managers' claims that they did not hear the engineers' concerns. The report concluded that not hearing the concerns was due in part to the managers not asking or listening. Managers created barriers against dissenting opinions by stating preconceived conclusions based on subjective knowledge and experience rather than on solid data. In the extreme, they listened to those who told them what they wanted to hear. Just one indication of the atmosphere existing at that time was statements in the 1995 Kraft report that dismissed concerns about Shuttle safety by labeling those who made them as partners in an unneeded "safety shield conspiracy" (Kraft 1995). This accusation against those expressing safety concerns as being part of a "conspiracy" is a powerful demonstration of the attitude toward system safety at the time, and the change from the Apollo era when dissent was encouraged and rewarded.

A partial explanation for this change was that schedule and launch pressures in the Shuttle program created a mind-set that dismissed all concerns, leading to overconfidence and complacency. This type of culture can be described as a *culture of denial* where risk assessment is unrealistic and credible risks and warnings are dismissed without appropriate investigation. Managers begin to listen only to those who provide confirming evidence that supports what they want to hear. Neither Thiokol nor NASA expected the rubber O-rings sealing the joints to be touched by hot gases during motor ignition, much less to be partially burned. As tests and then flights confirmed damage to the sealing rings, however, the reaction by both NASA and Thiokol was to increase the amount of damage considered "acceptable." At no time did management either recommend a redesign of the joint or call for the Shuttle's grounding until the problem was solved. The Rogers Commission found that the Space Shuttle's problems began with a faulty design of the joint and increased as both NASA and Thiokol management first failed to recognize the problem, then failed to fix it when it was recognized, and finally treated it as an acceptable risk.

NASA and Thiokol accepted escalating risk apparently because they "got away with it last time" (Rogers, et al. 1986). Morton Thiokol did not accept the implication of tests done early in the program that the design had a serious and unanticipated flaw. NASA management did not accept the judgment of its engineers that the design was unacceptable, and, as the joint problems grew in number and severity, they were minimized in management briefings and reports. Thiokol's stated position was that "the condition is not desirable, but it is acceptable" (Leveson 1995). As Feynman observed, the decision-making was

> a kind of Russian roulette. . . . [The Shuttle] flies [with O-ring erosion] and nothing happens. Then it is suggested, therefore, that the risk is no longer so high for the next flights. We can lower our standards a little bit because we got away with it last time. (Rogers, et al. 1986, 148)

Every time an incident occurred that was a narrow escape, it confirmed for many the idea that NASA was a tough, can-do organization with high intact standards that precluded accidents (Leveson, et al. 2003). The exact same phenomenon occurred with the foam shedding, which had occurred during the life of the Shuttle but had never, prior to the *Columbia* loss, caused serious damage.

A NASA study report in 1999 concluded that the Space Shuttle program was using previous success as a justification for accepting increased risk (McDonald, et al. 2000). The practice continued despite this and other alarm signals. William Readdy, head of the NASA Manned Space Program, for example, wrote in 2001 that "The safety of the Space Shuttle has been dramatically improved by reducing risk by more than a factor of five" (Gehman, et al. 2003, 101). It is difficult to imagine where this number came from, as safety upgrades and improvements had been deferred while, at the same time, the infrastructure continued to erode. The unrealistic risk assessment was also reflected in the 1995 Kraft report, which concluded that "the Shuttle is a mature and reliable system, about as safe as today's technology will provide." A recommendation of the Kraft report was that NASA should "restructure and reduce overall safety, reliability, and quality assurance elements."

The CAIB report identified a perception that NASA had overreacted to the Rogers Commission recommendations after the *Challenger* accident, for example, believing that the many layers of safety inspections involved in preparing a shuttle for flight had created a bloated and costly safety program. Reliance on past success became a substitute for sound engineering practices and a reason to accept increasing risk. Either the decision makers did not have or they did not use inputs from system safety engineering. "Program management made erroneous assumptions about the robustness of a system based on prior success rather than on dependable engineering data and rigorous testing" (Gehman, et al. 2003, 184).

Many analysts have faulted NASA for missing the implications of the *Challenger* O-ring trend data. One sociologist, Diane Vaughan, went so far as to suggest that the risks had become seen as "normal" (Vaughan 1997). In fact, the engineers and scientists at NASA were tracking thousands of potential risk factors.[3] It was not the case that some risks had come to be perceived as normal (a term that Vaughan does not define), but that they had come to be seen as *acceptable* without adequate data to support that conclusion.

Edwin Tufte, famous for his visual displays of data, analyzed the way the O-ring temperature data were displayed at the meeting where the *Challenger* launch decision was made, arguing that they had minimal impact because of their physical appearance (Tufte 2003).

While the insights into the display of data are instructive, it is important to recognize that both the Vaughan and the Tufte analyses are easier to do in retrospect. In the field of cognitive engineering, this common mistake has been labeled "hindsight bias": it is easy to see what is important in hindsight (Woods and Cook 1999). It is much more difficult to achieve this goal before the important data have been identified as critical after the accident. Decisions need to be evaluated in the context of the information available at the time the decision is made, along with the organizational factors influencing the interpretation of the data and the decision-making process itself. Risk assessment is extremely difficult for complex, technically advanced systems such as the Space Shuttle. When this engineering reality is coupled with the social and political pressures existing at the time, the emergence of a culture of denial and overoptimistic risk assessment is not surprising.

Shuttle launches are anything but routine, so that new interpretations of old data or of new data will always be needed—that is, risk assessment for systems with new technology is a continual and iterative task that requires adjustment on the basis of experience. At the

[3]There are over 20,000 critical items on the Space Shuttle (i.e., items whose failure could lead to the loss of the Shuttle) and, at the time of the *Columbia* loss, over 3,000 waivers existed.

same time, it is important to understand the conditions at NASA that prevented an accurate analysis of the data and the risk and the types of safety culture flaws that contribute to unrealistic risk assessment.

Why would intelligent, highly educated, and highly motivated engineers engage in such poor decision-making processes and act in a way that seems irrational in retrospect? One view of culture provides an explanation. Social anthropologists conceive of culture as an ongoing, proactive process of reality construction (Morgan 1986). In this conception of culture, organizations are socially constructed realities that rest as much in the heads of members as in sets of rules and regulations. Organizations are sustained by belief systems that emphasize the importance of rationality. Morgan calls this the *myth of rationality* and it helps in understanding why, as in both the *Challenger* and *Columbia* accidents, leaders often appear to ignore what seems obvious in retrospect. The myth of rationality "helps us to see certain patterns of action as legitimate, credible, and normal, and hence to avoid the wrangling and debate that would arise if we were to recognize the basic uncertainty and ambiguity underlying many of our values and actions" (Morgan 1986, 134-135).

In both the *Challenger* and *Columbia* accidents, the decision makers saw their actions as rational at the time, although hindsight suggests otherwise. Understanding and preventing poor decision-making under conditions of uncertainty requires providing environments and tools that help to stretch our belief systems and overcome the constraints of our current mental models—i.e., to see patterns that we do not necessarily *want* to see (Leveson 2004).

A final common aspect of the dangerous complacency and overconfidence seen in the manned space program is related to the use of redundancy to increase reliability. One of the rationales used in deciding to go ahead with the disastrous *Challenger* flight despite engineering warnings was that there was a substantial safety margin (a factor of three) in the O-rings over the previous worst case of Shuttle O-ring erosion. Moreover, even if the primary O-ring did not seal, it was assumed that a second, redundant one, would. During the accident, the failure of the primary O-ring caused conditions that led to the failure of the backup O-ring. In fact, the design changes necessary to incorporate a second O-ring contributed to the loss of the primary O-ring.

The design of the Shuttle solid rocket booster (SRB) was based on the US Air Force *Titan III*, one of the most reliable ever produced. Significant design changes were made in an attempt to further increase that reliability, including changes in the placement of the O-rings. A second O-ring was added to the Shuttle solid rocket motor design to provide backup: If the primary O-ring did not seal, then the secondary one was supposed to pressurize and seal the joint. In order to accommodate the two O-rings, part of the Shuttle joint was designed to be longer than in the *Titan*. The longer length made the joint more susceptible to bending under combustion pressure, which led to the failure of both the primary and backup O-rings. In this case, and in a large number of other cases (Leveson 1995), the use of redundancy requires design choices that in fact defeat the redundancy at the same time that the redundancy is creating unjustified confidence and complacency.

In this case, the ineffectiveness of the added O-ring was actually known. An engineer at NASA Marshall Space Flight Center concluded, after tests in 1977 and 1978, that the second O-ring was ineffective as a backup seal. Nevertheless, in November 1980, the SRB joint design was classified as redundant—a status it retained until November 1982. Its classification was changed to nonredundant on December 17, 1982, after tests showed that the secondary O-ring was no longer functional after the joints rotated under 40% of the SRB maximum operating pressure. Why that information did not get to those making the *Challenger* launch decision is unclear, but communication and information system flaws may have contributed (see the section on safety engineering practices below).

Organizational Structure

Organizational change experts have long argued that structure drives behavior. Much of the dysfunctional behavior related to both accidents can be traced to flaws in the NASA organizational safety structure, including poorly designed independence, ill-defined responsibility and authority, a lack of influence and prestige leading to insufficient impact, and poor communication and oversight.

Independence, Responsibility, and Authority

Both accident reports criticized the lack of independence of the safety organization. After the *Challenger* loss, a new independent safety office was established at NASA Headquarters, as recommended in the Rogers's Commission report. This group is supposed to provide broad oversight, but its authority is limited and reporting relationships from the NASA Centers are vague. In essence, the new group was never given the authority necessary to implement their responsibilities effectively and nobody seems to have been assigned accountability. The CAIB report noted in 2003 that the management of safety at NASA involved "confused lines of responsibility, authority, and accountability in a manner that almost defies explanation" (Gehman, et al. 2003, 186).

The CAIB also noted that "NASA does not have a truly independent safety function with the authority to halt the progress of a critical mission element" (Gehman, et al. 2003, 180). In essence, the project manager "purchased" safety from the quality assurance organization. The amount of system safety applied was limited to what and how much the project manager wanted and could afford. "The Program now decides on its own how much safety and engineering oversight it needs" (Gehman, et al. 2003, 181).

The problems are exacerbated by the fact that the project manager also has authority over the safety standards applied on the project. NASA safety "standards" are not mandatory. In essence, they function more like guidelines than standards. Each program decides what standards are applied and can tailor them in any way it wants.

There are safety review panels and procedures *within* individual NASA programs, including the Shuttle program. Under various types of pressures, including budget and schedule constraints, however, the independent safety reviews and communication channels within the Shuttle program degraded over time and were taken over by the Shuttle Program office.

Independence of engineering decision-making also decreased over time. While in the Apollo and early Shuttle programs the engineering organization had a great deal of independence from the program manager, it gradually lost its authority to the project managers, who again were driven by schedule and budget concerns.

In the Shuttle program, all aspects of system safety are in the mission assurance organization. This means that the same group doing the system safety engineering is also doing the system safety assurance—effectively eliminating an independent assurance activity.

In addition, putting the system safety engineering (e.g., hazard analysis) within the assurance group has established the expectation that system safety is an after-the-fact or auditing activity only. In fact, the most important aspects of system safety involve core engineering activities such as building safety into the basic design and proactively eliminating or mitigating hazards. By treating safety as an assurance activity only, safety concerns are guaranteed to come too late in the process to have an impact on the critical design decisions. Necessary information may not be available to the engineers when they are making decisions and instead potential safety problems are raised at reviews, when doing something about the poor decisions is costly and likely to be resisted.

This problem results from a basic dilemma: either the system safety engineers work closely with the design engineers and lose their independence or the safety efforts remain an independent assurance effort but safety becomes divorced from the engineering and design efforts. The solution to this dilemma, which other groups use, is to separate the safety engineering and the safety assurance efforts, placing safety engineering within the engineering organization and the safety assurance function within the assurance groups. NASA attempted to accomplish this after *Columbia* by creating an Independent Technical Authority within engineering that is responsible for bringing a disciplined, systematic approach to identifying, analyzing, and controlling hazards. The design of this independent authority is already undergoing changes with the result unclear at this time.

Influence and Prestige

The Rogers Commission report on the *Challenger* accident observed that the safety program had become "silent" and undervalued. A chapter in the report titled "The Silent Safety Program" concludes that a properly staffed, supported, and robust safety organization might well have avoided the communication and organizational problems that influenced the infamous *Challenger* launch decision.

After the *Challenger* accident, as noted above, system safety was placed at NASA Headquarters in a separate organization that included mission assurance and other quality assurance programs. For a short period thereafter this safety group had some influence, but it quickly reverted to a position of even less influence and prestige than before the *Challenger* loss. Placing system safety in the quality assurance organization, often one of the lower prestige groups in the engineering pecking order, separated it from mainstream engineering and limited its influence on engineering decisions. System safety engineering, for all practical purposes, began to disappear or became irrelevant to the engineering and operations organizations. Note that the problem here is different from that before *Challenger*, where system safety became silent because it was considered to be less important in an operational program. After *Challenger*, the attempt to solve the problem of the lack of independence of system safety oversight quickly led to loss of its credibility and influence and was ineffective in providing lasting independence (as noted above).

In the testimony to the Rogers Commission, NASA safety staff, curiously, is never mentioned. No one thought to invite a safety representative to the hearings or to the infamous teleconference between Marshall and Thiokol. No representative of safety was on the mission management team that made key decisions during the countdown to the *Challenger* flight.

The *Columbia* accident report concludes that, once again, system safety engineers were not involved in the important safety-related decisions, although they were ostensibly added to the mission management team after the *Challenger* loss. The isolation of system safety from the mainstream design engineers added to the problem:

> Structure and process places Shuttle safety programs in the unenviable position of having to choose between rubber-stamping engineering analyses, technical errors, and Shuttle program decisions, or trying to carry the day during a committee meeting in which the other side always has more information and analytical ability. (Gehman, et al. 2003, 187)

The CAIB report notes that "We expected to find the [Safety and Mission Assurance] organization deeply engaged at every level of Shuttle management, but that was not the case" (Gehman, et al. 2003, 177).

One of the reasons for the lack of influence of the system safety engineers was the stigma associated with the group, partly resulting from the placement of an engineering activity in the quality assurance organization.

Safety was originally identified as a separate responsibility by the Air Force during the ballistic missile programs of the 1950s to solve exactly the problems seen here—to make sure that safety is given due consideration in decision-making involving conflicting pressures and that safety issues are visible at all levels of decision-making. Having an effective safety program cannot prevent errors in judgment in balancing conflicting requirements of safety and schedule or cost, but it can at least make sure that decisions are informed and that safety is given due consideration. To be effective, however, the system safety engineers must have the prestige necessary to have the influence on decision-making that safety requires. The CAIB report addresses this issue when it says that:

> Organizations that successfully operate high-risk technologies have a major characteristic in common: they place a premium on safety and reliability by structuring their programs so that technical and safety engineering organizations own the process of determining, maintaining, and waiving technical requirements with a voice that is equal to yet independent of Program Managers, who are governed by cost, schedule, and mission-accomplishment goals. (Gehman, et al. 2003, 184)

Both accident reports note that system safety engineers were often stigmatized, ignored, and sometimes actively ostracized. "Safety and mission assurance personnel have been eliminated [and] careers in safety have lost organizational prestige" (Gehman, et al. 2003, 181). The author has received personal communications from NASA engineers who write that they would like to work in system safety but will not because of the negative stigma that surrounds most of the safety and mission assurance personnel. Losing prestige has created a vicious circle of lowered prestige leading to stigma, which limits influence and leads to further lowered prestige and influence and lowered quality due to the most qualified engineers not wanting to be part of the group. Both accident reports comment on the quality of the system safety engineers and the SIAT report in 2000 also sounded a warning about the quality of NASA's Safety and Mission Assurance efforts (McDonald, et al. 2000).

Communication and Oversight

Proper and safe engineering decision-making depends not only on a lack of complacency—the desire and willingness to examine problems—but also on the communication and information structure that provides the information required. For a complex and technically challenging system like the Shuttle with multiple NASA centers and contractors all making decisions influencing safety, some person or group is required to integrate the information and make sure it is available for all decision makers.

Both the Rogers Commission and the CAIB found serious deficiencies in communication and oversight. The Rogers Commission report noted miscommunication of technical uncertainties and failure to use information from past near misses. Relevant concerns were not being reported to management. For example, the top levels of NASA management responsible for the launch of *Challenger* never heard about the concerns raised by the Morton Thiokol engineers on the eve of the launch, nor did they know about the degree of concern raised by the erosion of the O-rings in prior flights. The Rogers Commission noted that memoranda and analyses raising concerns about performance and safety issues were subject to many delays in transmittal up the organizational chain and could be edited or stopped from further transmittal by some individual or group along the chain (Zebroski 1989, 443-454).

A report written before the *Columbia* accident notes a "general failure to communicate requirements and changes across organizations" (Gehman, et al. 2003, 179). The CAIB found that "organizational barriers . . . prevented effective communication of critical safety information and stifled professional differences of opinion." It was "difficult for minority and dissenting opinions to percolate up through the agency's hierarchy" (Gehman, et al. 2003, 183).

As contracting of Shuttle engineering has increased over time, safety oversight by NASA civil servants has diminished and basic system safety activities have been delegated to contractors. According to the CAIB report, the operating assumption that NASA could turn over increased responsibility for Shuttle safety and reduce its direct involvement was based on the 1995 Kraft report that concluded the Shuttle was a mature and reliable system and that therefore NASA could change to a new mode of management with less NASA oversight. A single NASA contractor was given responsibility for Shuttle safety (as well as reliability and quality assurance), while NASA was to maintain "insight" into safety and quality assurance through reviews and metrics. In fact, increased reliance on contracting necessitates *more* effective communication and more extensive safety oversight processes, not less (Gehman, et al. 2003).

Many aerospace accidents have occurred after the organization transitioned from oversight to "insight" (Leveson 2004). The contractors have a conflict of interest with respect to safety and their own goals and cannot be assigned the responsibility that is properly that of the contracting Agency. In addition, years of workforce reductions and outsourcing had "culled from NASA's workforce the layers of experience and hands-on systems knowledge that once provided a capacity for safety oversight" (Gehman, et al. 2003, 181).

System Safety Engineering Practices

After the *Apollo* fire in 1967 in which three astronauts were killed, Jerome Lederer (a renowned aircraft safety expert) created what was considered at the time to be a world-class system safety program at NASA. Over time, that program declined for a variety of reasons, many of which were described earlier. After the *Challenger* loss, there was an attempt to strengthen it, but that attempt did not last long due to failure to change the conditions that were causing the drift to ineffectiveness. The CAIB report describes system safety engineering at NASA at the time of the *Columbia* accident as "the vestiges of a once robust safety program" (Gehman, et al. 2003, 177). The changes that occurred over the years include:

- *Reliability engineering was substituted for system safety.* Safety is a system property and needs to be handled from a system perspective. NASA, in the recent past, however, has treated safety primarily at the component level, with a focus on component reliability. For example, the CAIB report notes that there was no one office or person responsible for developing an integrated risk analysis above the subsystem level that would provide a comprehensive picture of total program hazards and risks. Failure Modes and Effects Analysis (FMEA), a bottom-up reliability engineering technique, became the primary analysis method. Hazard analyses were performed but rarely used. NASA delegated safety oversight to its operations contractor USA, and USA delegated hazard analysis to Boeing, but as of 2001, "the Shuttle program no longer required Boeing to conduct integrated hazard analyses" (Gehman, et al. 2003, 188). Instead, Boeing performed analysis only on the failure of individual components and elements and was not required to consider the Shuttle as a whole—i.e., system hazard analysis was not being performed. The CAIB report notes: "Since the FMEA/CIL process is designed for bottom-up analysis at the component level, it cannot effectively support the kind of 'top-down' hazard analysis that is needed. . .to identify potentially harmful interactions between systems" (Gehman, et al. 2003, 188) (like foam from the external tank hitting the forward edge of the orbiter wing).

- *Standards were watered down and not mandatory* (as noted earlier).
- *The safety information system was ineffective.* Good decision-making about risk is dependent on having appropriate information. Without it, decisions are often made on the basis of past success and unrealistic risk assessment, as was the case for the Shuttle. Lots of data was collected and stored in multiple databases, but there was no convenient way to integrate and use the data for management, engineering, or safety decisions (Aerospace Safety Advisory Panel 2003b).

Creating and sustaining a successful safety information system requires a culture that values the sharing of knowledge learned from experience. Several reports have found that such a learning culture is not widespread at NASA and that the information systems are inadequate to meet the requirements for effective risk management and decision-making (Aerospace Safety Advisory Panel 2003a; Aerospace Safety Advisory Panel 2003b; Government Accounting Office 2001; McDonald, et al. 2000; Gehman, et al. 2003). Sharing information across centers is sometimes problematic and getting information from the various types of lessons-learned databases situated at different NASA centers and facilities ranges from difficult to impossible. Necessary data is not collected and what is collected is often filtered and inaccurate or tucked away in multiple databases without a convenient way to integrate the information to assist in management, engineering, and safety decisions; methods are lacking for the analysis and summarization of causal data; and information is not provided to decision makers in a way that is meaningful and useful to them. In lieu of such a comprehensive information system, past success and unrealistic risk assessment are being used as the basis for decision-making.

- *Inadequate safety analysis was performed when there were deviations from expected performance.* The Shuttle standard for hazard analyses (NSTS 22254, *Methodology for Conduct of Space Shuttle Program Hazard Analyses*) specifies that hazards be revisited only when there is a new design or the design is changed: there is no process for updating the hazard analyses when anomalies occur or even for determining whether an anomaly is related to a known hazard.

- *Hazard analysis, when it was performed, was not always adequate.* The CAIB report notes that a "large number of hazards reports contained subjective and qualitative judgments, such as 'believed' and 'based on experience from previous flights' this hazard is an accepted risk." The hazard report on debris shedding (the proximate event that led to the loss of the *Columbia*) was closed as an accepted risk and was not updated as a result of the continuing occurrences (Gehman, et al. 2003). The process laid out in the Shuttle standards allows a hazard to be closed when a mitigation is *planned*, not when the mitigation is actually implemented.
- *There was evidence of "cosmetic system safety."* Cosmetic system safety is characterized by superficial safety efforts and perfunctory bookkeeping: hazard logs may be meticulously kept, with each item supporting and justifying the decisions made by project managers and engineers (Leveson 1995). The CAIB report notes that "Over time, slowly and unintentionally, independent checks and balances intended to increase safety have been eroded in favor of detailed processes that produce massive amounts of data and unwarranted consensus, but little effective communication" (Gehman, et al. 2003, 180).

Conclusions

Space exploration is inherently risky. There are just too many unknowns and requirements to push the technological envelope to be able to reduce the risk level to that of other aerospace endeavors such as commercial aircraft. At the same time, the known and preventable risks can and should be managed effectively.

The investigation of accidents creates a window into an organization and the opportunity to examine and fix unsafe elements. The repetition of the same factors in the *Columbia* accident implies that NASA was unsuccessful in permanently eliminating those factors after the *Challenger* loss. The same external pressures and inadequate responses to them, flaws in the safety culture, dysfunctional organizational safety structure, and inadequate safety engineering practices will continue to contribute to the migration of the NASA manned space program to states of continually increasing risk until changes are made and safeguards are put into place to prevent that drift in the future. The current NASA administrator, Michael Griffin, and others at NASA are sincerely trying to make the necessary changes that will ensure the safety of the remaining Shuttle flights and the success of the new manned space program missions to the Moon and to Mars. It remains to be seen whether these efforts will be successful.

The lessons learned from the Shuttle losses are applicable to the design and operation of complex systems in many industries. Learning these lessons and altering the dynamics of organizations that create the drift toward states of increasing risk will be required before we can eliminate unnecessary accidents.

Acknowledgments

Many of the ideas in this essay were formulated during discussions of the "Columbia Group," an informal, multidisciplinary group of faculty and students at MIT that started meeting after the release of the CAIB report to discuss that report and to understand the accident. The group continues to meet regularly but with broader goals involving accidents in general. The author would particularly like to acknowledge the contributions of Joel Cutcher-Gershenfeld, John Carroll, Betty Barrett, David Mindell, Nicolas Dulac, and Alexander Brown.

References

Aerospace Safety Advisory Panel. 2003a. *Annual report*. Washington, DC: National Aeronautics and Space Administration.

Aerospace Safety Advisory Panel. 2003b. *The use of leading indicators and safety information systems at NASA*. Washington, DC: National Aeronautics and Space Administration.

Gehman, Harold et al. 2003. *Columbia accident investigation report*. Washington, DC: National Aeronautics and Space Administration and Government Printing Office.

Government Accounting Office. 2001. *Survey of NASA's lessons learned process*. GAO-01-1015R. Washington, DC: United States General Accounting Office.

Kraft, Christopher. 1995. *Report of the Space Shuttle Management Independent Review*. http://www.fas.org/spp/kraft.htm.

Leveson, Nancy G. 1995. *Safeware*. Boston: Addison-Wesley Publishers.

Leveson, Nancy G. 2004. The role of software in spacecraft accidents. *AIAA Journal of Spacecraft and Rockets* 41:4:1–27.

Leveson, Nancy G. 2006. *System safety: Back to the future*. Unpublished book draft. http://www.sunnyday.mit.edu/book2.html.

Leveson, Nancy et al. 2006. Engineering resiliance into safety critical systems. In *Resiliance engineering*, ed. Molnagel et al. London: Ashgate Publishing.

Leveson, Nancy G. et al. 2003. Risk analysis of NASA Independent Technical Authority. http://www.sunnyday.mit.edu/ITA-Risk-Analysis.doc.

Leveson, Nancy G. et al. 2004. Effectively addressing NASA's organizational and safety culture: Insights from systems safety and engineering systems. Engineering Systems Division Symposium, MIT, March 29–31.

McCurdy, Howard. 1994. *Inside NASA: High technology and organizational change in the U.S. space program.* Baltimore: Johns Hopkins University Press.

McDonald, Henry et al. 2000. *Shuttle Independent Assessment Team (SIAT) Report to Associate Administrator*. Washington, DC: National Aeronautics and Space Administration.

Morgan, Gareth. 1986. *Images of organizations*. Thousand Oaks, CA: Sage Publications.

Rogers, William P. et al. 1986. *Report of the Presidential Commission on the Space Shuttle Challenger Accident.* Washington, DC: US Government Accounting Office.

Santayana, George. 1905–6. *The life of reason: Or, the phases of human progress.* New York: Scribner.

Shein, Edgar. 1986. *Organizational culture and leadership*, 2nd ed. Safe Publications.

Sterman, John. 2000. *Business dynamics: Systems thinking and modeling for a complex world.* Boston: McGraw-Hill.

Tufte, Edward. 2003. *The cognitive style of PowerPoint*. Cheshire, CT: Graphics Press.

Woods, D. D., and R. I. Cook. 1999. Perspectives on human error: Hindsight bias and local rationality. In *Handbook of applied cognition*, ed. F. Durso, 141–171. New York: Wiley.

Vaughan, Diane. 1997. *The Challenger launch decision*. Chicago: University of Chicago Press.

Zebroski, Edwin. 1989. Sources of common cause failures in decision making involved in man-made catastrophes. In *Risk assessment in setting national priorities*, ed. James Bonin and Donald Stevenson, 443–454. New York: Plenum Press.

13

On Slippery Slopes, Repeating Negative Patterns, and Learning from Mistake: NASA's Space Shuttle Disasters

Diane Vaughan

> Accident: nonessential quality or circumstance; 1. an event or circumstance occurring by chance or arising from unknown or remote causes; lack of intention or necessity; an unforeseen or unplanned event or condition; 2. sudden event or change occurring without intent or volition through carelessness, unawareness, ignorance, or a combination of causes and producing an unfortunate result; 3. an adventitious characteristic that is either inseparable from the individual and the species or separable from the individual but not the species; broadly, any fortuitous or nonessential property, fact or circumstance (~ of appearance) (~ of reputation) (~ of situation).
>
> —*Webster's Third New International Dictionary.*

If we adhere to a strict dictionary definition, accidents do not arise from the innate nature of things. They occur unpredictably from chance alone. Accidents, therefore, are events with no forewarning and thus are not preventable. However, a comparison of NASA's two space shuttle accidents, *Challenger* in 1986 and *Columbia* in 2003, contradicts this definition. Both events were unfortunate and unforeseen, but their origins were not random or chance occurrences, and both disasters might have been prevented.[1]

In a press conference a few days after the *Columbia* tragedy, Ron Dittemore, NASA's space shuttle program manager, held up a large piece of foam approximately the size of the one that fatally struck *Columbia* and discounted it as a probable cause of the accident, expressing confidence in the integrity of the foam. NASA officials could have made similar claims about O-ring erosion after the *Challenger* disaster. Yet the O-ring erosion that caused the loss of *Challenger* and the foam debris that took *Columbia* out of the sky both had long histories. Both accidents were the result of a series of decisions that led to a gradual slide into disaster. The histories of decisions about the risk of O-ring erosion and the foam debris were littered with early warning signs that NASA either misinterpreted or ignored. For years preceding both accidents, technical experts dismissed the possibility of risk by repeatedly normalizing technical anomalies that deviated from expected performance. By the "normalization" of technical deviations I mean that anomalies on returning flights that were first viewed as signals of potential danger and risk were, upon engineering analysis, officially categorized as "acceptable risks." Thus their recurrence was considered a normal and acceptable aspect of shuttle performance, not a threat to flight safety. That these disasters were the result of an extended decision-making process is significant, since a long incubation period provided greater opportunity for intervention that could have avoided the catastrophes. But that did not happen in either situation. How—and why—was this possible?

[1]The *Challenger* and *Columbia* disasters have attracted critical attention from scholars in many different disciplines, including history, sociology, communications, sociology, and engineering—to name only a few. This chapter, a condensation from a more complex analysis (Vaughan 1996), cannot accommodate the full spectrum of arguments and analyses, but the chapter bibliography and the following works will provide interested readers with a strong starting point for further reading: Stephen B. Johnson, "Revisiting the *Challenger*," QUEST 7:2 (1999):18–24 and *The Secret of Apollo: Systems Management in American and European Space Programs* (Baltimore, MD: The Johns Hopkins University Press, 2002); Phillip K. Tompkins, *Organizational Communication Imperatives: Lessons of the Space Program* (Los Angeles, CA: Roxbury, 1993) and *Apollo, Challenger, Columbia: The Decline of the Space Program* (Los Angeles, CA: Roxbury, 2005); Randy Avera, *The Truth About Challenger* (Good Hope, PA: Randolph Publishing, 2003).

Within weeks of beginning its official investigation, the *Columbia* Accident Investigation Board (CAIB) noted strong parallels between *Columbia* and *Challenger*. The CAIB concluded that NASA's second shuttle accident resulted from an "organizational system failure"—or, the complex set of factors that affected how NASA engineers defined and redefined risk. These factors, which included NASA's political and economic environment, its organizational structure and culture, had contributed to the *Challenger* explosion but had not been fixed. The CAIB claimed that the causes of both accidents lay within three specific parts of NASA's organizational system (CAIB 2003, Ch. 8).

First, managers and engineers made each decision to launch under circumstances that made sense to them at the time, even if they seemed risky in hindsight. They normalized the technical deviations that were occurring. The question was why. Although NASA treated the shuttle as if it were an operational vehicle, it was experimental: alterations of design and unpredictable flight conditions led to anomalies on many parts on every mission. Because anomalies were common on the shuttle, they were expected, and therefore having anomalies was, itself, normal. Consequently, neither O-ring erosion nor foam debris was the signal of danger it seemed in retrospect. Engineers interpreting the anomalies saw mixed signs of danger; an incident would be followed by a mission with a lesser degree of damage or no damage at all, convincing them that they had fixed the problem and understood the parameters of cause and effect. In other cases, the signals were weak: incidents that were outside what had become defined as acceptable were not alarming because their circumstances were so unprecedented that they were viewed as unlikely to repeat. Or, finally, anomalies became normal since they occurred so frequently; the repeating pattern then became a sign that the shuttle was operating as predicted. The result was an overall assumption that the problems were not a threat to flight safety—a belief repeatedly reinforced by the safe return of each mission. Flying with these flaws became normal and acceptable to NASA insiders, not irresponsible, as it appeared to outsiders after the accidents.

Second, production pressure permeated the culture. NASA employees were under both internal and external pressure not to delay launches. Meeting deadlines and following schedules were important to NASA's scientific goals and also for securing annual Congressional funding. Flights were always halted to correct permanently problems that were a clear threat to the safety of the shuttle and its astronauts (a cracked fuel duct to the main engine, for example), but the schedule and resources could not give way for a thorough hazard analysis of ambiguous, low-lying problems that seemed to pose no threat to the vehicle.

Finally, structural secrecy masked potential threats. The decisions to fly with O-ring erosion and foam debris had gone on for years. Why had no one responsible for safety oversight acted to halt NASA's two transitions into disaster? Individual secrecy—the classic explanation that individuals tried to keep bad news from top management—does not work here. Everyone at the agency knew about the two problems and their histories; the question was how did they define the risk? Structural secrecy, not individual secrecy, explained the failure of administrators and safety regulators to intervene. Structural secrecy refers to the obstacles to communication that are built into normal organizational structure: division of labor, hierarchy, and subunits, for example. Each time project managers and engineers assessed risk, finding anomalies safe to fly, their evidence and conclusions were passed up the hierarchy, forming the basis for further assessments. The organizational structure and specialization obscured the seriousness of such problems from people with responsibility for oversight. As a result, instead of reversing the pattern of flying with erosion and foam debris, NASA's Flight Readiness Review—the launch decision-making structure designed to identify and correct problems—ratified it.

This kind of structural secrecy also interfered with the ability of safety regulators at NASA to address and correct potential problems. NASA's internal safety organization did not have independent authority and funding. Consequently, safety units fell prey to personnel cuts and loss of expertise as responsibilities were shifted to contractors in order to cut operating costs, and NASA safety units had no ability to independently run tests that might challenge existing assessments. NASA's external safety panel, on the other hand, had the advantage of independence, but was handicapped by inspection at infrequent intervals. Unless NASA engineers defined something as a serious problem, it was not brought to the attention of safety personnel. As a result of structural secrecy, the belief that it was safe to fly with O-ring erosion and foam debris prevailed throughout the agency in the years prior to each of NASA's tragedies.

These three factors combined to produce both the *Challenger* and *Columbia* disasters: a decision-making pattern normalizing technical anomalies, which created the belief that it was safe to fly; production goals that encouraged launching over conducting a thorough hazard analysis; and structural secrecy, which prevented intervention that would have halted NASA's incremental descent into poor judgment. The amazing similarity between the organizational causes of these accidents, 17 years apart, raises two questions: Why do negative patterns persist? Why do organizations fail to learn from mistakes and accidents?

The Presidential Commission's 1986 report about the *Challenger* accident followed the traditional format of prioritizing the technical causes of the accident and identifying human factors as "contributing causes,"—according them less, not equal, importance. The report went well beyond the usual human factors focus on individual incompetence, poor training, negligence, mistake, and physical or mental impairment, to identify some relevant institutional causes, such as the weakened safety structure. However, the primary cause was seen as a "flawed decision-making process" by individuals. Managerial failures dominated the findings: managers in charge had a tendency to solve problems internally, rather than forwarding them to all hierarchical levels; inadequate testing was performed; neither the contractor nor NASA understood why the O-ring anomalies were happening; and escalated risk-taking was endemic, apparently "because they got away with it the last time" (Presidential Commission report 1986, 148). Furthermore, managers and engineers failed to analyze flight history carefully, so data were not available on the eve of *Challenger*'s launch to evaluate the risks properly and the anomaly tracking system permitted flight to continue despite erosion, with no record of waivers or launch constraints.

The Commission's recommendations for change were consistent with the human factors/individual failure model of causes identified in their findings (1986, 198–201). To correct NASA's "flawed decision making," the report called for changes in individual behavior and procedures. It issued a series of mandates to NASA, including eliminating the tendency of managers not to report upward, "whether by changes of personnel, organization, indoctrination, or all three" (1986, 200); developing rules regarding launch constraints; recording Flight Readiness Reviews and Mission Management Team meetings. The Commission called for centralizing safety oversight in part by introducing astronauts to management to lend a practical awareness of risk and safety. Some organizational changes were recommended. A new Shuttle Safety Panel would report to the shuttle program manager, and an independent Office of Safety, Reliability and Quality Assurance would be established with independent funding and direct authority over all safety bodies throughout the agency. This office was to report to the NASA administrator, rather than program management, thus keeping safety structurally separate from the part of NASA responsible for budget and efficiency in operations. Finally, to address scheduling pressures, the Commission recommended that NASA establish a flight rate that was consistent with its resources and recognize in its policies that the space shuttle would always be experimental, not operational.

How did the space agency respond? NASA implemented most of the changes suggested by the Commission. Managers responsible for flawed decisions were either transferred or took early retirement. NASA also addressed its flawed decision-making by changing its policies, procedures, and processes so that the probability that anomalies would be recognized early and problems corrected would increase. NASA went further, and took the opportunity to "scrub the system totally." NASA established data systems and trend analysis, recording all anomalies so that problems could be tracked over time. Rules were changed for Flight Readiness Review so that engineers, formerly included only in the lower-level reviews, could participate in the entire process. The astronaut managers participated in final prelaunch Flight Readiness Review and the signing of the authorization for the final mission "go."

At the organizational level, NASA made several structural changes that centralized control of operations and safety (CAIB 2003, 101). The agency shifted control for the space shuttle program from Johnson Space Center in Houston to NASA headquarters in an attempt to replicate the management structure at the time of *Apollo* and restore communication to a former level of excellence. NASA also established the Headquarters Office of Safety, Reliability and Quality Assurance (renamed Safety and Mission Assurance) as recommended, but instead of the direct authority over all safety operations, as the Commission recommended, each of the centers had its own safety organization and reported to the center director (CAIB 2003, 101). Finally, NASA repeatedly acknowledged in press conferences that the space shuttle was and always would be treated as an experimental vehicle and vowed that henceforth safety would take priority over schedule in launch decisions.

Given that each of these changes targeted the causes identified in the report, why did the negative pattern repeat, leading to the *Columbia* disaster? First, from our post-*Columbia* hindsight, we can see that the Commission did not identify all the layers of NASA's organizational system as targets for change. NASA took the blame for safety cuts (Presidential Commission report 1986, 160), while the external budgetary actions by the White House and Congress that forced NASA leaders to make the cuts in the first place were not mentioned. Further, the Commission did not name organizational culture as a culprit, although the pressure to launch on schedule was the subject of an entire chapter. The report ultimately places responsibility for "communication failures" not with NASA's organizational structure, but with the individual middle managers responsible for key decisions, inadequate rules, and procedures. The obstacles to communication inherent in the organization's hierarchy and the consequent power that managers wielded over engineers, stifling their input in crucial decisions, are not mentioned.

Second, the CAIB found that many of NASA's initial changes were implemented as the Commission directed, but the changes to the safety structure were not, thus impacting *Columbia* decisions. NASA's new Headquarters Office of Safety, Reliability and Quality Assurance did not have direct authority, as the Commission mandated; further, the various safety offices in its domain remained dependent because their funds came from the very activities that they were overseeing (CAIB 2003, 101, 178–9). Thus, cost, schedule, and safety all remained intertwined.

Third, the CAIB found that other positive changes were undone over time. Although NASA's own leaders played a role in determining goals and how to achieve them, the institutional environment was not under their control. NASA remained essentially powerless as a government agency dependent upon political winds and budgetary decisions made elsewhere. Thus, NASA had little recourse but to try to achieve its ambitious goals—politically necessary to keep the agency a national budgetary priority—with limited resources. As a consequence, production expectations were unchanged, yet the organizational structure became more complex. Also, the post-*Challenger* surge in funding was followed by budget cuts that reproduced conditions at the time of *Challenger*. As a result of the budgetary squeeze, NASA Administrator, Daniel Goldin, introduced new efficiencies and smaller programs with the slogan, "faster, better, cheaper." The initial increase in NASA safety personnel was followed by a repeat of preaccident reductions that again cut safety staff and placed even more responsibility for safety with contractors. The accumulation of successful missions (defined as flights returned without accident) also reinvigorated belief in the status quo, thus legitimating these cuts. Tracking post-*Challenger* decisions shows the effects of financial pressures on NASA's organization: the persistence of NASA's legendary can-do attitude, excessive allegiance to bureaucratic proceduralism and hierarchy due to increased contracting out, and the squeeze produced by "an agency trying to do too much with too little" (CAIB 2003, 101–120) as funding dropped so that downsizing and sticking to the schedule became the means to all ends.

Thus, the CAIB found that the conditions contributing to *Challenger* persisted. In the CAIB report, in contrast to that of the 1986 commission, recommendations for change targeted the organizational system that produced both accidents. The political environment led to ongoing pressure for the shuttle to fly and NASA complied. The CAIB found that "the White House and Congress must recognize the role of their decisions in this accident and take responsibility for safety in the future. . . .Leaders create culture. It is their responsibility to change it. . . .The past decisions of national leaders—the White House, Congress, and NASA headquarters—set the *Columbia* accident in motion by creating resource and schedule strains that compromised the principles of a high-risk technology organization" (CAIB 2003, 196, 203).

What about the normalization of deviance? Because the shuttle is and always will be an experimental vehicle, technical problems will proliferate. In such a setting, categorizing risk will always be difficult, especially with ambiguous problems like foam debris and O-ring erosion, where the threat to flight safety is not readily apparent and the absence of serious incident constitutes success. In order to make early warning signs about low-lying problems more salient, NASA created a new Engineering and Safety Center as a resource for engineering decisions made throughout the agency. The Center was required to review recurring anomalies determined by agency engineers to be inconsequential for flight safety, to see if engineering assessments were correct (Morring 2004a). From this revelation, a common database was to be created to identify early warning signs that may have been misinterpreted or ignored. However, as we have seen from *Columbia* and *Challenger*, what happens at the level of everyday interaction, interpretation, and decision-making does not occur in a vacuum, but in an organizational system in which other factors affect problem definition, corrective actions, and problem dispositions. A second consideration is clout: how will the National Engineering and Safety Center reverse long-standing institutionalized definitions of risk of specific problems, and how will it deal with the continuing pressures of the culture of production?

NASA remains a politically vulnerable agency, dependent on the White House and Congress for its share of the budget and approval of its goals. After *Columbia,* the Bush administration supported continuing the space shuttle program and supplied the mission vision that the CAIB report concluded was missing: the space program would explore Mars. However, the initial funds required for the shuttle to return to flight and simultaneously accomplish this new goal were insufficient (Morring 2004b). Thus, NASA, following the CAIB mandate, attempted to adjust its goals to account for available resources by phasing out the Hubble telescope program and, eventually, planning to phase out the shuttle itself. Meanwhile, however, the International Space Station is still operating and remains dependent upon the shuttle to ferry astronaut crews, materials, and experiments back and forth in space. How the conflict between NASA's goals and the constraints upon achieving them will unfold is uncertain, but one lesson from *Challenger* is that external pressures can perpetuate problems. The board mandated independence and resources for the safety system, but when goals, schedule, efficiency, and safety conflicted post-*Challenger*, NASA goals were reigned in, but safety oversight was compromised.

Whereas the CAIB recommendations for changing structure were specific, its directions for changing culture were vague. Trained in engineering and accustomed to human factors analysis, NASA leaders did not understand how to go about changing culture. Further, many NASA personnel believed that the report's conclusion about agency-wide cultural failures wrongly indicted parts of NASA that were working well. Finally, and more fundamentally, they had a problem translating the contents of the report to identify what cultural changes were necessary and what actions they implied.

So how to do it? NASA's approach was this. On December 16, 2003, NASA headquarters posted a Request for Proposals on its Web site for a cultural analysis followed by the elimination of cultural problems detrimental to safety. Verifying the CAIB's conclusions about NASA's deadline-oriented culture, proposals first were due January 6, then the deadline was extended by a meager ten days. NASA, furthermore, required data on culture change in six months and a transformed culture in 36 months.

The contractor who got the bid proceeded to undertake a survey to gather data on the extent and location of cultural problems in the agency. The ability of any survey to tap into cultural problems is questionable, however, because it relies solely on insiders who can be blind to certain aspects of their culture. A better assessment results when insider information is complemented by data collected by outside observers who become temporary members, spending sufficient time there to be able to identify cultural patterns, examine documents, participate in meetings and casual conversations, and interview, asking open-ended questions. Not surprisingly, under the circumstances, the contractor's strategy for change was individually oriented, training managers to listen and decentralize, and encouraging engineers to speak up. NASA, in turn, to date, has not confronted cultural beliefs about risk, goals, schedule pressures, structure, and power distribution.

Organization System Failures: Lessons Learned

The dilemmas of slippery slopes, repeating negative patterns, and learning from mistakes are not unique to NASA. Think of the incursion of drug use into professional athletics, US military abuse of prisoners in Iraq, and Enron—to name some sensational cases in which incrementalism, commitment, feedback, cultural persistence, and structural secrecy have created a "blind spot" that allowed individuals to see their actions as acceptable, thus perpetuating a collective incremental descent into poor judgment. Knowing the conditions that cause a gradual downward slide, regardless of the immediate cause of such disasters, does give us some insight into how it happens that may be helpful to managers and administrators hoping to avoid these problems.

Although such mistakes often appear to be sudden and may have devastating public outcomes, they can have a long incubation period. How do early warning signs become normalized? A first decision, once taken and met with either outright success or no obvious failure, sets a precedent upon which future decisions are based. Or, if initially viewed as problematic, the positive outcome may override perceptions of risk and harm; thus, what was originally defined as an anomaly becomes normal and acceptable, as decisions that build upon the precedent accumulate. Patterns of information bury early warning signs amidst subsequent indicators that all is well. As decisions and their positive results become known to others in the organization, those making decisions become committed to their chosen line of action. Reversing direction—even in the face of contradictory information—becomes more difficult (Salancik 1997).

The accumulating actions become so ingrained that newcomers may take over from others without questioning the status quo, or their objections are overridden by claims that "this is the way we do it here." Cultural beliefs persist because people tend to make the problematic nonproblematic by defining a situation in a way that makes sense of it in cultural terms. NASA's gradual slide continued unchecked because the decisions made conformed to the mandates of the dominating culture of production and because NASA's organizational structure impeded the ability of those with regulatory responsibilities—top administrators and safety representatives included—to critically question decisions and intervene.

Why do negative patterns repeat? Was it true, as the press concluded after *Columbia*, that the lessons of *Challenger* weren't learned? The findings and recommendations of the Commission's 1986 report on *Challenger* identified its cause primarily in individual mistakes, misjudgments, flawed analysis and decision-making, and communication failures. The findings about schedule pressures and safety structure were attributed also to flawed decision-making—not by middle managers, but by NASA leaders. In response, the Commission recommended adjusting decision-making processes, creating structural change in safety systems, and bringing goals and resources into alignment. NASA acted on each of those recommendations; thus, we could say that the lessons were learned. The *Columbia* accident and the CAIB report that followed taught different lessons, however. They showed that an organizational system failure, not individual failure, was behind both accidents, and NASA had not connected its strategies for control with the full social causes of the first accident.

Since *Columbia*, NASA has tried to address the organizational system problems that the CAIB identified. There have been practical difficulties, however, in correcting the second organizational system failure. NASA leaders had difficulty integrating new functions into existing parts of the operation; cultural change and how to go about it eluded them. Some of the CAIB recommendations were puzzling to NASA personnel who believed their system was working well under most circumstances. Further, understanding how social circumstances affect individual actions is not easy to grasp, especially in an American ethos in which both success and failure are seen as the result of individual action. Finally, negative patterns can repeat because causing change has widespread effects that can produce unintended consequences. Changing organizational structure can increase complexity and therefore the probability of mistake; it can also change culture in unpredictable ways (Jervis 1997; Perrow 1984; Sagan 1993; Vaughan 1999).

Even under the best of circumstances, negative patterns can still repeat. External forces are often beyond a single organization's ability to control. Cultures of production are a product of larger historical, cultural, political, ideological, and economic institutions. Making organizational change that contradicts them is difficult to implement, but in the face of continuing and consistent institutional forces, it is even more difficult to sustain as time passes. Although NASA seems like a powerful governmental agency, its share of the federal budget is small in comparison to other agencies. In the aftermath of both shuttle accidents, external political and budgetary decisions that altered goals and resources made it difficult to create and sustain a different NASA. It may be argued that, under the circumstances, NASA's space shuttle program has had a remarkable safety record.

But even when everything possible is done, we cannot have mistake-free organizations. As Jervis argues (1997) and the NASA case verifies, system effects will produce unanticipated consequences. Further, not all contingencies can be predicted; most people don't understand how social context affects individual action, and that organizational changes that correct one problem, may, in fact, have a dark side, creating other unpredictable problems. External environments are likewise difficult to control. Although not all mishaps, mistakes, and accidents can be prevented, both of NASA's accidents had long incubation periods; thus, they *were preventable.* By addressing the systemic organizational causes of gradual slides and repeating negative patterns, organizations can reduce the probability that these kinds of harmful outcomes will occur.

Acknowledgments

This essay was modified from "System Effects: On Slippery Slopes, Negative Patterns, and Learning from Mistake," in *Organization at the Limit: NASA and the Columbia Disaster*, ed. William Starbuck and Moshe Farjoun (Oxford: Blackwell 2005) and is reprinted with permission of the publisher.

References

Behavioral Science Technology, Inc. 2004. *Status report: NASA culture change*. Ojai, CA.

Cabbage, M., and W. Harwood. 2004. *COMM check: The final flight of the shuttle Columbia.* New York: Free Press.

Clarke, L. 1999. *Mission improbable: Using fantasy documents to tame disaster.* Chicago: University of Chicago Press.

Clarke, L. 2005. *Worst Cases*. Chicago: University of Chicago Press.

CAIB (*Columbia* Accident Investigation Board). 2003. *Report,* vol. 1. Washington, DC: Government Printing Office. www.caib.us/news/report/default.html.

Deal, D. W. 2004. Beyond the widget: *Columbia* accident lessons affirmed. *Air and Space Power Journal* (Summer):29–48.

Edmondson, A. C. et al. 2005. *The Columbia's last flight,* MultiMedia Business Case, 305–332. Cambridge, MA: Harvard Business School.

Gove, P. B., ed. 1971. *Webster's third new international dictionary.* Springfield, MA: G. and C. Merriam Company.

Hutter, Bridget, and M. Power, eds. 2006. *Organizational encounters with risk.* Cambridge: Cambridge University Press.

Jervis, R. 1997. *System effects: Complexity in political and social life*. Princeton: Princeton University Press.

Kanter, R. M. 1983. *The Changemasters.* New York: Simon & Schuster.

Kanter, R. M. 2004. *Confidence: How winning streaks and losing streaks begin and end.* New York: Simon & Schuster.

Miller, D. 1990. *The Icarus paradox: How exceptional companies bring about their own downfall.* New York: Harper.

Morring, F., Jr. 2004a. Anomaly analysis: NASA's engineering and safety center checks recurring shuttle glitches. *Aviation Week and Space Technology* (August) 2:53.

Morring, F., Jr. 2004b. Reality Bites: Cost growth on shuttle return-to-flight job eats NASA's lunch on Moon/Mars exploration. *Aviation Week and Space Technology* (July) 26:52.

Perrow, C. B. 1984. *Normal accidents: Living with high-risk technologies.* New York: Basic Books.

Presidential Commission. 1986. *Report to the President by the Presidential Commission on the Space Shuttle Challenger Accident*, 5 vols. *(the Rogers report)*. Washington, DC: Government Printing Office.

Pressman, J. L., and A. Wildavsky. 1984. *Implementation: How great expectations in Washington are dashed in Oakland, or Why it's amazing that federal programs work at all*. Los Angeles: University of California Press.

Roberts, K. H. 1990. Managing high reliability organizations. *California Management Review* 32(4):101–114.

Sagan, S. D. 1993. *The limits of safety: Organizations, accidents, and nuclear weapons*. Princeton: Princeton University Press.

Salancik, G. R. 1977. Commitment and the control of organizational behavior and belief. In *New Directions in Organizational Behavior*, ed. B. M. Staw and G. R. Salancik. Malabar, FL: Krieger.

Sietzen, F., Jr., and K. L. Cowing. 2004. *New moon rising: The making of America's new space vision and the remaking of NASA*. New York: Apogee Books.

Snook, S. A. 2000. *Friendly fire: The accidental shootdown of U. S. black hawks over northern Iraq*. Princeton: Princeton University Press.

Starbuck, W. H. 1988. Executives' perceptual filters: What they notice and how they make sense. In *The executive effect*, ed. D. C. Hambrick. Greenwich, CT: JAI.

Tucker, A. L., and A. C. Edmondson. 2003. Why hospitals don't learn from failures: Organizational and psychological dynamics that inhibit system change. *California Management Review* 45:55–72.

Turner, B. 1978. *Man-made disasters*. London: Wykeham.

Vaughan, D. 1996. *The Challenger launch decision: Risky technology, culture, and deviance at NASA*. Chicago: University of Chicago Press.

Vaughan, D. 1997. The trickle-down effect: Policy decisions, risky work, and the *Challenger* accident. *California Management Review* 39:1–23.

Vaughan, D. 1999. The dark side of organizations: Mistake, misconduct, and disaster. *Annual Review of Sociology* 25:271–305.

Vaughan, D. 2003. History as cause: *Columbia* and *Challenger*. In *Columbia Accident Investigation Board. Report*, vol. 1. Washington, DC: US Government Printing Office.

Weick, K. E. 1993. The collapse of sensemaking in organizations: The Mann Gulch disaster. *Administrative Science Quarterly* 38:628–652.

Weick, K. E. 2006. *Managing the Unexpected*. San Francisco: Jossey-Bass.

Weick, K. E., and K. H. Roberts. 1993. Collective mind in organizations: Heedful interrelating on flight decks. *Administrative Science Quarterly* 38:357–381.

Weick, K. E. et al. 1990. Organizing for high reliability: Processes of collective mindfulness. *Research in Organizational Behavior* 21:81–123.

14

Justifiably within the Risk

James Lattis

On the morning of February 1, 2003, the Space Shuttle *Columbia* reentered the atmosphere after 16 days in low Earth orbit. Without the knowledge of the crew, reinforced carbon-carbon tiles on the leading edge of the orbiter's left wing had been damaged during launch. The damage rendered the left wing unable to withstand the heat and aerodynamic stress of reentry, with the result that the wing and the rest of the spacecraft disintegrated, raining debris over Texas and Louisiana. *Columbia*'s entire crew of seven was killed. Like the *Challenger* accident of January 1986, the loss of vehicle and crew was highly visible, carried great consequences for NASA's plans in space, and received deep and well-deserved scrutiny. During the extensive search operations to collect and analyze debris from the *Columbia* accident, two more people, US Forest Service personnel, were killed in the crash of a search helicopter (*New York Times*, 2003, A7).

Among the many issues raised in the wake of *Columbia*'s disaster was that of the Space Shuttle system's role in maintaining the Hubble Space Telescope (HST), which, in the course of the debate, would be judged "arguably the most important telescope in history" in a report of the National Research Council (2004, 3). Although HST had been successfully serviced by three earlier Space Shuttle missions, then NASA administrator Sean O'Keefe announced in January 2004 that the shuttle could not be made safe enough to send it again on a servicing mission, although the urgent need and great potential benefits of such a mission were undisputed. Missions to the International Space Station (ISS) were, on the other hand, judged acceptable in terms of risk because the shuttle can be thoroughly examined for tile or other damage from the ISS and because, in the event of damage to the shuttle, the ISS can serve as a refuge for a crew awaiting rescue. O'Keefe's decision, although controversial even within NASA and eventually reversed in October 2006, was justified in significant measure by considerations of risk to astronauts. It therefore generated an active and public debate about the roles of human beings in space and whether the benefits of space exploration are worth the perils experienced by those who choose to explore (Carreau 2004; *Wisconsin State Journal* 2004).

Is the study of space important enough to risk human life in the endeavor? We might first take a step back and ask just what is worth the risk of human life? We risk human life constantly, of course. We appear to accept that the transport of citrus fruits from Florida to Wisconsin is worth the risk to (and indeed the steady loss of) human life caused by heavy trucks on our highways. So apparently the question is of the cost/benefit type. With a commercial activity like interstate trucking, we could arrive at a reasonable figure for how many human lives are lost (the cost) per dollar of economic activity (the benefit). But the benefits are really somewhat more slippery than that, because we could argue that among the benefits should be included the health benefits enjoyed by northerners of a diet including fresh citrus. The costs, too, might be vague if we decide (as we probably should) to consider the environmental degradation that follows from heavy truck traffic and the various infrastructures that support it. So perhaps it is not so obvious that citrus commerce passes a cost/benefit test, and yet it is not widely controversial. Why not? Perhaps because the costs are geographically widespread in time and space, it is not easy to balance them against the benefits that pile up in our grocery stores. If a year's worth of trucking accidents occurred all on one day within a few miles of each other, there would probably be a public outcry to evaluate whether such traffic was worth allowing and even subsidizing (e.g., through highway construction and maintenance), and whether it was worth the loss of human life. We tolerate and even subsidize interstate trucking without requiring that it justify itself in terms of costs vs. benefits. (And if we did, we would ship citrus by rail.)

Even when losses are sudden, dramatic, and highly visible, we, as a society, are still sometimes reluctant to submit to the logic of a cost/benefit analysis. The human costs of war, for example, are often justified by the alleged need to protect or establish such abstract entities as freedom, democracy, or justice. Even when a war drags on and demands for a cost/benefit analysis finally gain political traction, the difficulty of defining, much less measuring, the benefits remains an obstacle.

Acts of altruism, in society's view, also resist a cost/benefit analysis. Firefighters considering entry into a burning building in order to save lives use their experience of such situations to evaluate their chances of success, which is a kind of cost/benefit analysis. But when they are unsuccessful in the attempt, they are not faulted for trying. And when the rescue attempt ends in the loss of a firefighter's life, the would-be rescuer is treated as a hero—not as having miscalculated the cost/benefit ratio. There were those two fatalities in a helicopter crew involved in the search for debris from the *Columbia* accident, so one could (and some did) reasonably ask whether that search for debris was worth the risk of human life. Others viewed the goals of the search (including both investigation of the accident itself and protection of the public from hazardous debris) as worthy of the risks associated with helicopter operations.

In fact, we think of risk in at least two senses. There are the risks we all encounter every day by getting out of bed in the morning, which are basically the risks faced by the citrus truck drivers and by the commuters who share the roads with the trucks. These are the risks we know are there, think we understand, and unreflectively (for the most part) defy because in our experience the bad outcomes are rare (i.e., widely scattered in time and space) and usually follow from coincidences of multiple contributing mistakes or multiple technical failures, or both. Of course, a misunderstanding of this kind of risk can contribute, for example, to a fear of scheduled commercial air travel, which, by the usual cost/benefit analysis, is the safest way to travel per passenger mile. But in general, these mundane risks (both commercial trucking and commercial aviation) are acceptably low because technology (e.g., vehicle and highway construction) and management techniques (e.g., informational road signs and air traffic control systems) are good enough to reduce the frequency of accidents to a low level—low enough that we don't feel threatened by our participation in them.

Then there are activities that are truly more risky than truck driving, although the reasons for the risk vary. A soldier in active combat is at very high risk, as are firefighters and police officers at certain points in their jobs. Combat is intrinsically risky because soldiers struggle with other soldiers, humans who can bring as much wile, courage, resourcefulness, and determination to the battle field as anyone else. The combatants constantly adjust and redeploy their technology and techniques in a deliberate attempt to frustrate their opponent; gaining control over the situation is the goal, and the risk is intrinsic to the struggle for control. Law enforcement can be a similar situation because lawbreakers employ the full range of human creativity in the frustration of the police. Why do people choose such risky careers? They don't always, of course: many a soldier has been conscripted. But many others do choose to go to war, to become police officers, to serve as firefighters. They do so for many reasons, among which might be a sense of duty, the influence of a role model, a desire to be of service, a hope to improve the world, or an interest in the techniques, equipment, and locales involved in a risky profession. A cold cost/benefit analysis is not the kind of selling point found on an Army recruiting poster. Society generally approves of and encourages career choices such as the military and law enforcement because, in the end, someone must do those jobs; and we seem to think that the consequent risks to human life are acceptable and worth the often unquantifiable benefits.

Other kinds of risky activities are not, strictly speaking, necessary. Attempting to climb Mt. Everest, riding a rocket into space, and other such endeavors are risky because the maximum control that human technology can muster is only barely enough to allow the successful completion of the task. This is the essence of exploration, whether deep sea diving, wintering over in Antarctica, or any number of other examples: the risk is inherent in pushing human activities into realms that are only barely accessible at our current state of knowledge and technology. Nobody is required to do these jobs, they are not (yet) economically important, and life could go on pretty much as most of us expect even if no one ever again reached the summit of Everest.

When the South Pole was first reached in 1911, it strained the technology of the day to its limits to complete the trip in the summer. Today we can fly to Earth's South Pole fairly reliably during the summer, but reaching that point during the winter is not undertaken lightly, primarily because of the limits of contemporary aircraft to operate in such harsh weather conditions. Because of that isolation, wintering over at the South Pole is done only with careful planning and a certain degree of risk—risk stemming largely from the fallibility of critical life-support systems or the onset of an acute medical condition.

Why, then, do we engage in the risky business of working at the South Pole? There are no immediate economic benefits. It satisfies no military or law enforcement needs. Yet many nations invest in research stations in Antarctica. They send geologists, paleontologists, biologists, climatologists, and other scientists to study that unique part of our planet. Meteoriticists harvest meteorites, fallen to Earth from various places in the solar system, which would be so much harder to find and identify elsewhere that it is worth the trip to the Antarctic. Astronomers build huge and innovative telescopes that can take advantage of the unique conditions of the Antarctic to observe infrared light and identify astronomical sources of elusive neutrinos. Why are those activities worth the risk to human life? Because wherever we probe into the unknown, there are inevitably, and almost by definition, new knowledge and opportunities to be won. People in general seem to accept, without a cost/benefit rationale, that exploration of the unknown is an activity worth their support, even their participation, and sometimes even great loss. In January 1912, Robert F. Scott, having been beaten to the South Pole by Roald Amundsen, strove to return with his men, in the face of a fierce storm, to their base camp and safety. But they failed to find an essential supply depot, and Scott, knowing that they were doomed, penned a poignant acceptance of the hazard in one of his last letters: "We have missed getting through by a narrow margin which was justifiably within the risk of such a journey" (Scott 1949, 475).

It should not surprise us that humans have a general desire to explore. Modern humans originated in Africa and migrated into Europe and Asia and ultimately to Australia, the Americas, the widely spaced islands of Oceania—indeed, settling nearly everywhere, from sea level to the Himalayas, that Neolithic technology could make habitable. The exploratory drive has contributed to forging a species of considerable diversity, with adaptations to a wide range of habitats comprising various climates, diets, and contagious disease environments. Those prehistoric explorations, undertaken for motivations at which we can only guess, also resulted in great cultural developments: specialized technologies, agricultural practices, domestication of animals, varieties of social organization, forms of information storage and retrieval, and so on. The best evidence of anthropology and paleontology demonstrates that some subset of human populations have consistently found success by moving into previously unexplored and unexploited environments. Whether genetic tendency, cultural trait, or some of each, exploration has been a beneficial practice for human populations.

The key to this success is due in some part to human biological adaptations (Tibetan women, for example, exhibit adaptations allowing them to bear healthy babies at altitudes where other women cannot (Willis 1998, 64-68)), but mainly to the adaptive power of human culture, which has allowed us to establish self-sustaining populations on every continent, except Antarctica, on time scales that are vanishingly small compared to those over which natural selection occurs. Antarctica, although not discovered until the 19th century, could probably not be rendered habitable by the technology of our distant ancestors, and a Neolithic settlement at the South Pole itself is almost unimaginable. But we can survive there now. Given our current technological capabilities, which make permanent occupation there quite possible (if not quite desirable), the absence of a self-sufficient Antarctic human population is more owing to a lack of readily exploitable economic resources than to anything else. Nations accept the costs and take the risks of going and staying there now almost entirely for the scientific research opportunities. Other reasons for Antarctic exploration might emerge, and other motivations have certainly been significant, e.g., wealth, nationalistic pride, the thrill of novelty, fascination with the unknown. They will remain in the mix, but always subject to the winds of politics and the whims of fashion.

It is scientific research that possesses the best claim—a claim backed by history and tangible results—on society's attention and resources as a reason for exploration. So although far from being the only reason to explore, scientific research can act as a wedge that motivates exploration even in the absence of other drivers and, indeed, helps bring other drivers to light. Furthermore, in cases such as space exploration, in which the large costs require some kind of societal consensus, the near certainty of improving our understanding of nature through exploration helps the public understand the value of such projects. The use of science in this sense may be essentially rhetorical, but that does not make it less compelling or valid.

This is true of our current forays into space: the enduring rationale is science. Longer missions beyond Earth, such as explorations of our moon or Mars, will take place (at least at first) for the scientific research opportunities. Although economic rationales (such as the mining of Helium-3 on the Moon) that could conceivably lead to self-sufficient extraterrestrial settlements have been explored, their actual feasibility will only be tested during scientific expeditions. It will be harder to do than reaching Antarctica, but unlike the hypothetical Neolithic explorer who would have been incapable of surviving a year at the South Pole, we have the minimal technologies for reaching and surviving on the lunar surface. Working on Mars, while more demanding than working in Antarctica, will probably be easier than working on the Moon, but the long trip will be a significant challenge. Scientific discovery and exploration are most exciting and inspirational when poised on this cusp of technical feasibility.

The risks will be there, on the Moon and Mars, just as they are in Antarctica, and they will be worse. Will the enterprise of space exploration be worth the risks? I think so, but not because of a favorable cost/benefit analysis. Cost/benefit analysis will be appropriate when costs and benefits can be quantified, i.e., when space commerce is a reality: will the profits be worth the costs to take tourists on suborbital flights to 100 kilometers and higher? That particular experiment is well under way. Will harvesting Helium-3 from the lunar soil result in profitable energy production? Perhaps. The quantifiable nature of those sorts of costs and benefits is why we can leave the sorting out of alternative business models, within appropriate regulatory contexts, to market forces. Risk assessment is, of course, a highly sophisticated endeavor with a vast literature.[1]

Like the Antarctic exploration (and, indeed, most exploration and pure scientific research in general), space exploration is beyond the cost/benefit realm. There are benefits, of course; otherwise we wouldn't be interested in such things at all. The benefits, all difficult to quantify, fall into four broad categories, all with profound roots in the history and success of human culture and society. Like other human exploration, space exploration 1. expands the ranges of alternatives, 2. inspires and motivates learning, 3. opens up transformational possibilities, and 4. enhances the survival of life.

1. New ranges of alternatives open up in many areas, but most clearly in availability of physical environments, natural resources, and economic possibilities. Access to vacuum, microgravity, and continuous solar energy, for example, has opened up scientific and industrial options unobtainable on Earth. Other extraterrestrial environments will offer more possibilities. Examples of extraterrestrial life, should

[1]For a starting point in thinking about risk in the context of specific spaceflight operations, see Michael L. Gernhardt (2005).

we find any, as well as nonliving but nonetheless novel chemistry, will expand our spectrum of alternatives via the results of nature's myriad environmental experiments on other worlds—experiments that have been running for billions of years. This is conceptual breadth we can gain by no other means than exploring new worlds. Human experience of states such as weightlessness, or sights such as the unfiltered view of the stars, will also have their effects on the range of arts, literature, and other modes of human expression. As alternative environments, processes, substances, experiences, and so on, arise, we will sort them out according to interest, danger, utility, beauty, and whatever categories become useful, but it is only through exploration that the hopper of alternatives remains stocked.

2. Space exploration, like other exploration, has an inspirational and motivational effect on people that can benefit society. Solving the problems associated with exploration of new environments motivates scientific, medical, and technological advances. To the extent that the exploratory enterprise is seen as vital to a vibrant society (in the same sense that art, music, and literature enrich society), we, and especially our youth, will see participation in it and support of it as praiseworthy and deserving of study and effort. This is often, sad to say, why young people will willingly march off to war: because such an act of personal commitment and potential sacrifice is treated as praiseworthy. How much better it is to enlist them in the expansion of knowledge and the cause of peaceful exploration while, at the same time, engaging human curiosity and testing the intellect to the utmost by exploring the unknown.

3. Space exploration can also be a transformational activity with the potential to change both individual outlooks and societal structures. At the personal level, exploration of the world serves as a transformational activity in which new belief systems can evolve in response to novel information and circumstances that are inadequately addressed by older belief systems. For example, most traditional belief systems are ill-equipped to cope with modern information and circumstances such as the advent of scientific knowledge of nature, the increasing power and rate of change of technology, the finite nature of Earth, and the emergence of global communities (Lemke 1996, 7). Exploration beyond Earth will fuel the furnace of change. The transformations are possible no matter how the exploration of space happens; that is to say, knowledge gained from robotic explorers will certainly contribute to transformative pressures. But the sight

of human explorers and their accounts of their experiences will have much greater impact than mere data and images on how people choose to see their relationship to the world.

As other frontiers in human history have done, the unique conditions in actual human settlements or on long-duration missions will create pressures and opportunities for socioeconomic experimentation, and consequent transformation, that could not arise on "old world" Earth. This "catalytic" function of frontiers, or "frontier shock," has arguably helped drive innovation and development in societies that expand into new environments, and the space frontier will be the most powerful agent for social transformation yet (Zubrin 1999, 7).

4. Perhaps most important of all, the exploration of space is the only way to ensure the long-term (and perhaps the short-term) survival of human society, human life, and perhaps even life itself on Earth. As has become clear in recent decades, the history of life on Earth is one of recurrent mass extinctions, die-offs that follow from major impacts of asteroids and comet nuclei or from enormous volcanic episodes (or perhaps both). The subsequent repopulation of the planet necessarily emerges from those species that were able to survive the cataclysm. Given that collisions between Earth and any of the many "Near Earth Objects" (NEOs) are inevitable, it is obviously desirable to have the capabilities to survive, mitigate, and even avoid such a catastrophe. Carl Sagan presents a different view that nevertheless arrives at the same ultimate conclusion (1994 320-326). Abilities to detect, investigate, and predict the danger of NEOs are clearly prerequisites to coping with the problem. Space exploration is needed to locate and ascertain the physical nature of these objects. Once the threats are identified, one possibility is to alter the course of the NEO that threatens us so as to avoid the collision. This possibility is in itself a good reason to develop interplanetary travel capability. William Burrows makes this case forcefully in *The survival imperative: Using space to protect Earth* (2006).

Of course, it might be the case that NEOs cannot be deflected or that any advance warning will be too short to allow preventive action. So, another approach is to establish a repository of information, technology, genetics, art, music, etc. to serve as a "backup" for human civilization. If established on the Moon, or in a lower but stable orbit, such a repository would be secure against many potential calamities, including nuclear war, epidemic disease, massive vulcanism, cosmic impacts, environmental disasters, or anything else that might cause the destruction of civilization on the surface of Earth. The survivors would have the possibility of establishing communication with the repository, downloading its information, eventually retrieving its stored materials, and thus mitigating the loss and hastening the recovery (Burrows 2006; http://www.lifeboat.com). But the best insurance of human life and civilization would be to have self-sufficient human populations on such places as the Moon and Mars. Then, even if Earth were rendered uninhabitable, the adventure of human consciousness would continue.

The first generation of experimenters to put modern rocketry to practical use deployed their rockets to rain terror on England from across the Channel. But their ambitions for the new technology of rocket propulsion had started with interplanetary travel, and after the war their vision of the expansion of humanity into the universe regained its status as the ultimate justification for space travel:

> And what would be the purpose of all this? For those who have never known the relentless urge to explore and discover, there is no answer. For those who have felt this urge, the answer is self-evident. For the latter there is no solution but to investigate every possible means of gaining knowledge of the universe. This is the goal: to make available for life every place where life is possible. To make inhabitable all worlds as yet uninhabitable, and all life purposeful. (Oberth 1957, 166–167)

Many other writers, from Konstantin Tsiolkovsky and Werner Von Braun to Robert Zubrin, have expressed similar sentiments.[2]

In the end, the expansion, development, progress, and even survival of human life and culture depend on an eventual diffusion beyond the single planet that now sustains them. The compelling reasons include scientific investigation as well as the preservation and vitality of society, but while they are reasons that defy cost accounting, their benefits are manifest. It is hard to imagine what could be more worthy of the risk of individual human lives than ensuring the long-term survival and propagation of humanity itself and deepening our understanding of our relationship to the natural universe.

[2]A classic introduction to this line of thinking can be found in Arthur Clarke's *The exploration of space* (1959). For more current assessments, see the works of Harrison (2001) and Dick.

References

Burrows, William E. 2006. *The survival imperative: Using space to protect Earth*. New York: Tom Doherty Associates.

Carreau, Mark. 2004. Hubble's fate stirs debate on space. *Houston Chronicle*, April 12.

Clarke, Arthur C. 1959. *The exploration of space*. New York: Harper and Row.

Committee on the Assessment of Options for Extending the Life of the Hubble Space Telescope, National Research Council. 2004. *Assessment of options for extending the life of the Hubble Space Telescope*. Washington, DC: National Academies Press.

Dick, Steven J. Risk and exploration revisited. http://www.nasa.gov/mission_pages/exploration/whyweexplore/Why_We_14.html.

Gernhardt, Michael L. 2005. Exploration and the risk-reward equation. In *Risk and exploration: Earth, sea, and the stars*, ed. Steven J. Dick and Keith L. Cowing. Washington, DC: NASA History Division. NASA SP-2005-4701. http://history.nasa.gov/SP-4701/riskandexploration.pdf.

Harrison, Albert A. 2001. *Spacefaring: The human dimension*. Berkeley: University of California Press.

Helicopters are grounded after crash in debris hunt. 2003. *New York Times*, March 29.

Hubble's priceless value justifies repair mission. 2004. *Wisconsin State Journal*, August 29.

Lemke, Lawrence G. 1996. Why should humans explore space? In *Strategies for Mars: A guide to human exploration*, vol. 86, *science and technology series*, ed. Carol R. Stoker and Carter Emmart. San Diego: American Astronautical Society.

NASA on defensive about suspending Hubble servicing missions. http://www.space.com/spacenews/archive04/hubblearch_022404.html.

Oberth, Herman. 1957. *Man into space: New projects for rocket and space travel*. G. P. H. de Freville, trans. New York: Harper and Brothers.

Sagan, Carl. 1994. *Pale blue dot*. New York: Random House.

Scott, Robert Falcon. 1949. *Scott's last expedition: The personal journals of Captain R. F. Scott, C. V. O., R. N., on his journey to the South Pole*. London: John Murray.

Willis, Christopher. 1998. *Children of Prometheus*. Reading, MA: Perseus Books.

Zubrin, Robert. 1999. *Entering space: Creating a spacefaring civilization*. New York: Tarcher/Putman Books.

PART 4

Global Warming: Scientific Data, Social Impacts, and Political Debate

15

The Politics of Climate Change

Clark A. Miller

Climate change is arguably one of the most important and controversial arenas of US and global policy-making. Is the profligate use of fossil fuels to support energy-intensive economic and industrial growth putting the planet at risk of major, disruptive climatic shifts? If so, how should humanity respond? Should dramatic investments be made in transforming the energy infrastructure of the globe to noncarbon-based fuels? Is a global regulatory framework necessary to force countries to reduce greenhouse gas emissions? Do we need to be prepared to engineer planetary ecosystems in order to avoid the consequences of our own excesses as a species?

These are difficult and contentious questions facing humanity in the 21st century. I make no pretense of answering them here. Rather, this essay seeks to clarify the issues at stake, as well as to lay out a bit of history regarding policy debates over climate change and to describe some of the ways in which various groups have tried to reason their ways to answers to some of these questions. I open with a short trip in time to the outset of the climate change debates in a set of experiments carried out in the late 1950s. I then turn, in subsequent sections, to an overview of some of the key elements of debates about climate change and what I see as at stake.

In 1957, two scientists set out from the Scripps Institution of Oceanography in San Diego to conduct a simple scientific study.[1] Their objective was to measure the atmospheric concentration of carbon dioxide and to contribute to the emerging field of atmospheric chemistry and to the larger study of the Earth's geophysics known as the International Geophysical Year.

Roger Revelle and Charles Keeling knew that industrial and other human activities, including automobiles, were responsible for significant amounts of carbon dioxide being poured into the atmosphere. Hence, they would need to sample the atmosphere at long distances from major industrial centers, if they were to make undisturbed measurements. So they chose Hawaii, isolated in the middle of the Pacific Ocean, and set up their equipment at the top of Mauna Loa. They took their first measurements in 1958, starting a continuous series of measurements of atmospheric carbon dioxide that runs to this day. Initially, what they found was not terribly surprising. Atmospheric carbon dioxide concentrations fluctuated from spring to fall, as a result of seasonal changes in plant growth, decay, and respiration. While plants respire constantly, the net balance of intake to outflow between respiration, growth, and decay varies seasonally. Since the seasons are reversed in the two hemispheres, seasonal variations would balance out, except that the Northern hemisphere has much greater biomass. Hence, during the Northern hemisphere growing season, there is a net loss of carbon dioxide from the atmosphere, while, when plant material in the Northern hemisphere is primarily decaying, there is a net addition of carbon dioxide to the atmosphere. More surprising was what they found over the next year and a half.

By the end of the eighteen months of the International Geophysical Year, Revelle and Keeling had established a new fact. From one year to the next, carbon dioxide concentrations in the atmosphere cycled, but they didn't return to where they started. Rather, the base level increased over time. Two more years of data confirmed the finding. Atmospheric carbon dioxide levels were rising at the slow but deliberate pace of a little less than 1% per year. Humans were changing the atmosphere, everywhere. Revelle and Keeling wrote up their results, calling the increase a "geophysical experiment" of a kind never carried out previously. What would changing the atmospheric carbon dioxide concentration do to the Earth's climate? No one knew, but it was now a pressing question.

---

[1]For readers who are interested in a deeper account of the early history of carbon dioxide research, and especially the work of Revelle and Keeling, please see Weart 1997.

Today, nearly half a century after Revelle and Keeling published their initial findings, scientists know a great deal more about the geophysical experiment humans are conducting on the climate. For one thing, atmospheric concentrations of carbon dioxide have risen over one third from their levels in the late 1950s, from 281 to 380 parts per million. For another, carbon dioxide is a significant greenhouse gas, trapping solar radiation in the atmosphere and maintaining livable environmental conditions in the biosphere. Other chemicals in the atmosphere are also important greenhouse gases, and they, too, are increasing due to human activities: methane (largely from cattle and rice agriculture), nitrous oxide (from fertilizer use), and chlorofluorocarbons (from industrial processes).

Since 1960, a number of scientific laboratories around the world have also created sophisticated computer models of the Earth's climate system. One of the models' major objectives has been to project the future consequences for the climate system if greenhouse gases continue to increase in the atmosphere. Their results are not comforting. The models all agree that the mean global temperature will rise between 1.5 and 4.5 degrees Celsius by 2100. The specific implications of this rise on the ground are less clear, but a range of studies suggests dramatic shifts in ecological zones for natural systems and also for crops and agricultural production, changes in the frequency and intensity of hurricanes and other dangerous storms and natural hazards, and also increased melting in the polar ice caps and glacial regions of the planet. Much of this melting is now well documented. In the past few years, in fact, signs of dangerously rapid melting have become visible in the Arctic, Antarctic, and many of the world's glaciers. This melting is expected to accelerate sea-level rise and potentially put at risk many of the world's low-lying cities. Also feared are shifts in the ocean's circulation patterns, which could cause enormous fluctuations in worldwide weather patterns.[2]

Scientists are almost unanimous in their concern about future global warming—the more popular name for the phenomenon. A recent study by Naomi Oreskes at the University of California–San Diego randomly sampled 10% of peer-reviewed publications on climate change and found none that expressed skepticism about the basic theoretical or em-

---

[2]The most exhaustive, up-to-date survey of climate research can currently be found in the Intergovernmental Panel on Climate Change's Fourth Assessment Report (IPCC 2007), which was being released as this volume went to press.

pirical research supporting the claim that greenhouse gases are driving and will continue to drive changes in the climate system (Oreskes 2004). Less certain are issues such as the precise timing and magnitude of climatic change, as well as what climatic changes will mean for human communities and natural ecosystems in the coming decades. Even here, however, there is a growing conviction within the climate science community that the changes are coming, perhaps even faster than anticipated, and that they will be significant. Some climate scientists are, in fact, concerned that the climate system may soon encounter dangerous "tipping points," such as melting of glaciers, that in turn provide such significant feedbacks that they create runaway warming.

Shifting one's gaze away from the scientific literature to more public contexts brings a quite different view, however. While it has achieved considerable prominence on the public agenda, global warming appears much more contentious there than in the scientific literature. Throughout the 1990s, conservative think tanks, politicians, and radio talk-show hosts—many funded by the oil and coal industries—pursued a vigorous public relations campaign seeking to deny the reality of climate change. Republicans in Congress made significant efforts to derail funding for research on climate change and passed rules forbidding the Environmental Protection Agency from addressing the problem. These events culminated in March 2001, when President George W. Bush declared that the most recent report of the Intergovernmental Panel on Climate Change was junk science, propagated by that un-American institution, the United Nations.

Looking beyond political institutions to the media and the public reveals similar skepticism about the reality of climate change. Major newspapers covering climate change almost always include coverage of climate skeptics in stories on climate change. Public opinion polls reveal similar expressions of uncertainty and disagreement. A recent poll of US adults by NBC News and the *Wall Street Journal* revealed, for example, that of just over 1,000 respondents, only 29% felt that climate change had been established as a serious problem demanding immediate action. Twenty-eight percent believed that we don't know enough yet about climate change and that more research is required before we take any action. Nine percent said that concern about climate change is unwarranted, and a further four percent were unsure.[3] Skeptical, too,

[3]Reported on Pollingreport.com. NBC News/*Wall Street Journal* poll conducted by the polling organizations of Peter Hart (D) and Bill McInturff (R). June 9–12, 2006. N = 1,002 adults nationwide. MoE ± 3.1 (for all adults). http://pollingreport.com/enviro.htm. Accessed on 7/18/06.

is Michael Crichton's *State of Fear*, which depicts climate change as an elaborate hoax by extremist environmental groups, and which rose to become a *New York Times* bestseller. Despite being exceedingly poorly written (at least measured by the criteria my English teachers taught me), the novel has attracted significant attention and has apparently been taken by many of its readers to be a sincere and believable reading on the author's part of the scientific evidence (a view that the author very much encourages in the book's preface). Despite the widespread media attention in February 2007 to the most recent report of the Intergovernmental Panel on Climate Change, and especially its finding that the human influence on climate change is "unequivocal," skeptics in the public and among politicians remain.

What explains this persistent divergence between scientists and the public, the media, and the politicians? One theory is that corrupt politicians are carrying out a prolonged campaign to distort science and that the media is simply parroting their propaganda. The result is widespread perplexity in the public about the truth of global warming. In his recent film and book, for example, Al Gore calls climate change "an inconvenient truth":

> The truth about global warming is especially inconvenient and unwelcome to some powerful people and companies making enormous sums of money from activities they know full well will have to change dramatically in order to ensure the planet's livability. These people . . . have been spending many millions of dollars every year in figuring out ways of sowing public confusion about global warming. (Gore 2006, 284)

Numerous scientists similarly accuse the media of inadvertent bias in its reporting. Media reports, they argue, too uncritically follow a common journalistic practice of seeking out divergent opinions on any story. The practice, which is designed to avoid the bias of basing a story on only a single viewpoint, nonetheless distorts science by empowering minority views with the same apparent weight as majority expert opinion. Thus, climate scientists argue, the perception of scientific dissent is magnified out of true proportion, and reporters and editors should modify their practices to stop reporting on skepticism about the reality of climate change. From this perspective, if the public simply heard less dissent in the media, they would be less skeptical about the reality of climate change and support stronger action to protect the global environment.

There are several problems with this explanation, however. First, there is little evidence that distortion in public opinion results from how the media reports stories. In fact, considerable evidence suggests that the public is quite smart about the media and that news stories do not directly control perceptions on controversial issues. If this is true, then even a significant reduction in the reporting of scientific dissent in the mainstream press may have little impact on public perceptions of the climate debate. Indeed, given the rapidly growing importance of alternative media, via cable news and the Internet, it seems unlikely that altering the practices of the mainstream media would do much, in the first instance, to reduce the expression of dissent about climate change in the media overall.

It also seems politically problematic, at least to me, to suggest radical changes in the practices of investigative journalism at this moment in history. A free press has long been seen as a thorn in the side of powerful institutions and a hallmark and an essential element of democracy in the United States. The practice of seeking at least two perspectives on any story is part of a commitment in journalism to avoid capture by any single viewpoint on what is important—no matter how persuasive or powerful. It is a form of institutionalized skepticism that helps serve multiple communities in a pluralist society by providing insights into the actions and beliefs of diverse social groups. In doing so, it lowers the barriers to entry into public debate, providing a platform for minority voices and the voices of those skeptical of dominant positions to be heard, thus contributing to resistance to the emergence of hegemonic elements and the preservation of minority rights. If it also contributes to giving more weight to minority viewpoints than perhaps they deserve, that has long been a well-known trade-off of the right to free speech. The option, of course, is to rein in speech deemed unacceptable, thus raising the barriers to entry into public discourse, so that only those who are appointed experts are allowed to speak on a relevant issue. In a time of growing concentrations and inequalities of power and wealth, this seems a dangerous precedent.

One alternative way to think about the persistent gap between scientific and other views of climate change is through the lens of credibility. Climate scientists appear to have a credibility problem. This is a significant liability in democratic politics, where reasoned deliberation is an essential element of achieving policy change. What should we make of this credibility problem? One aspect of it, I believe, is that climate scientists have tended to leave participation in public debate to others, including environmental organizations, the media, and politicians, most of whom have even bigger credibility problems vis-à-vis US publics than scientists. I do not mean by this that scientists have not been active trying to raise the issue. For example, many scientists have participated in a collaboration between the Union of Concerned Scientists and the Ecological Society of America, which brought scientists together to develop state-of-the-science climate change reports for several regions of the United States A major component of these efforts has been to engage politicians and policy makers on issues of climate science. The Californian effort is credited with facilitating the passage of new vehicle-emission standards several years ago.

However, scientists have been less active in efforts to mobilize broad public opinion on the issue of climate change. It seems somewhat surprising, given the popularity that some scientific personalities (such as Stephen Hawking) achieve, that no major campaign to persuade the public about the importance of climate change was launched in the United States until 2006. Even then, this effort was led by a politician (Al Gore), working with a group of Hollywood film directors. Where are the scientific voices who, as they did in 1945 and 1946 after the dropping of the atomic bomb, could transform the popular debate on climate change?[4] I believe such voices could be significant, if aimed at working with grassroots communities to persuade them of the seriousness of the issue. Policy makers listen to their constituents, yet much of the focus of scientists has been on taking their message to Washington or Sacramento. Sometimes this works, but other times it does not.

Part of the challenge is also that scientists have lost ground in terms of public trust (Ezrahi 1990). What sociologists know about credibility is that it is a product of social norms that lend credence and trust to claims, and it often requires a long period of time and considerable investment in social capital to establish (Shapin 1996). For much of the 20th century, scientists had a considerable reservoir of trust among American publics, but that trust has worn away over time. Some of the reasons for this stem from a loss of faith in narratives of technological progress, as disasters and environmental crises have worn away at a belief that scientific and technological innovation will lead inevitably to enhanced human welfare. Other reasons include the drawing of sci-

---

[4]Naomi Oreskes offers one perspective on this question in her forthcoming book, *Science on a Mission: American Oceanography in the Cold War and Beyond* (Chicago: University of Chicago Press, 2007).

ence into political conflict and contestation (of which, more later). Still others include growing skepticism toward all forms of authority in a culture increasingly committed to a narrow individualism.

One result of this is that scientific claims no longer command unquestioned deference from skeptical publics. I believe that persistent opposition among some citizens to certain kinds of research, such as stem cells and genetic engineering, reflects, in part, this overall shift in public attitudes toward science. So, too, does persistent skepticism about climate change. Nor will trust easily be regained, as any observer of arms control policies or relations between war-torn communities will quickly recognize. Efforts must be made to persuade multiple publics that the scientific community has their best interests in mind and is committed to truth-telling. Rebuilding credibility may require significant commitments of time and energy on the part of not only scientific leaders but also the rank-and-file of the scientific community to reengage citizens in the places where they live and work.

That said, however, it would be a mistake to overestimate the degree to which the issue of public and political dissent is, fundamentally, a failure of credibility. Climate change has been, at least until recently, a largely invisible problem. An untrained eye cannot simply look out the window and see climate change or its implications for society. The only knowledge we have had about the potential risks of climate change have come from incredibly sophisticated computer models whose inner workings are understood by at most a few hundred individuals worldwide. And yet, climate scientists have persuaded much of the world's population that global warming is a major public concern and that it demands considerable attention and resources directed toward its solution. The fact that international diplomats have inked two major international treaties that regulate global warming—the UN Framework Convention on Climate Change and the Kyoto Protocol—signals the enormous influence of climate scientists on world opinion. That a few pockets of resistance remain—even if one of them is the world's largest economy and emitter of greenhouse gases—cannot really be seen as a crisis of credibility.

How, then, should we view the fact that skepticism of climate science remains rampant in the United States, if not entirely as a problem of credibility or the distortion of truth? Why does it remain possible for US politicians to assert credibly that global warming is simply a theory—on a par with the theory of intelligent design—without concern for their political support?

I think the most persuasive answer is rooted in political culture. In particular, I see climate change as a product of the way that US political culture reasons about risk. What do I mean by that? Put simply, there are many kinds of risks in the world, and for various reasons, different communities understand and respond differently to different kinds of risks (Jasanoff and Wynne 1998; Thompson and Rayner 1998). Social scientists who have studied this issue highlight a number of different factors involved in giving rise to this selectivity about risk perception and judgment, including the constitutional organization of politics, the kinds of evidence, expertise, and models that acquire cultural credibility, the ways cultures factor uncertainty into decisions, and many others.

In US political culture, one of the most significant factors in thinking about climate change is its emphasis on the concept of harm. American environmental regulation is built on the backbone of tort law and the concept of liability. In US tort law, legal standing—the right to bring a claim before a court—is obtained by demonstrating that a harm has taken place. A recent book with the title *Evidence of Harm* exemplifies the issue (Kirby 2006). The account, written by a journalist, follows the efforts of hundreds of parents-turned-activists in the United States to document evidence that their children have been harmed by mercury contained in childhood vaccines. Like climate scientists, these men and women have faced significant opposition from well-funded industries and politicians—in this case, pharmaceutical giants and, interestingly, key Democrats—many of whom deploy quite sophisticated technical arguments and evidence suggesting that there are no causal links between mercury and autism or other health risks. It is a sobering account of US political culture, and it highlights throughout that the concept of harm is central to both scientific and legal discourse about risk in the United States. Without harm, there is no risk.

Disputes about climate change offer a similar tale. The principal tactic of opponents of global warming has been to argue that global warming is just a theory. They attack climate models as inherently uncertain efforts to project *possible future* risks, not *present day* harms. They note that risk, as described by Langdon Winner, is something that one takes and that is often offset by other risks. Climate opponents often argue, for example, that potential risks to the economy from premature regulation are more dangerous than the possible risk of climate change. Absent demonstration of a serious harm, therefore, climate change remains politically weak. The paradigm of harm-based environmental law also helps explain why the concept of precaution has achieved no traction in the United States, even among scientists who deeply believe that something must be done to address climate change. Scientists in the United States by and large accept the legal norm that action to rectify a complaint must be based on solid (read "scientific") evidence of injury.

One of my favorite facts about *An Inconvenient Truth* is that it contains exactly zero references to climate models. Al Gore is no stranger to US political culture, having participated in innumerable battles for environmental protection in the 1970s and 1980s. Book and film are chock full of stunning visuals of glaciers melting, oil rigs smashed into bridges, bodies floating in the flood waters left after hurricanes, cars smashed by a river of mud—not quite babies with cancer, but pretty darn close. In the book, too, are projections of future flooding, not from climate models, but from simple extrapolations of sea-level rise along the coastlines of Florida, San Francisco, Manhattan, Beijing, Shanghai, and Calcutta. Here is the extremism lambasted by Republicans (the book's subtitle is "The planetary emergency of global warming . . .") but grounded in visible evidence of harm. These are the kind of facts that sell well in US politics, and Gore puts them to good effect.

Other recent studies, too, have focused on evidence of harm. A recent review in *Nature* estimates, using figures from the World Health Organization, that an excess of 150,000 individuals *already* die each year in poorer parts of the world from global warming (Patz, et al. 2005). The authors stress that this is a conservative estimate, based on studies of only a small range of potential health-related impacts of climate change, such as changes in the number and ecological ranges of mosquitoes that serve as malaria vectors. Many of the major cities of Africa, for example, which were built in elevated regions above mosquito ranges, now lie well within them. The authors also stress that this is a significant indication of harm that raises serious questions about US opposition to strong action to prevent additional climate change.

That such evidence is becoming available has led some climate scientists to warn that it may already be too late to stop significant amounts of global warming. The climate system has enormous momentum—as does the carbon-based energy infrastructure that drives the global economy. Neither will be stopped anytime soon. But in the United States, this is the kind of evidence that can motivate action. While much of the evidence deployed is quite recent, it builds on earlier warnings that harm was beginning to occur. Within a month of the President's specious attack on the Intergovernmental Panel on Climate Change in March 2001, for example, Bush had been forced to back down. The prestigious National Academy of Sciences had weighed in, supporting what it viewed as good science and noting, in passing, that a majority of the panel's scientists were from the United States. Although I doubt the conflict over the reality of global warming is entirely dead, the end is near. The release of the 2007 IPCC report and its assertion that human influence on the climate was "unequivocal" has created a groundswell of public and political support for climate change policies. Also, President Bush acknowledged in his 2007 State of the Union address that "global climate change" was "a serious challenge." It is not an accident, however, that the *New York Times* article announcing the IPCC findings was accompanied by a photograph of two polar bears isolated on a shrinking iceberg. What has changed people's minds about the credibility of climate change is the growing evidence of its *harm* to things we care about as a society. Indeed, it is quite possible that both candidates for US President in 2008 may be active supporters of strong action to protect the Earth's climate.

There is one final reason that US political culture has been persistently skeptical of the reality of climate change. An important element of US political culture is that political controversies are almost always fought on technical grounds, even when such technical debates are simply proxies for deeper ideological and policy disagreements. For decades, for example, tobacco companies fought efforts to regulate smoking through massive efforts to discredit the scientific findings linking smoking to health risks. Whatever the reason for opposition to policy action, good tactical strategy in the United States involves extensive criticism of scientific claims about risk. It is striking that, even where political opposition has formed to global warming policies in other countries, they have not experienced the extensive public debates over science present in the United States. Opposition takes other forms, in other settings. This unique feature of American politics has been described at length by scholars (Nelkin 1992). In turn, others have explained how such conflicts arise as US democracy seeks to strip power of its sometimes ambiguous and ad hoc application by insisting that officials justify their behaviors with reference to science (Ezrahi 1990; Porter 1995). Technical argumentation has become the central arena of political dispute, especially in the politics of risk, and a wide range of interests in society have built up the necessary resources to compete on scientific grounds.

A key consequence of this politicization of science in US political culture is that no appeal to science, no matter how well grounded, is likely to shift policy debate unless it is also accompanied by some form of social or political realignment. In the case of climate change, I believe the end of the debate about the reality of global warming is near, partly because of the shifting terrain of the technical debate, away from climate models to evidence of harm, where new resources come into play for persuading skeptical publics, such as the visual imagery so pervasive in *An Inconvenient Truth*. At the same time, a fundamental convergence has occurred, at least temporarily, regarding climate policy in the United States. Despite the appearance of ongoing rifts, the climate policies pursued by the Bush Administration have converged significantly with those advocated by key climate activists, such as Gore. At the end of *An Inconvenient Truth*, the film highlights the need for voluntary actions by individuals and companies and for investments in new technologies that will reduce the nation's dependence on fossil fuels. These suggestions are not too far off from Congressional energy debates. Indeed, in his 2007 State of the Union address, President Bush announced a major initiative to promote biofuels. What distinguishes these positions is essentially how much money they allocate to green technologies, and how much of a role should be allocated to nuclear power.

I believe this convergence is temporary, however. It masks a set of emerging policy debates that are likely to re-sunder US politics over how the country should respond to climate change. One such debate is likely to emerge around the problem of installed infrastructure in the energy industry. Power plants have considerably longer lives than automobiles, for example, meaning plans that move forward in the short-term will have enormous impact on long-term emissions profiles. Governments are likely, therefore, to need to signal to energy producers about the dates by which they expect carbon-based fuels to be phased out of the nation's energy portfolio. States have begun making medium-term initiatives, but the real benefits will come from setting long-term goals that the energy industry can plan for over decades. Setting policy for the second half of the

21st century has certain advantages, in that it is unlikely any of the politicians in question will still be in office, but it will, nonetheless, remain highly controversial. Governments will also need to provide incentives for energy companies to invest in alternative forms of energy production, should the lifetimes of soon-to-be-installed plants begin to impinge on long-term policy objectives, but should existing technologies not yet prove competitive in the energy markets.

A second set of controversies is likely to surround the societal impacts of large-scale technological transformation in the energy sector. There is a myth, I think, that switching from fossil fuels to hydrogen can be done simply through technology substitution, with no associated social changes. Anyone with the least introduction to the history or sociology of technological change will instantly recognize the ludicrousness of this myth. Large-scale technological transformations inevitably create significant societal dislocations that communities are forced to work through, and there is perhaps no single technological system that is more deeply embedded in contemporary society than fossil fuel–based energy sources. Already, we are witnessing early dislocations as Ford and General Motors get hammered by Japanese automakers, who appear far more technically adept at making vehicles that run more efficiently and consume fewer hydrocarbons. Entire sectors of jobs, such as gas station attendants or refinery workers, are likely to disappear, necessitating new programs of social insurance to cushion the transformation to a noncarbon energy economy. Social protests in Mexico City offer another illustration of the potential societal dislocations that may stem from changes in energy technologies. The protests, driven by higher tortilla prices, are an outgrowth of a rapid spike in global commodity markets, which is, in turn, a consequence of rapidly increasing demand for biofuels as an alternative fuel source in the US energy market.

A third set of difficult policy choices is likely to involve whether, and if so in what capacity, the United States should participate in future international climate treaties. Continued US intransigence is likely to scuttle any such talks, as countries are unlikely to commit themselves to significant reductions in greenhouse gas emissions so long as the world's largest carbon emitter and largest and most ruthlessly competitive economy refuses to play by common rules. Yet, as with so many other arenas of international policy-making, US politicians are likely to remain reluctant to place limits on US sovereignty in a domain as highly critical to the nation's economic production as energy. Nor will negotiating an appropriate and equitable set of long-term policies be easy, given the already significant chasms between rich and poor nations in general, and in climate politics specifically. Put simply, most developing countries see the current politics of climate change as an explicit attempt to limit their economic development and so retain the current global inequities in wealth and well-being. Given the context in which climate negotiations are likely to take place—e.g., US policies on free trade, intellectual property, and investors' rights in recent decades—it would appear that they are largely right in concluding that the United States is likely to take every opportunity to protect its own economic future at the expense of others. Bridging that chasm will be extremely difficult—and controversial here in the United States.

Finally, emergent debates about geoengineering are worth a brief note. If climate scientists are right, by midcentury humanity is likely to face dramatic changes in the distribution of rainfall and in temperature patterns, leading to considerable social and ecological dislocation. Many people argue that the United States should be beginning to prepare now, and should be helping poorer nations to prepare as well, for radical changes in flood and drought regimes, hurricane and storm intensity, as well as considerable rises in sea level. I think this focus on adaptation is likely to remain mostly uncontroversial, aside from the scale of foreign assistance that may be necessary in a few decades. For the past 20 years or so, however, a smaller group of individuals has argued for the need to invest in technologies that might be used to directly counter global warming in the atmosphere. They call these technologies geoengineering, and it is a field that is receiving new attention from an odd assortment of people and places, including climate scientists and the US national laboratories. So far, public sentiment and activist opposition have kept discussions of geoengineering out of the mainstream of climate policy. Indeed, it does seem peculiarly wrong-headed to think about handling the unexpected by-products of the failure of one massive technological system with another massive technological intervention into nature. Nor, I should say, does weather modification have a good history of success in its few limited trials to date. Nonetheless, proponents of geoengineering suggest that we may change our minds in 20 or 30 years, if we see we have no other option but to suffer through decades of dangerous climatic transformation. They argue, therefore, that we should invest now in researching possible technologies, just in case.

What is at stake in continuing scientific and policy debates about climate changes? *At its most basic, the human future.* What will the technological infrastructures of the second half of the 21st century look like? Which industrial sectors will have vanished? Which new sectors will have arisen? How will wealth have been redistributed? What will happen to managed and natural ecosystems as climatic patterns shift? What will happen to those who depend on those ecosystems for survival or livelihood? How will patterns of health and disease have changed? Will we have a world government—to impose sanctions on profligate users of fossil fuels, to salvage communities ravaged by climatic disasters, to fight off hordes of refugees from impoverished nations thrown into disarray, or to secure the resources necessary to engineer the planet to fix global warming?

How we reason our way through these problems as a human community will say much about our capacity to make good on the promises of the Enlightenment in a divided world. It will demand clear thinking and an inexorable commitment to improving the human condition for everyone on the planet. It will also demand a new faith in the ability of experts and governments to plan long-term infrastructural changes in an equitable fashion. Or, if not that, it will require markets finding mechanisms to address the public good, in addition to private interests, and careful scrutiny from attentive citizens. It will certainly involve international cooperation. Solving the coming climate crisis will not be possible for the United States or any other nation alone. Whether we, here in the United States, can overcome our parochial attitudes toward our fellow inhabitants of Earth, and pitch in to help prevent and respond to global warming is, I think, a deeply troubling and open question. Of course, the naysayers may yet turn out to be right. Uncertainties in scientists' knowledge about the climate system may mean that we have misjudged the extent of the problem. Perhaps our luck will hold out. . . . Somehow, I doubt it.

Acknowledgments

I thank the reviewer for his suggestions, including the addition of Californian vehicle emissions standards as an example of climate scientists' involvement in political debates.

References

Ezrahi, Yaron. 1990. *The descent of Icarus: Science and the transformation of contemporary democracy.* Cambridge, MA: Harvard University Press.

Gore, Al. 2006. *An inconvenient truth: The planetary emergency of global warming and what we can do about it.* Emmaus, PA: Rodale.

Intergovernmental Panel on Climate Change. 2001. *Climate change 2001.* Cambridge, UK: Cambridge University Press.

Jasanoff, Sheila, and Brian Wynne. 1998. Science and Decisionmaking. In *Human choice and climate change: The societal framework,* ed. S. Rayner and E. Malone, 1–88. Columbus, OH: Battelle Press.

Kirby, David. 2006. *Evidence of harm: Mercury vaccines and the autism epidemic: A medical controversy.* New York: St. Martin's Press.

Nelkin, Dorothy. 1992. *Controversy: Politics of technical decisions.* 3rd ed. Newbury Park, CA: Sage Press.

Oreskes, Naomi. 2004. Beyond the ivory tower: The scientific consensus on climate change. *Science* 306:1686.

Patz, J.A. et al. 2005. Impact of regional climate change on human health. *Nature* 438:310–317.

Porter, Theodore. 1995. *Trust in numbers: The pursuit of objectivity in science and public life.* Princeton, NJ: Princeton University Press.

Shapin, Steven. 1996. Cordelia's love: Credibility and the social studies of science. *Perspectives on Science: Historical, Philosophical, Social* 3:255–275.

Thompson, Michael, and Steve Rayner. 1998. Cultural discourses. In *Human choice and climate change: The societal framework,* ed. S. Rayner and E. Malone, 265–344. Columbus, OH: Battelle Press.

Weart, Spencer. 1997. The discovery of the risk of global warming. *Physics Today* 50(1):34–40.

16

Climate Change: Impacts and Implications for Justice

Stephen H. Schneider
and Patricia R. Mastrandrea

Future climate change impacts are highly likely to be significant, affecting human populations, as well as populations of plants and animals globally. These impacts are likely to disproportionately harm developing nations, which contain roughly 80% of the world's population. Despite rapid economic and carbon dioxide emissions growth in some large developing countries, when viewed historically, these nations have contributed little to cumulative greenhouse gas (GHG) emissions (about 20% of the cumulative global total to the year 2000). Compounding this inequity, most cost-benefit analyses (CBAs) weighing the costs of climate policies and the damages from unmitigated climate change typically undervalue the extent of climate change–induced damages in these countries, since CBAs traditionally assess damages easily quantified in monetary terms, while mostly ignoring—or treating via very crude methods—issues such as distributional inequity and the value of Nature or quality of life. We discuss a system proposed by Stephen Schneider and others that positions market-system impacts along with other metrics of harm and incorporates nonmarket impacts for a more equitable approach to damage assessment (Schneider, et al. 2000b). Mitigation, in both developed and developing countries, albeit with developed countries leading, is necessary to relieve pressure on the climate system. Even with stringent mitigation efforts, however, some increased level of climate change is inevitable and adaptation will continue to be required. Therefore, developed countries, with the largest per capita emissions and responsible for the lion's share of cumulative emissions, must help developing countries and marginalized people in both developed and developing countries to put adaptive strategies in place to reduce damages already occurring and projected to increase sharply. Implementing a combination of mitigation and adaptation policies will be necessary to meet the challenges ahead.

Climate Change Impacts

Climate has varied substantially on geological time-scales. Changes of about 5–7°C over many thousands of years have occurred over the past half-million years, triggering ecological responses including large shifts in vegetation and animal patterns and some extinctions. More recently, human activities that clear land or burn fossil fuels have been injecting greenhouse gases such as carbon dioxide (CO_2) and methane (CH_4) into the atmosphere: CO_2 has increased by ~35% and CH_4 has increased by ~150% since the beginning of the Industrial Revolution. The net effect of these perturbations has been to warm the global climate by ~0.75°C since 1850. The 2007 IPCC Fourth Assessment Report (AR4) states that this warming is "unequivocal" and that it is very likely (greater than a 90% chance) that most of this observed increase in globally averaged temperature over the past 4 decades is due to the observed increases in anthropogenic greenhouse gas concentrations, rather than known natural causes (IPCC 2007a). Human activities are the main driving force of the recent warming trend. In addition, climatic theory—supported by climate models—suggests that several more degrees Celsius of global warming will likely occur during this century. The IPCC AR4 projects a "likely" range of warming between 1.1 and 6.4°C by 2090 (compared with a range between 1.4° and 5.8°C by 2100 projected six years ago by IPCC in its Third Assessment Report [IPCC 2001a]). Although warming of, say, 1.5°C, would have many associated "key impacts" (Schneider, et al. 2007), changes of this magnitde would very likely create much less damage than warming at the higher end of the range. Even another 1°C warming above current levels would still be significant for some "unique and valuable systems" (IPCC, 2001b; IPCC 2007b). Mastrandrea and Schneider have noted (2004) that the

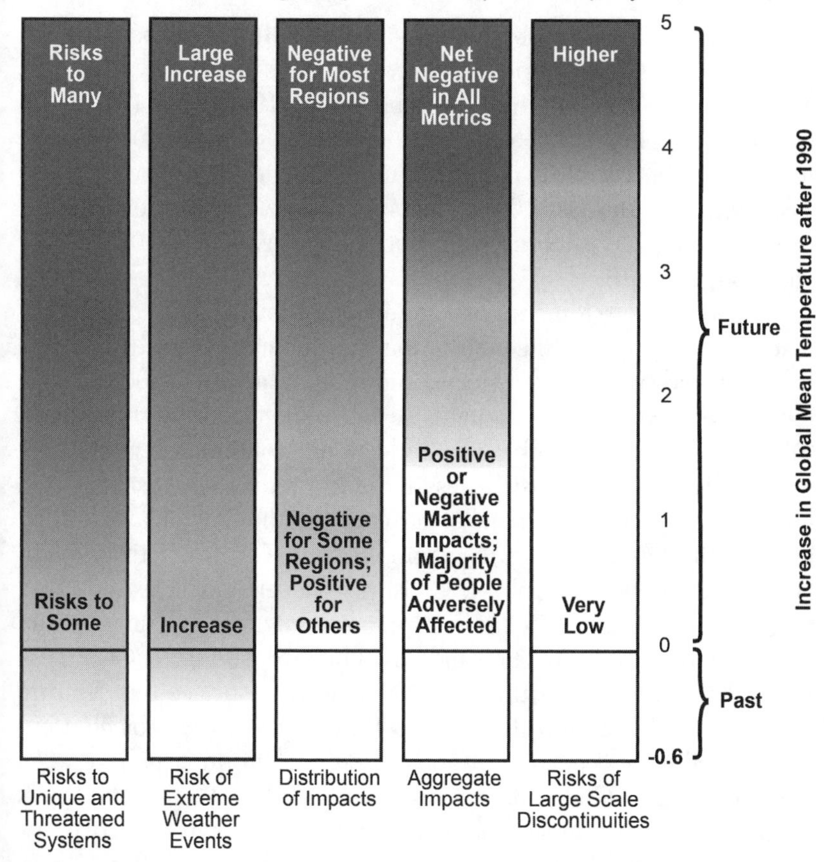

Figure 16.1 Reasons for concern about climate change impacts. The risks of adverse impacts from climate change increase with the magnitude of the climate change itself (represented here by degrees Celsius increase in global mean temperature after 1990). Figure 16.1 displays conceptualizations of "five reasons for concern" regarding climate change risks evolving through 2100. White indicates neutral or minimal negative or positive impacts/risks, light gray shading indicates negative impacts for some systems, and dark gray shading represents widespread adverse impacts. The transitions from medium to dark gray shading on the figure have major implications for defining percentile thresholds of "dangerous anthropogenic interference with the climate system," as Mastrandrea and Schneider (2004) demonstrated quantitatively (source IPCC 2001b).

high end of the projection in the Third Assessment Report—based on the IPCC's figure, "Reasons for Concern" (see Figure 16.1)—is very likely to exceed "dangerous" climate thresholds. (This point has also been made in IPCC 2007a [see Table 1] and Schneider, et al. 2007 [see Table 19.1]). Warming of nearly 6°C (or more at the upper end of the 2007 IPCC "likely" projections) would be very likely to have catastrophic consequences, since the differences between ice ages and interglacial periods involve a change of 5° to 7°C in the global average temperature (Azar and Rodhe 1997). The IPCC projections would also entail a global average rate of temperature change that, for the next century or two, would dramatically exceed the rates sustained over the last 10,000 years with potential for very serious ecological impacts such as plant and animal extinctions (Schneider and Root 2001, 6).

Based on these temperature projections, the IPCC listed likely climate change impacts, including: more frequent heat waves and less frequent cold spells; bigger storms and more weather-related damage; more intense floods and droughts (with droughts expected mostly in subtropical drylands and midlatitude, inner-continent regions); expanded disease transmission; less farming productivity, especially in hotter places, and movement of farming to other regions (mostly to higher latitudes/altitudes) where some increases in productivity might be expected for warming of 1–3°C; rising sea levels and increased storm surges, which could inundate low-lying, coastal areas and small island nations; and species extinction and loss of biodiversity (see IPCC 2007a; IPCC 2007b). In fact, with only a global average surface temperature increase of ~0.7°C since the mid-1800s, many of these projected impacts are already occurring. Among these, certain impacts have been identified as key vulnerabilities in numerous studies. These include loss of glaciers, adverse impacts on biodiversity, increases in severity of extreme events, and loss of cultural amenities (Schneider, et al. 2007). As glaciers melt, sea level rises and water becomes scarcer in regions that depend heavily on glacier-derived water for their main dry season water supply. In South America, a significant fraction of the population west of the Andes Mountains could be at risk due to shrinking glaciers. According to Barnett, et al. (2005), glacier-covered areas in Peru have experienced a 25% reduction in the past three decades. The authors note that "at current rates some of the glaciers may disappear in a few decades, if not sooner," and warn that fossil water lost through glacial melting will not be replaced in the foreseeable future.

China, India, and other parts of Asia are even more vulnerable. As glaciers vanish, water supplies will be more impacted in the coming decades, affecting vast populations throughout this region. The ice mass in the region's mountainous area is the third largest on Earth, following Arctic-Greenland and Antarctica. The Chinese Academy of Sciences has announced that the glaciers of the Tibetan plateau are vanishing so fast that they will be reduced by 50% every decade. Each year, enough water permanently melts from them to fill the entire Yellow River (Lean 2006).

With regard to climate impacts on biodiversity, there has been an increase in rainfall in mid- to high-latitude regions in the Northern Hemisphere, and the behavior of certain plants and animals has changed in response to climate warming. Root and Schneider (2001) reported that the ranges of the vast majority of species observed to have been changing over the past decades were moving poleward, up mountain slopes, or both. They also found that many of the species they investigated had consistently shifted some important biological traits to occur earlier in the spring. These findings have been supported by more recent, greatly expanded results (Root, et al. 2003). After analyzing over 1700 species, Parmesan and Yohe (2003) also detected habitat shifts toward the poles (or higher altitudes) averaging about 6.1 km per decade, as well as the advancement of typical spring activities by 2.3 days per decade (Root, et al. find a greater spring shift, but the differences are largely due to their exclusion of species reporting no change, since that number is such an abstraction as to have no clear meaning—a methodological issue beyond the scope of this chapter). Furthermore, scientists have identified "geoboundaries" such as coastlines and mountaintops that are more prone than other areas to irreversible losses. If these areas become incapable of supporting their present occupants, many of the plant and animal species that dwell there will be unable to find suitable alternative habitats, making extinction much more probable (e.g., Thomas, et al. 2004). The severity of climate change impacts will vary among species and may result in the dismantling of existing plant and animal communities as individual species' responses to climate unfold. This could create disruptions in what Daily (1997) calls ecosystem goods (seafood, fodder, fuel wood, timber, pharmaceutical products) and services (air and water purification, waste detoxification and decomposition, climate moderation, and soil fertility regeneration). The more rapidly climate changes, the more likely there will be communities with no analogue in the current world (see Overpeck, et al. 1992; Schneider and Root 1998).

Cultural amenities are also already being disturbed by climate impacts in several regions. This is particularly true in the Arctic. For several decades, surface air temperatures in the Arctic region have warmed at approximately twice the global rate (McBean et al. 2005), and the region is being impacted at a faster rate of change. Humans have lived in the Arctic for thousands of years (Pavlov, et al. 2001), and many communities are already being substantially affected. In the past, many Arctic peoples moved between settlements according to the seasons, as well as moving seasonally between farming and fishing, and more nomadic activities like following game animals and herding. Now, most Arctic residents live in permanent communities. Many communities are located in low-lying coastal areas and retain a strong relationship with the land and sea, despite the socioeconomic changes taking place (Duhaime 2004). Arctic communities, particularly coastal indigenous communities, are especially vulnerable to climate change because of their close relationship with the land, their geographic location and reliance on the local environment for their diet and economy, and the current state of social, cultural, economic, and political change taking place in these regions.

The island village of Shishmaref located off the coast of northern Alaska has been inhabited for 4,000 years, and now the 600 residents are facing the very real possibility of evacuation. Rising temperatures are melting sea ice (which allows higher storm surges to reach the shore) and thawing permafrost along the coast (which increases shoreline erosion). Homes, water systems, and other essential structures are being undermined. One end of the village has been literally eaten away, losing 15 meters of land overnight after one storm. Plus, the absence of sea ice in the fall makes traveling to the mainland to hunt moose and caribou more difficult.

According to elder Clifford Weyiouanna, "The currents have changed, ice conditions have changed, and the freeze-up of the Chukchi Sea has really changed, too. Where we used to freeze up in the last part of October, now we don't freeze up 'til around Christmas time. Under normal conditions, the sea ice out there should be four feet [1.2 meters] thick. I went out, and the sea ice was only one foot [0.3 meter] thick" (quoted in Hasol 2004).

The Arctic is home to numerous indigenous peoples who have interacted with their environment over countless generations through careful observations and skillful adjustments in traditional harvesting activities and ways of life, developing uniquely insightful ways of observing, interpreting, and responding to the impacts of environmental change. Arctic resources both sustain indigenous peoples economically and nutritionally through hunting, fishing, and gathering and also provide a fundamental basis for social identity, spiritual life, and cultural survival. Rich cultural traditions embody the essential social, economic, and spiritual relationships that indigenous peoples have with their environment. The rapid climate change of recent decades, however, combined with other ongoing alterations in the region, present new challenges.

Indigenous people are reporting impacts: Inuit hunters in Canada's Nunavut Territory cite thinning sea ice, reduced numbers of ringed seals, and insects and birds not usually found in their region. In the western Canadian Arctic, Inuvialuit are observing more thunderstorms and lightning—formerly very rare in this region. Norwegian Saami reindeer herders report that prevailing winds that they rely on for navigation have shifted and become more variable, forcing them to change their traditional travel routes. Unpredictable weather, snow, and ice conditions make travel hazardous, endangering lives. Impacts of climate change on wildlife, including caribou, freshwater fish, and seals and polar bears are having enormous effects on both the diets and cultures of indigenous peoples; change is occurring faster than indigenous knowledge can adapt. "The periods of weather are no longer the norm. We had certain stable decisive periods of the year that formed the traditional norms. These are no longer at their places. Nowadays the traditional weather forecasting cannot be done anymore as I could before. For the markers in the sky we look now in vain. . . ." Heikki Hivasvuopio, Kalslauttanen, Finland, 2002 (quoted in Hasol 2004).

In addition to climate change impacts like the loss of glaciers, adverse impacts on biodiversity, increases in severity of extreme events, and loss of cultural amenities that are already being observed, the IPCC also suggested that climate change could trigger "surprises." These are fast, nonlinear climate responses, thought to occur when environmental thresholds are crossed and new, potentially harmful equilibriums are reached. Some of these surprises could be anticipated, but others may be truly unexpected. Schneider, et al. (1998) identified "imaginable surprises," such as the collapse of the North Atlantic thermohaline circulation (THC), which could cause significant cooling in parts of the North Atlantic region, and deglaciation of Greenland or the West Antarctic, which would, over many centuries, cause many meters of additional sea-level rise (Schneider 2004). There is also the possibility of true surprises not currently imagined (Schneider, et al. 1998). Global environmental change abounds with both types of surprises because of the enormous complexities of the processes and interrelationships involved (for example, between oceanic, atmospheric, and terrestrial systems) and our insufficient understanding of them individually and collectively.

Figure 16.1, also known as the "burning embers diagram," graphically represents the range of climate change impacts measured by five different metrics. The figure shows that the potentially most dangerous climate change impacts (the darker sections on the right of the figure) typically occur after only a degree or two Celsius of warming. What it does not fully depict, except in general terms in the third "reason for concern," is how those impacts will be distributed across specific countries and societies. Mastrandrea and Schneider (2004) have noted that some who focus on equity and justice will consider the third reason for concern, distribution of impacts, to be most important and want to weigh it most heavily, whereas those focusing on market-tradable commodities will put the largest weight on the fourth reason, aggregate impacts. Conservationists worried about species extinctions and ecosystems disruptions might weight most heavily the first, "burning ember," risks to unique and threatened systems. These value judgments are inherent to analyses that reflect different conceptions of justice.

Damage Assessment

In order to better measure climate damages, Schneider, et al. (2000b) described a different set of metrics for impact assessment—the "five numeraires" or measures of impacts. These measures have not been as well analyzed as the IPCC "reasons for concern," which were formulated by dozens of authors and hundreds of reviewers, but they have an advantage in being more explicit about market, nonmarket, and justice issues (see Table 16.1). The five numeraires are: conventional market system costs in dollars per ton of carbon emitted; human lives lost per ton of carbon; species lost per ton of carbon; distributional effects in changes in income differentials between rich and poor per ton of carbon; and quality of life changes, such as heritage sites lost or refugees created per ton of carbon.

Considering such factors provides a more accurate assessment of climate change damages than conventional market-damage–oriented CBAs. One must consider all of these factors to arrive at a fair and accurate damage assessment; it is difficult, however, to assign a monetary value to nonmarket categories of damages. Can we, for example, place a dollar value on a human life and the quality of that life?

Table 16.1 Five Numeraires for Judging the Significance of Climate Change Impacts

Vulnerability to Climate Change	Numeraire
Market impact	$ per ton C emitted
Human lives lost	Persons per ton C
Biodiversity loss	Species per ton C
Distributional impacts	Income redistribution per ton C
Quality of life	Loss of heritage sites; forced migration; disturbed cultural amenities; etc., per ton C

Note: Multiple metrics for the valuation of climatic impacts are suggested. Typically in economic cost-benefit calculations, only the first numeraire—market sector elements—is included. Different individuals, cultures, and governments might put very different weights on these five—or other—numeraires, and thus it is suggested that analysis of climatic impacts be first disaggregated into such dimensions and that any reaggregation provide a traceable account of the aggregation process so that decision makers can apply their own valuations to various components of analysis (Schneider, et al. 2000a).

How do we value ecosystem goods and services (Daily 1997)? In addition to the absolute costs in each of the five numeraires, relative costs should also be considered in some categories. For example, we should consider market-system costs relative to a country's Gross Domestic Product (GDP), or species lost relative to the total number of species in that family. This will better account for potential damages and could help merge the often disparate values of different groups when gauging damage severity. In other cases, such as human lives lost, the absolute measure remains more appropriate.

Traditional CBAs, which many economists and governments use, usually consider a sole measure (market values) to assess damages. Market values, however, rarely capture the full costs of climate impacts. For example, under a traditional CBA, an industrialized country with a large economy that suffered the same relative climate damages as a less-industrialized nation with a less robust economy would be viewed as suffering more (due to its larger absolute dollar loss) and would be more important to "rescue." Even more problematic, CBAs often ignore distributional issues. For example, what if an industrialized country experienced a monetary gain from global warming due to longer growing seasons and other factors, while poor countries suffered a monetary loss equivalent to that gained in the industrialized country? This is not a "neutral" outcome for human well-being, despite the zero net change in monetary welfare. Because it is difficult to assign monetary values to nonmarket damage categories (such as reduced quality of human life), this does not mean these damages should be ignored. Using multiple measures helps focus on "soft" issues that might be ignored under traditional CBAs.

Under the five numeraires, developing countries' interests have more weighting as threats to nonmarket entities are counted. For example, if rising sea levels in Bangladesh cause losses equivalent to about 80% of national GDP, these losses would be catastrophic for Bangladesh but would only be 0.1% of global GDP (IPCC 2001b, 97). A market-aggregation–only analysis would classify the damage as relatively insignificant from a global perspective. Those who consider multiple measures, however, would say this is unfair, given the severity of the aggregate market impacts in relation to GDP, in addition to loss of life, degraded quality of life, and potential loss of biodiversity.

Coping Capacity

Socioeconomic conditions driving emissions also influence adaptive and mitigative capacities (Yohe 2000), and thus there is an imbalance between rich and poor nations' ability to cope with climate change impacts. National GHG emission levels are highly correlated with wealth. Wealthier societies also have the institutional infrastructure needed to build adaptive and mitigative capacities. Thus, richer countries, that have historically contributed most to global emissions, can usually cope better with climate change impacts—such events as the 2003 European heat wave (which resulted in over 52,000 deaths) and Hurricane Katrina (which caused more that 1,800 deaths and $100 billion in damage), notwithstanding. In the latter examples, poorer people and the elderly were disproportionate victims. Poorer countries and poorer groups within countries have lower adaptive capacities due to financial, technological, and institutional constraints (IPCC 2001b; IPCC 2007b). The need to develop adaptive capacity in vulnerable countries is magnified by the possibility that many such countries may face particularly severe impacts. As evidenced by the high costs (in lives, fraction of GDP, quality of life, etc.) paid by developing countries when disasters such as cyclones occur, these countries are poorly equipped to deal with such catastrophes. Within poor nations, the most marginalized groups (including small-scale subsistence farmers and fishermen, small island inhabitants, victims of war/conflict, refugees, nomads, those without access to clean water or healthcare, the elderly, children, and victims of AIDS and other diseases) will be most vulnerable and therefore need particular attention. It will be difficult enough for these countries to cope with gradually occurring climate change, but "surprises" could be devastating in both an absolute sense and relative to the impacts on developed nations.

The Developed Versus Developing Country Divide

There are three main questions to ask when examining regional climate change impacts: 1. Have a country/region's climate change vulnerabilities been fairly assessed, and if not, how do we do this? 2. What is the country/region's capacity for coping with climate change? 3. How do we allocate responsibility for causing the problem, and therefore for paying for responses to it?

The greatest attention, in our value system, should be given to those who will suffer the greatest impacts from climate change, have the lowest capacity to cope with damages, and have contributed least to the problem. The need for resources to build appropriate capacity in developing countries to minimize damages and the allocation of this burden among industrialized countries must also be reemphasized.

Allocating Responsibility

In practice, the question of who will provide the financial, technical, and institution-building resources needed to help developing countries build their capacity to mitigate GHG emissions and adapt to climate impacts remains mostly unresolved. This issue is intimately linked with aspects of fairness and justice. Most developing countries have typically contributed little to atmospheric GHG build-up. For example, in 2004, the United States emitted 75 times more metric tons of carbon dioxide per capita than Bangladesh (Energy Information Administration 2006). Responsibility for helping these countries, and the most vulnerable groups within them, cope with climate change rests, at least initially, with the major polluters. The United Nations Framework Convention on Climate Change (UNFCCC) reiterates this: "the developed country Parties should take the lead in combating climate change and the adverse effects thereof." The UNFCCC also refers to developing country Parties that are particularly vulnerable to the adverse effects of climate change (UNFCCC 1992), but industrialized countries have made little effort to meet these obligations. One concrete action would be to assess the resources required to build adaptive capacity in developing countries. Another could be exploring ways to allocate responsibility among industrialized nations with a view to equitable burden-sharing of necessary GHG emissions reductions between these countries and developing nations.

The Policy Challenge

Adaptation

Unlike mitigation, adaptation is a response to, rather than a means of slowing, global warming. The IPCC has identified two types of adaptation: autonomous and planned. An autonomous adaptation is a non–policy-driven reactive response to a climatic stimulus that occurs after the initial impacts of climate change are felt (IPCC 2001a, 88). Planned adaptation, which is more promising for climate change, can be passive or anticipatory (Schneider and Thompson 1985; Adger et al. 2007). Passive adaptation, such as buying additional water rights to offset the impacts of a drying climate, is a reactive response. Schneider, et al. (2000a) offer farming examples to illustrate that passive adaptation cannot be assumed to occur automatically and question whether farmers will invest heavily in new crops or improved irrigation to adapt before demonstrable climate change occurs. Although some have argued that farmers do adapt to changes in markets, technology, and climatic conditions, others have contended that this view neglects such problems as resistance to unfamiliar practices, difficulties with new technologies, unexpected pest outbreaks, and the high degree of natural variability of weather (Schneider, et al. 2000a). Unlike autonomous adaptation, which is often assumed to be smooth or instantaneous, any passive adaptation will almost certainly not be fast or easy. Rather, adaptations to slowly evolving trends embedded in a noisy background of inherent variability are likely to be delayed by decades, as farmers attempt to sort out true climate change from random climatic fluctuations. In fact, if by dint of poor luck, a sequence of weather anomalies ensued which was the opposite of slowly building climatic trends, such a sequence could easily be mistaken for a new climatic regime and actually lead to maladaptive practices. Even if policies are instituted to facilitate passive adjustment, for example, regulations concerning sharing losses, changes in land use, changes in location, and retreat from rising sea levels (West, et al. 1998), maladaptation can still occur. Such actions can be more damaging to developing countries and marginalized groups, who have limited financial and other resources, than not adapting at all. For them, even one round of successful adaptations will be taxing, let alone multiple rounds when the early measures prove to be maladaptive.

Anticipatory or proactive adaptation has considerable policy potential. It could include such technical actions as purchasing more efficient irrigation equipment, building higher bridges and dams, and engineering seeds so that they are able to cope better with altered climates. It could also include political actions such as setting up networks to disseminate climate information and suggest potential adaptive actions, and creating insurance mechanisms or transferring payments to disadvantaged groups. Some of these actions would also be a step toward sustainable development. Good policy coordination on a range of anticipatory adaptation actions can help avoid maladaptation.

Most anticipatory adaptation studies assume that countries and groups can afford it, which is not universally true. Several funds have therefore been established to help developing countries pursue adaptation measures, the most well-known being the Marrakech Funds (established at the seventh Conference of the Parties—COP 7—in Marrakech in 2001) and the Global Environmental Facility's (GEF) Climate Change Operational Programme. These funds are a promising development, but guidelines are lacking for determining which adaptation projects deserve funding. The GEF requires such projects to show "global environmental benefits" and the Marrakech Funds try to assure funding of adaptation to long-term climate change rather than to short-term climate variability. Yet it is difficult to assess adaptation projects on these grounds because they are local (and therefore bring local, rather than global, benefits) and will likely improve an area's ability to adapt to climate change and climate variability (Huq and Burton 2003). The rules for eligibility need clarification and standardization. Huq and Burton (2003) suggest that, once an adaptation effort has been selected for funding, both the GEF and Marrakech resources should serve as incremental financing for adaptation activities that are already underway, activities that preferably have other benefits (so-called "co-benefits" or "ancillary benefits"), as well, such as biodiversity conservation.

Mitigation

Many assume, particularly under cost-benefit analyses (CBAs), that mitigation (GHG abatement) and adaptation are competing strategies for dealing with climate change, but this "trade-off" requires us to consider justice implications. Suppose it were cheaper for a rich, high-emitting industrialized nation to adapt than to mitigate—at least at first. If that nation chose only to adapt, its unabated emissions could damage a poorer, less adaptable country. Comparing mitigation and adaptation costs across all nations in a "one dollar, one vote" system therefore has serious equity implications. Low-cost options for one country are often high-cost options for other countries or groups within countries. In this situation, emission cuts below standard projected baselines become paramount. Industrialized nations should therefore reduce their emissions and help developing nations to do the same. At the same time, they should help developing countries build their adaptive capacities, as even with stringent global mitigation efforts, some level of future climate change is inevitable due to historical emissions.

The most comprehensive agreement on emissions abatement is the Kyoto Protocol that "entered into force" (an unfortunate expression, since there are no real consequences for nations exceeding their targets) February 16, 2005, despite the Bush Administration's refusal to accept it. Under the Protocol, the developed (Annex I) signatory countries, which account for about 80% of current anthropogenic atmospheric CO_2, agreed to reduce their overall emissions by 5.2% below 1990 levels between 2008 and 2012. Developing countries have no such emissions targets or timetables. However, if developing nations insist on equal global per capita emissions, CO_2 levels could more than triple beyond the 21st century, almost certainly causing "dangerous" warming. The Protocol is therefore only a starting point for international climate policy. To slow down the rate of climatic change, the international community has recognized and begun to address the need for stronger emissions reductions than envisioned in the first commitment period of the Kyoto Protocol.

Eventually, *all* major emitters must participate, developed and developing countries alike. Decisions on abatement levels and other actions could be aided by better information on the risks of "dangerous anthropogenic interference" (Mastrandrea and Schneider 2004; Schneider and Mastrandrea 2005). From a justice perspective, however, it is inappropriate that the costs of abatement or adaptation activities be shared equally, especially in the initial phases of global agreements. The current situation favors the prime burden falling to the developed coun-

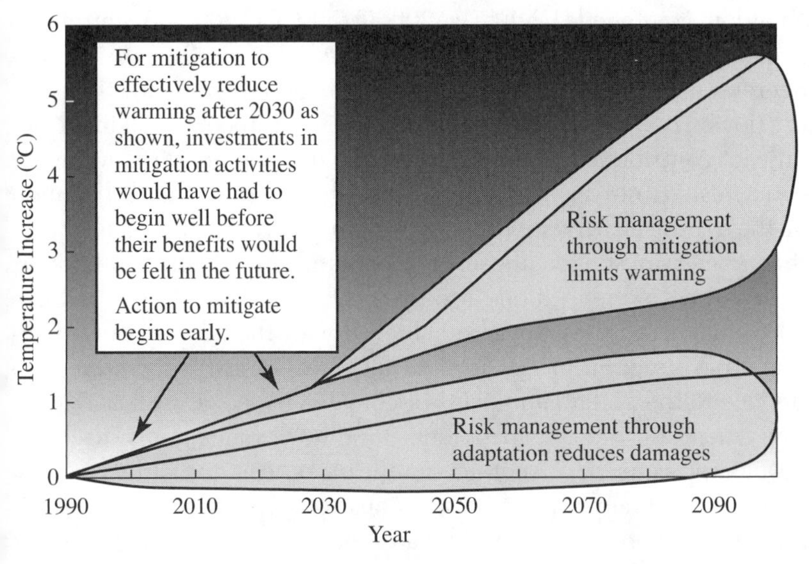

Figure 16.2 Adaptation and mitigation. This figure displays the envelope of temperature increase over the 21st century for the highest and lowest projected future GHG emissions scenarios (upper and lower lines) described in the Special Report on Emissions Scenarios IPCC 2000. The A1FI scenario, the most fossil fuel–intensive case, projects a CO_2 tripling by 2100. The B1 scenario, the lowest emissions projection, represents a world of egalitarian sharing of new technologies, resources, and information. As the figure indicates, mitigation can limit warming to a lower level, but adaptation is necessary to respond to the level of temperature increase associated even with low emissions scenarios. Thus it is necessary to ensure that both adaptation and mitigation strategies are put in place for developed and developing nations, with developed nations taking the lead, at least initially (adapted from Jones 2003).

tries, which typically enjoy a tenfold advantage in emissions per capita. The distribution of "who plays" can differ greatly from "who pays"—the latter being what the COP process calls "common, but differentiated responsibilities" for burden sharing. The Kyoto Protocol's Clean Development Mechanism (CDM) and Joint Implementation (JI) initiatives, formally approved at the United Nations Climate Change Conference 2005, embrace this idea. They help rich nations reach their Protocol targets more cost-effectively, while helping developing nations become venues for low-cost, clean technologies for emissions abatement. This bypassing of older technologies and pursuing more efficient, high-technology solutions is known as "technology leapfrogging."

The number of Clean Development Mechanism projects is increasing dramatically. The United Nations Climate Change Secretariat estimates that registered CDM projects will generate more than one billion tons of emission reductions by the end of 2012, which is equivalent to the present annual emissions of Spain and the United

Kingdom combined (UNFCCC 2006; UNFCCC 2007). According to Richard Kinley, acting head of the United Nations Climate Change Secretariat, an important threshold has been crossed with these reductions. "It is now evident that the Kyoto Protocol is making a significant contribution towards sustainable development in developing countries." However, there are issues of fairness in the distribution of these projects. Even with very strong growth in the mechanism, the projects are still too unevenly distributed among different regions.

At the United Nations Montreal Climate Change Conference 2005, major progress was also achieved with the adoption of a five-year work program to develop the technical basis for increasing resilience to the potential impacts of climate change, as well as the advancement of special adaptation funding arrangements to assist developing countries. Another significant outcome was the opening of a two-track approach to discussions to determine the future directions of global action on climate change: one involving the 157 Parties to the Kyoto Protocol; the other involving all 189 Parties to the UNFCCC. There was also a greater openness on the part of developing nations to discuss actions they might take to limit the growth in emissions in their economies, linked to financial assistance, technology transfer, and market mechanisms (Kinley 2006).

One hopes that the efforts in the last decade to develop a cooperative international negotiation process based on cost-effectiveness and fairness can be extended to further initiate, expand, and strengthen climatically "safe" agreements in the future. Using a "sustainability approach" rather than a traditional CBA method will help achieve more equitable outcomes. Using multiple measures to evaluate vulnerability will also support climate policy that helps avoid in the distant future climate surprises that are "built into" the system by behaviors of this generation—which CBAs downplay due to discounting. Such discounting makes future catastrophes insignificant in present cost considerations, which is inherently unfair. Why should a life today be worth 10 times more than a life in 40 years? For business investments, discounting at expected return on investment rates makes sense, but for sustainability practices, perhaps the most appropriate pure rate of time preference for the present over the future is simply zero. This is a political value judgment, not a professional analytic determination.

The Kyoto Protocol allocates emission quotas according to 1990 emission levels. Such a system may help smooth the transition, but eventually, a scheme based on fairness and justice may be the only way to build consensus for global participation in the UNFCCC. Climate policy decision makers should aim for a long-term goal of equitably allocating "rights to the atmosphere" (Sagar 2000), or at least agreeing on some common principles for determining acceptable national emissions trajectories and a mechanism to help out those with lesser emissions opportunities.

The Numbers Game

Finally, it is common for some opposed to climate policies to cite frightening absolute numbers: trillions of dollars of annual costs for climate mitigation policies; or a few percent of GDP lost. But there is a wide variance across economic models on how much mitigation might cost—and some estimates suggest that it could actually improve the economy at first by promoting the implementation of cost-effective efficiency actions sooner. But even if one accepts some of the seemingly staggering estimates like trillions of dollars of costs, some perspective is needed. Figure 16.3 shows the results that Azar and Schneider (2002) produced, based on conventional economic models that estimate the costs of climate policy. They found that a typical shadow price on carbon (a carbon fee or tax, for example) to prevent the concentrations of CO_2 from more than doubling was around $200 per ton of carbon emitted. A fee twice that high could eventually keep concentrations near present values (though an overshoot of concentrations above present in the next half century seems unavoidable—see Schneider and Mastrandrea 2005). Azar and Schneider used typical economic models estimates of the costs of such policies, although they personally believed them to be too pessimistic. These models estimate between a half a percent and several percent GDP lost annually by century's end. This range was reaffirmed by the Working Group III of the AR4 (IPCC 2007c).

Let us reframe this for perspective. If the annual costs in the future were indeed a few trillion dollars lost from climate policies, and one compared that to today's level of GDP, it would indeed seem astronomically high—equivalent to a depression—some tens of percent loss of economic product of some countries. But that comparison would be totally misleading, if not pernicious. We can't legitimately compare potential *future* costs of climate mitigation policies in 2100 to the *present* size of the economy. Nearly all mainstream economic analyses typically project GDP growth rates of some 2% per year—barring pandemics, world wars, or other unforeseeable catastrophes we all work so hard to prevent. A few numbers to illustrate this follow.

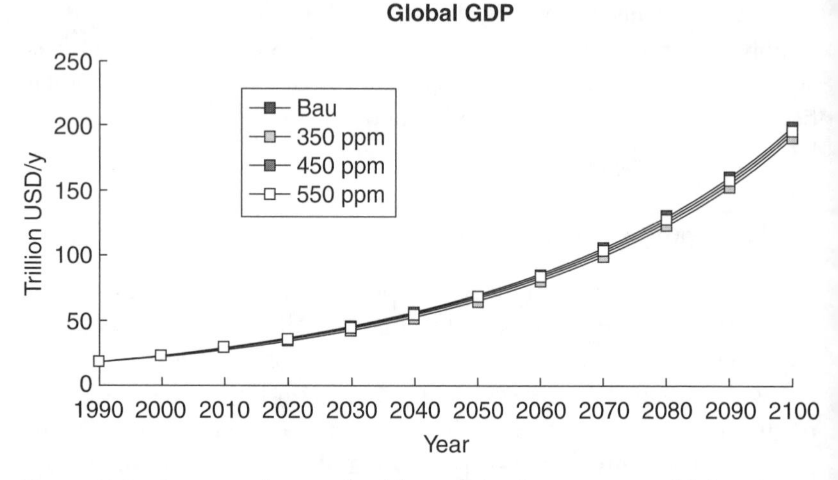

Figure 16.3 Comparing the cost of stabilization. This figure compares global income trajectories under "business as usual" (top curve) and for the case of stabilizing the atmosphere at 350 (bottom curve), 450, and 550 ppm. Note that rather pessimistic estimates were assumed of the cost of atmospheric stabilization (average costs to the economy assumed here are $200/tC for 550 ppm target, $300/tC for 450 ppm, and $400/tC for 350 ppm), and that the environmental benefits (in terms of climate change avoidance and reduction of local air pollution) of meeting various stabilization targets have not been included. (Source: Azar and Schneider 2002).

If the current economy of the world now were about $40 trillion and it grew at 2% per year, then in 100 years it would be about eight times bigger—about $320 trillion annually. So indeed, a 2% loss in 2100 from a century of shadow prices on carbon that reduced most of the climate change risks would be a seemingly very daunting figure: about $6.4 trillion in 2001—a major fraction of the economy today. But in 2100, that loss would be made up in only one year by economic growth! In other words, if our economy continues to grow as typically projected, that growth will swamp the costs of mitigation. In this simple demonstration, we would be about 500% per capita richer on average in 2101, with major climate policies to reduce risks, versus being 500% per capita richer in 2100, having taken no climate policy action and thus being faced with the full risk of dangerous climate change. In the language of risk-management, such an investment in mitigation is a cheap insurance ("deterence" is actually a better metaphor) policy or hedging strategy to avoid significant threat to our planetary life-support system. It is unacceptable to compare future costs to the present scale of the economy. Framing costs in terms of the delay time it would take to be x% richer is much more understandable than frightening, but largely out of perspective in absolute dollar costs.

But just because overall costs of climate mitigation may not be a large number relative to projected growth in the global economy, there will still be, as mentioned earlier, individuals and groups with more than average difficulties (e.g., coal miners and manufacturers of monster vehicles like Hummers and over 6,000-pound SUVs that would have to be curtailed if emissions targets are to be met). Thus, the critical challenge to governance is to protect the planetary commons for our posterity and the conservation of Nature, *while at the same time* fashioning solutions to deal fairly with those particularly hard hit by both the impacts of climate change (via adaptation programs) and by climate policies (perhaps via job retraining, incentives for relocation of industries, side payments, etc.).

Policy Winners and Losers

Even if future climate policies produce "optimal" combinations of abatement and adaptation measures, the results may still be unfair. Climate change policy decisions don't always benefit the most vulnerable countries or groups within countries. The most marginalized groups often have limited access to information and communications, and little political or economic power. They therefore have little influence over the decision-making process. Conversely, people with political and economic power who are less vulnerable to climate change damages tend to control policy-making. Hence, policies cater to special interest groups like the coal or auto industries or countries like the United States, at the expense of more needy groups or nations.

Marginalized groups are often excluded from climate debates because they are not major emitters. Most climate change aid goes to current or future polluters in developing nations, while people conducting relatively climate-friendly practices are ignored. This has been sarcastically termed the "polluters get paid" principle (Sagar 1999). For example, two billion people globally depend on biomass (animal dung, crop residues, and wood) for energy and need improved energy services. Recent research has shown that using biomass is not carbon neutral (Smith 2000; Balis 2003) and traditional biomass use is not sustainable—for example, the indoor air pollution it causes has major health effects, but policy makers and nongovernment organizations dealing with climate change policy often forget traditional biomass users, who constitute the poorest third of humanity and have very low per capita energy use and emissions.

To avoid overlooking or further subjugating already-marginalized groups when forming local, national, and international climate policy, decision makers should consider assessing climate change costs and benefits using a more equitable framework like the five numeraires. They should also consider the effects of actions (and inactions) on the distribution of people's well-being and the sustainability of other species. In a framework of "distributed justice," disadvantaged countries and groups should be prioritized. In the case of traditional biomass users, some suggest that petroleum should have an energy-poverty alleviation levy, with funds used to provide such biomass users with cleaner-burning fuels like kerosene. This levy would lead to a decrease in GHG emissions, as higher petroleum costs would reduce demand enough to more than offset emissions from increased kerosene use. In addition, eliminating the products of incomplete biomass combustion, which are also GHGs, will provide climate and health benefits. Changing from traditional uses of biomass to fossil fuel use may seem heretical, but moving climate debates forward justly requires unconventional thinking (Sagar 2005). Of course, such schemes are essentially transitional devices, as over the long term there simply has to be a charge for dumping greenhouse gases in the atmosphere, and that charge must be imposed on *all* emitters. If some techniques for energy supply like cellulostic ethanol or solar thermal emerge as socially acceptable and cost effective after a few decades of learning-by-doing deployments, then we may well "invent our way out of the problem." But such learning requires investments, so the key is to fashion an incentives strategy to foster such investments in alternative technologies for cleaner supply and demand energy emissions management. Of course, in addition to production and efficiency incentives there also may need to be side payments to those who caused less of the problem and have less ability to fix it; these can be fashioned for fairness and political cooperation.

Conclusions

Despite the well-established, mainstream science that warming since 1850 is unequivocal and recent decades of warming trends are very likely a result of human emissions of greenhouse gases, there are still many detailed uncertainties in climate science and impacts estimates. But because some warming is clearly already "in the pipeline" and up to 6 degrees more possible in the 21st century (with very likely dangerous outcomes resulting), we need to slow down the rate at which we add to atmospheric GHG levels as a risk-management response. This will give us more time to understand climate risks and to help develop and deploy lower-cost mitigation options, while making climate "surprises" less likely. GHG abatement policies need to provide incentives to invent cleaner, decreasing cost technologies, and developed countries should aggressively lead the effort, given their historical contribution to the problem and greater capacity to help. Simultaneously, the needs of developing countries and marginalized groups should be considered and integrated into international policies through adaptation and abatement actions coordinated within and among countries. Developed countries should shoulder this burden, (at least initially for a few decades) as well, as required by the UNFCCC. Developing countries need assurances that national GHG abatement actions will not jeopardize a fair resolution of long-term emissions allocation. Discussions and movement on this issue are urgently needed. Slowing down pressure on the climate system and addressing marginalized country and group needs are the main "insurance policies" we have against potentially dangerous irreversible climate events and associated injustices. The policy most likely to make sustainability unattainable is inaction and the maintenance of the status quo.

Acknowledgments

This chapter is modified from the following papers: A. Sager, et al., "Equity in climate change." *Tiempo* 55 (2005):9–14, and S.H. Schneider, and J. Lane, "Dangers and thresholds in climate change and the implications for justice," in *Politics, science, and law in justice debates,* ed. N. Adger, 23–51 (Cambridge, MA: MIT Press, 2006). The authors would like to give a special thanks to the Winslow Foundation for partial support.

References

Adger, W.N. et al. 2007. Assessment of adaptation practices, options, constraints, and capacity. In *Climate Change 2007: Impacts, Adaptation, and Vulnerability—Contribution of Working Group II to the Fourth Assessment Report*, ed. M. Parry, et al. Cambridge, UK: Cambridge University Press.

Azar, C., and H. Rodhe. 1997. Targets for stabilization of atmospheric CO_2. *Science* 276:1818–1819.

Azar, C., and S. H. Schneider. 2002. Are the economic costs of stabilizing the atmosphere prohibitive? *Ecological Economics* 42:73–80.

Bailis, R. et al. 2003. Greenhouse gas implications of household energy technology in Kenya. *Environmental Science and Technology* 37:2051–2059.

Barnett, T. P. et al. 2005. Potential impacts of a warming climate on water availability in snow-dominated regions. *Nature* 438:303–309.

Daily, G. C. 1997. *Nature's services: Societal dependence on natural ecosystems*. Washington, DC: Island Press.

Duhaime, G. 2004. Economic systems. In *The Arctic human development report*, ed. N. Einarsson et al., 69–84. Akureyri, Iceland: Stefansson Arctic Institute.

Energy Information Administration. 2006. World per capita carbon dioxide emissions from the consumption and flaring of fossil fuels, 1980–2004. *International Energy Annual 2004*, Table H.1co2. http://www.eia.doe.gov/pub/international/iealf/tableh1co2.xls.

Hasol, Susan Joy. 2004. *Impacts of a warming Arctic: Artic climate impact assessment*. Cambridge, UK: Cambridge University Press.

Huq, S., and I. Burton. 2003. *Funding adaptation to climate change: What, who and how to fund*? Sustainable development opinion paper. London: International Institute for Environment and Development (IIED). http://www.iied.org.

Intergovernmental Panel on Climate Change (IPCC). 2000. *Emissions scenarios—A special report of Working Group III of the Intergovernmental Panel on Climate Change*, ed. N. Nakicenovic, et al. Cambridge, UK: Cambridge University Press.

Intergovernmental Panel on Climate Change (IPCC). 2001a. *The scientific basis, Summary for policy makers—Contribution of Working Group I to the third assessment report of the Intergovernmental Panel on Climate Change*, ed. J. T. Houghton, et al. Cambridge, UK: Cambridge University Press.

Intergovernmental Panel on Climate Change (IPCC). 2001b. *Impacts, adaptation, and vulnerability—Contribution of Working Group II to the third assessment report of the Intergovernmental Panel on Climate Change*, ed. J. J. McCarthy, et al. Cambridge, UK: Cambridge University Press.

Intergovernmental Panel on Climate Change (IPCC). 2007a. *Climate change 2007: The Scientific basis, summary for policy makers—Contribution of Working Group I to the fourth assessment report of the Intergovernmental Panel on Climate Change*, ed. S. Solomon, et al. Cambridge, UK: Cambridge University Press.

Intergovernmental Panel on Climate Change (IPCC). 2007b. *Climate change 2007: Impacts, adaptation, and vulnerability, summary for policy makers—Contribution of Working Group II to the fourth assessment report of the Intergovernmental Panel on Climate Change*, ed. M.L. Parry, et al. Cambridge, UK: Cambridge University Press.

Intergovernmental Panel on Climate Change (IPCC), 2007c. *Climate change 2007: Mitigation of climate change, summary for policy makers—Contribution of Working Group III to the fourth assessment report of the Intergovernmental Panel on Climate Change*, ed. B. Metz, et al. Cambridge, UK: Cambridge University Press.

Kinley, Richard. 2006. The world "can take heart" from the Montreal Climate Change Conference. *United Nations Framework Convention on Climate Change (UNFCCC) Newsletter #1*. http://www.un.org/Pubs/chronicle/2005/issue4/0405p74.html.

Lean, Geoffrey. 2006. Ice-capped roof of world turns to desert. *The Independent*, UK, May 7.

Lucon, O. et al. 2004. LPG in Brazil: Lessons and challenges. *Energy for Sustainable Development* VIII (3):82–90.

Mastrandrea, M., and S. H. Schneider. 2004. Probabilistic integrated assessment of "dangerous" climate change. *Science* 304:571–575.

McBean, G. et al. 2005. Arctic climate—Past and present. In *Arctic climate impacts assessment* (ACIA), ed. C. Symon et al., 22–60. Cambridge, UK: Cambridge University Press.

Modi, V. 2004. Energy and transport for the poor. Commissioned paper for the Millennium Project Task Force 1. http://www.unmillenniumproject.org/documents/EnergyandtheMDGs_Sept1.pdf.

Overpeck, J. T. et al. 1992. Mapping Eastern North America vegetation change over the past 18,000 years: No-analogs and the future. *Geology* 20:1071–1074.

Parmesan, C., and G. Yohe. 2003. A globally coherent fingerprint of climate change impacts across natural systems. *Nature* 421:37–42.

Pavlov, P. et al. 2001. Human presence in the European Arctic nearly 40,000 years ago. *Nature* 413:64–67.

Root, T. L. et al. 2003. Fingerprints of global warming on wild animals and plants. *Nature* 421:57–60.

Sagar, A. D. 2000. Wealth, responsibility, and equity: Exploring an allocation framework for global GHG emissions. *Climatic Change* 45:511–527.

Sagar, A. D. 2005. Alleviating energy poverty for the world's poor. *Energy Policy* 33:1367–1372.

Sager, A., et al. 2005. Equity in climate change. *Tiempo* 55:9–14.

Schneider, S. H. 2004. Abrupt non-linear climate change, irreversibility and surprise. *Global Environmental Change* 14:245–258.

Schneider, S. H., and J. Lane. 2006. Dangers and thresholds in climate change and the implications for justice. In *Politics, science, and law in justice debates,* ed. N. Adger, 23–51. Cambridge, MA: MIT Press.

Schneider, S. H., and M. D. Mastrandrea. 2005. Probabilistic assessment of "dangerous" climate change and emissions pathways. In *Proceedings of the National Academy of Sciences USA* 102:15728–15735.

Schneider, S. H., and T. L. Root. 1998. Climate change. In *Status and trends of the nation's biological resources,* ed. M. J. Mac et al., 89–105. Reston, VA: U.S. Department of the Interior, U.S. Geological Survey.

Schneider, S. H., and T. L. Root, eds. 2001. *Wildlife responses to climate change: North American case studies*. Washington, DC: Island Press.

Schneider, S. H., and S. L. Thompson. 1985. Future changes in the atmosphere. In *The global possible,* ed. R. Repetto, 363–430. New Haven, CT: Yale University Press.

Schneider, S. H. et al. 1998. Imaginable surprise in global change science. *Journal of Risk Research* 1:165–185.

Schneider, S. H. et al. 2000a. Adaptation: Sensitivity to natural variability, agent assumptions and dynamic climate changes. *Climatic Change* 45:203–221.

Schneider, S. H. et al. 2000b. Costing nonlinearities, surprises, and irreversible events. *Pacific and Asian Journal of Energy* 10:81–106.

Schneider, S.H. et al. 2007. Key vulnerabilities and the risk from climate change. *Climate change 2007: Impacts, adaptation, and vulnerability—Contribution of Working Group II to the fourth assessment report*, ed. M. Parry, et al. Cambridge, UK: Cambridge University Press.

Smith, K. R. et al. 2000. "Greenhouse implications of household stoves: An analysis for India." *Annual Review of Energy and the Environment* 25:741–763.

Thomas, C. D. et al. 2004. Extinction risk from climate change. *Nature* 427:145–148.

United Nations Framework Convention on Climate Change (UNFCCC). 1992. United Nations Framework Convention on Climate Change: Convention text. Bonn: UNFCCC. http://www.unfccc.int.

United Nations Framework Convention on Climate Change (UNFCCC). 1997. Proposed elements of a protocol to the United Nations Framework Convention on Climate Change, presented by Brazil in response to the Berlin Mandate. UNFCCC/AGBM/1997/MISC.1/Add.3.

United Nations Framework Convention on Climate Change (UNFCCC). 2003. Review of adaptation activities under the convention. Working Paper No. 10. Bonn: United Nations Framework Convention on Climate Change.

United Nations Framework Convention on Climate Change (UNFCCC). 2006. Emission reductions from Kyoto Protocol's Clean Development Mechanism pass the one billion tonnes mark. Climate Change Secretariat, June 9, 2006 press release. Bonn: United Nations Framework Convention on Climate Change. http://unfccc.int/press/items/2794.php.

United Nations Framework Convention on Climate Change (UNFCCC). 2007. CDM Statistics. Bonn: United Nations Framework Convention on Climate Change. http://cdm.unfccc.int/Statistics/index.html

Yohe, G. 2000. Assessing the role of adaptation in evaluating vulnerability to climate change. *Climatic Change* 46:371–390.

17

The Steps Not Yet Taken

Daniel Sarewitz and Roger Pielke, Jr.

Introduction

The climate system of the planet Earth, and the energy system built by those who inhabit the Earth, are today seen as the integrated elements of a single problem: global warming. In turn, scientific inquiry, public concern, and policy prescription have given rise to an international regime for controlling the behavior of the climate through management of the global energy system. In this chapter, we explain why this regime, and in particular its codification through the Kyoto Protocol, is a failure. Our central point is simple: protecting people and the environment from the impacts of climate is a different problem from reducing greenhouse gas emissions to combat global warming. The policies that have resulted from combining these two problems are, as a consequence, failing to address either problem meaningfully. Policies to reduce global warming must be pursued independently of policies to reduce climate impacts.

First, we explain why the Kyoto Protocol is not achieving its environmentally modest goals, a failure that has no connection to the refusal of the United States to sign onto the treaty, but rather reflects the complexity of energy systems and their management. We then consider the impacts of climate on society through the lens of Hurricane Katrina. Such impacts are unrelated to global warming, and cannot be addressed by emissions reductions. Instead, they require policies specifically focused on reduction of socioeconomic vulnerability to climate.

But emissions reductions are a key societal goal, and next we discuss the role of technological innovation in pursuing that goal. Current policies, embodied in Kyoto, are inappropriate and insufficient for making the necessary progress. A cornerstone of our argument is that much of the failure to date of climate change policy originates in a misunderstanding of the appropriate roles of science and technology in social and political change. Proponents of action on global warming have treated scientific evidence as the central catalyst for motivating necessary change, while technological advance has been viewed as a second-order consequence of such change. We argue that this reasoning is backwards, and that technological innovation is a much more effective scaffolding upon which to address energy policies than scientific knowledge.

The Kyoto Protocol is not effectively addressing the climate impacts problem or the energy technology problem. Although Kyoto is often portrayed as only a first step toward establishing an effective international climate change regime, we conclude that it is a step in the wrong direction.

Bush Saves Kyoto!

US President George W. Bush saved the Kyoto Protocol. By refusing even to consider bringing the treaty before the US Senate for a debate and vote over ratification, Bush ostentatiously ceded the moral high ground to those other nations—all 163 of them—that signed onto the treaty and brought it into force. In so doing, he shifted the world's attention onto America's scurrilous role in refusing to take seriously the problem of global climate change, and away from the inevitable failure of the international governance regime that led to Kyoto. Benedick made a related point in suggesting that Bush's rejection of Kyoto motivated pro-Kyoto forces not to give up on the treaty (2001).

This regime emerged from the famous 1992 "Earth Summit" in Rio de Janeiro, and was formally enshrined in the United Nations Framework Convention on Climate Change (UNFCCC). The Kyoto Protocol was negotiated under the terms of the UNFCCC to give teeth to the regime by requiring affluent nations (known as "Annex I Parties") to reduce greenhouse gas emissions to specific levels relative to their 1990 emissions by the year 2012. For example, Western European nations must reduce emissions to 8% below their 1990 baselines; the original US reduction target was 7%. Developing countries, in light of both the economic challenges they face, and the fact that the rich nations created the problem to begin with, were exempted from mandated reductions.

Even if all Annex I nations were to meet their reduction targets successfully, the effect on total greenhouse gas concentrations in the atmosphere would be inconsequential, and would have no discernible effect on the behavior of the climate. But supporters of Kyoto have always portrayed the treaty as only a first step toward a global regime for managing climate change.[1] This step would establish the international institutional and legal framework, and the economic incentives, necessary for making meaningful reductions over the longer term. Thus, if the direct environmental consequences of Kyoto were, even under the best of circumstances, trivial, in other ways this first step was a giant one, setting the initial conditions under which climate change would be debated and confronted in the 21st century, while mobilizing a global cadre of public servants, scientists, and activists for a period of fifteen years in pursuit of an international policy framework of unprecedented scope and complexity.

Most important, this first step has meant that action on climate change is framed in terms of Kyoto itself, where support for Kyoto means saving the world environment from catastrophe, and opposition means denial of the problem, perhaps combined with a preference for short-term profit over long-term sanity. This situation is perfectly suited for (and reinforced by) media coverage, which thrives on stark, simplistic dichotomies, but it means that a third option—that climate change is a serious problem, but that Kyoto and the broader UNFCCC regime are the wrong way to go about addressing it—has not been able to emerge into the marketplace of ideas. Independent analysts have been offering thoughtful, nonideological critiques of Kyoto since its earliest days, but with little effect on public debate (e.g. Rayner and Malone 1997; Laird 2000; Benedick 2001; Nordhaus 2001; Schelling 2002).

---

[1]E.g., "But Kyoto is, and has always been understood as, a first step, and a baby step at that." (Kolbert 2005); "Protection of the climate system will require substantial reductions in greenhouse gas emissions; hence, the Kyoto Protocol is recognized to be only the first step on a long journey to protect the climate system." (Watson 2003); "[A]s Bill Hare of Greenpeace describes it . . .'the Kyoto Protocol and the Bonn rules to implement it are the first step on what has to be a long climb.' " (Retallack 2001).

Had Bush and his advisors been a bit more clever, they would at the very least have voiced support for the treaty, and might even have pushed the Senate to ratify it. Besides fellow Kyoto-nonratifier Australia, many of the Annex I nations that actually have ratified Kyoto—including Canada, Japan, Spain, Portugal, Greece, Ireland, Italy, New Zealand, Finland, Norway, Denmark, and Austria—are failing to meet the treaty's requirements for greenhouse gas emissions reductions, and have little or no prospect of doing so before the treaty expires in 2012.[2] Had the United States, as the world's biggest energy user and greenhouse gas emitter, simply joined that group of scofflaws, the fact that Kyoto's emissions reduction goals were impossibly ambitious (even as they were, at the same time, environmentally inconsequential) would have become obvious to all the world, and the treaty could quickly have been pronounced dead, or at least abandoned to its death throes as the inertia stored up in the tens of thousands of people who depend on the international climate regime for a living and for ideological sustenance inexorably dissipated.

Instead, common wisdom has it that the failure of Kyoto can be laid at the feet of an organized right-wing political effort that has brought together conservative think tanks, politicians, corporate supporters (especially from the fossil fuel industry), and a few co-opted scientists, to oppose Kyoto specifically, and more generally to counter the notion that global warming is a problem worth worrying about (Austin 2002; McCright and Dunlap 2003). This effort, whose voice and power got a huge boost when Republicans gained control of the US Congress in 1994, made it futile for President Clinton to push for Senate ratification of Kyoto, and made it possible for President Bush to repudiate the treaty. The conservative strategy was, in fact, very effective in meeting its goals of making Kyoto a political liability in the United States. But if America's refusal to ratify Kyoto and cut its greenhouse gas emissions simply reflects the power of right-wing opponents, why are so many other countries that have ratified Kyoto failing to meet their emissions targets? The answer begins with an understanding that the question itself is incoherent: ratifying Kyoto and cutting emissions are not equivalent.

[2]The source for emissions data for all discussions in this chapter is the authoritative Web site of the United Nations Framework Convention on Climate Change; see: http://www.ghg.unfccc.int/index.html.

Energy systems are the metabolism of modernity; they cannot be managed in isolation from the political and economic conditions of a society. They include everything from automobiles and light bulbs (and the firms that manufacture them) to power plants, transmission lines, highways, buildings, research laboratories, regulatory bodies, and labor unions. A nation's geographic and climatic attributes, the structure of its workforce, the age of its manufacturing infrastructure, the organization of its tax code and regulatory structure, its patterns of immigration and demographics, the makeup of its transportation system and even its food system (among many other factors) constitute the complex conditions that determine both the structure of the energy system and the political context for modifying it. Changes in a nation's population, downturns or upturns in the economy, changes in political leadership, and changes in the price of various energy sources are among the unpredictable forces that can cause emissions to rise or decline independently of any actions taken to conform to Kyoto. For example, Australia, Canada, and the United States are all experiencing population growth rates of more than 10% per decade; in Germany, France, and the United Kingdom, the rates are between 3 and 4%. Over the past decade, the US GDP has increased an average of 3.4% per year; the UK rate was 1.85%; Germany, 1.39%. Yet, needless to say, Germany and the UK are not pursuing low population or economic growth rates as ways of lowering their carbon emissions.

Moreover, because of the huge complexity of energy systems, the connection between explicit energy policies and emissions levels is largely unpredictable. As Frank Laird (2000) noted shortly after Kyoto was negotiated, "no one knows in any precise sense how [policy options] are connected to emissions reductions . . . The instrument with the most direct effect on fossil fuel consumption—a carbon tax—is crude and imprecise. How big a tax would the United States need to achieve a 7% reduction in 1990 levels?" (Some economists think they can answer this question [e.g., Nordhaus 2001], but as 2005–06 gasoline price increases in the United States have shown, consumers are willing to absorb cost increases greater than any politically conceivable carbon tax, without appreciably reducing their fuel consumption, at least in the short term.) The policy cornerstone of the Kyoto approach is a market system to allow nations to establish emissions trading for greenhouse gases, but there is much disagreement about whether international trading can possibly work to reduce emissions significantly (Victor, et al. 2005; Bell 2006). The new European Emissions Trading Scheme already seems to be failing because the emissions allowances that nations have agreed upon are too high to provide incentives to actually cut emissions (Open Europe 2006). To make matters worse, this approach may stimulate profiteering, as utilities sell credits for emissions that they would not have released in the first place (Gow 2006).

Because of the historical contingencies embedded in national energy systems, and the significant costs and uncertainties entailed in overcoming those contingencies, most nations simply cannot identify and follow a particular path to meeting Kyoto's targets. Those countries that are on path toward meeting their targets were well on their way to doing so without Kyoto. The twelve nations that have reduced emissions the most since 1990 are all former Eastern Bloc or Soviet Union countries that have undergone radical economic restructuring and, in some cases, near-collapse. Germany, which has also managed considerable reductions, has done so largely because it absorbed a former Eastern Bloc country, and Great Britain's reductions are principally the consequence of actions by Conservative Prime Minister Margaret Thatcher in the 1980s to break the British coal union and move that nation's energy system away from coal and toward gas—for reasons that were economic and political, not environmental. In 1993, four years before Kyoto was negotiated, German and British emission levels were already 9.5 and 5.1%, respectively, below their 1990 levels. Meanwhile, Russia's cynical decision to ratify Kyoto in 2005, made in the face of years of official denial that global warming was a problem, was motivated by the fact that it is greatly exceeding its reduction targets due to its economic collapse in the 1990s, and the hope that it (along with other Annex I countries of the former-Soviet sphere) can therefore make money selling carbon credits to countries that are unable to reduce their own emissions.

As the experienced international environmental diplomat Richard Benedick (2001) has noted, for many European and post-Soviet countries, 1990 was a very fortunate choice as a base year for measuring emissions reductions. Put somewhat differently, the ability of many European nations to meet their Kyoto targets is largely fortuitous. Given these practical realities, Kyoto may indeed be only a first step, but the question one must ask is: toward what?

Global Warming and the Logic of Katrina[3]

Unlike, say, cancer-causing toxic waste dumps, or smog-filled air that makes throats burn and eyes water, global warming is an abstraction: no one actually feels an increase in the global average temperature that scientists calculate. Any effective strategy aimed at mobilizing political action to forestall climate change needs to connect the abstract global phenomenon to the local realities of felt human experience. Where is the evidence that people can see and feel, which can impart the necessary urgency to the cause?

The most poignant connections are those based on scientific predictions that a warming climate may bring greater human suffering due to increasingly severe weather events such as hurricanes, heat waves, and droughts. Such predictions can be found in scientific research articles and assessments (always modified by appropriate qualifying statements), in hyperbolic fictional accounts of a world afflicted by climate-gone-wild (like the movie *The Day After Tomorrow*), and everything in between.

The spectacle of real-world disasters has become powerful evidence for those making claims about the need to stop climate change. When Hurricane Mitch struck Central America and killed more than 10,000 people in October 1998, the delegates at the fourth Conference of Parties to the United Nations Framework Convention on Climate Change passed a resolution of "solidarity with Central America," expressing concern "that global warming may be contributing to the worsening of weather," and urging "governments. . .and society in general, to continue their efforts to find permanent solutions to the factors which cause or may cause climate events" (Conference of Parties 1998). (By "permanent solutions," of course, the delegates meant "the Kyoto approach to emissions reductions.") And despite the fact that the December 2004 Indian Ocean tsunami that killed more than 220,000 people was not a weather event, Britain's chief science advisor, Sir David King, made the connection to global warming: "What is happening in the Indian Ocean underlines the importance of the Earth's system to our ability to live safely. And what we are talking about in terms of climate change is something that is re-

[3]For more complete presentations of the data and arguments in this section, see: Pielke (2006); Pielke, et al. (2000); Sarewitz and Pielke (2000; 2005).

Figure 17.1 Climate change politics as a control fantasy: this image, used in a major promotional campaign by a leading environmental group, says that humans can dial in the weather they want by reducing greenhouse gas emissions.

ally driven by our own use of fossil fuels, so this is something we can manage" (Press Association 2004).

But the disaster that made the threat of a changing climate most palpable to all the world was of course Hurricane Katrina, which destroyed the city of New Orleans and unleashed social mayhem in the richest and apparently most environmentally profligate nation in the world—a cautionary event of almost biblical dimensions. While a few scientists and advocates directly blamed Katrina on human-caused global warming, many such voices upped the volume on the potential links between warming and future hurricanes, invoking Katrina as a harbinger of what was to come (compare Figure 17.1). Scientific claims and counterclaims began to get airtime, not just in the mass media but in technical journals—were hurricanes increasing in number and power? Was Katrina's severity exacerbated by a warming climate (Kerr 2005; Kerr 2006)?

The connection to Kyoto was irresistible. British Deputy Prime Minister John Prescott remarked in a speech that: "As a European negotiator at the Kyoto climate change convention, I was fully aware that climate change is changing weather patterns and raising sea levels. . . . The horrific flood of New Orleans brings home to us the concern of leaders of countries like the Maldives, whose nations are at risk of disappearing completely. There has been resistance by the US government to Kyoto—which I believe is wrong" (Jowit and Temko 2005). Similar comments were offered by Jürgen Trittin, Germany's Minister of the Environment, who also directly blamed the Bush Administration for Katrina (German Minister 2005).

The logic that connects Katrina and other disasters to Kyoto presumably goes something like this: human emission of greenhouse gases is changing the climate; these changes will cause more frequent and severe disasters; and, therefore, Kyoto is a first step toward protecting the world against ever-greater disasters. This logic depends on the idea that humans can control the behavior of climate to achieve particular social benefits by adjusting greenhouse gas emissions. The advocacy group Scientists and Engineers for Change apparently had this sort of control fantasy in mind when, in the midst of the 2004 US presidential election season and a disastrous Florida hurricane season, they posted billboards with the message, GLOBAL WARMING = WORSE HURRICANES. GEORGE BUSH JUST DOESN'T GET IT.

Katrina ripped the façade of civilization off a city—and a country—to expose a Hobbesian nightmare. While there was enough suffering and loss to ensure that most everyone living around New Orleans got a good dose, it was the poor, the disenfranchised, the infirm, and the historically discriminated-against who suffered most and were disproportionately left behind to fend for themselves. Of course, this concentration of suffering is to some extent tautological; it's no surprise that the most vulnerable members of society suffered the most. But the inescapable visibility, staring out from newspaper front pages and TV screens, of abject social and economic privation in the heart of the richest, most powerful nation that history has ever known, was a profoundly disheartening spectacle, to these authors at least.

Scientists have known—and warned for years—that the location of New Orleans on a rapidly subsiding river delta in the heart of the hurricane belt made some version of Katrina inevitable (Travis 2005). Indeed, the enormous destruction wrought by the disaster can mostly be blamed not on Katrina itself, which, by the time it struck the coast, was not an unusually powerful hurricane, but on the reality that New Orleans was built in harm's way, that the progressive development of the city and the environmental destruction of the surrounding wetlands rendered it increasingly vulnerable over time, and that the levees that were designed to protect the city under precisely the circumstances that Katrina presented were poorly designed and maintained (Schwartz 2006).

Was Katrina a harbinger of what global warming has in store for humanity if Kyoto is not implemented? The first and most obvious point is that Katrina was not a harbinger of anything—it was a real disaster, whose causes and consequences were fully anticipated, scientifically uncontroversial, and causally unrelated to "dangerous anthropogenic interference" with the climate. Transforming Katrina from an unequivocal indicator of persistent socioeconomic and racial inequality, developmental hubris, political ineptitude, and engineering incompetence into a symbol of global warming and the need for reduced carbon emissions demands some combination of factual ignorance, political cynicism, and moral nihilism.

Natural disasters have been rapidly increasing worldwide over the past century, in both number and severity, not because of increases in the frequency or severity of storms, earthquakes, or other extreme events, but because of growing populations, expanding economies, rapid urbanization, and migrations to coasts and other exposed regions. The number of disasters affecting at least 100 people or resulting in a call for international assistance has increased from an average of fewer than 100 per year in the early 1970s to more than 400 per year by the early 21st century. While economic losses from disasters are increasingly concentrated in the affluent world, as a percentage of GNP, the economic effects of natural disasters on poor countries can be hundreds of times greater (Guha-Sapir et al. 2004).

Disasters disproportionately harm poor people in poor countries because those countries typically have densely populated coastal regions, shoddily constructed buildings, sparse infrastructure, and inadequate public health capabilities. Poor land use leads to widespread environmental degradation, such as deforestation and wetlands destruction, which in turn exacerbates flooding and landslides. Emergency preparation and response capabilities are often inadequate, and hazard insurance is usually unavailable, further slowing recovery. New Orleans, with its stark juxtaposition of the affluent and the poor, provided a synoptic portrayal of this global situation.

Disparities in disaster vulnerability between rich and poor will continue to grow. About 97% of population growth is occurring in the developing world. This growth, in turn, helps drive urbanization and coastal migration. The result is that, in the next two decades, the population of urban areas in the developing world will increase by several billion people. This population is being added to cities that are commonly located on coastal or flood plains that are vulnerable to climate hazards but are unable to provide the quality of housing, services, infrastructure, and environmental protection that can help reduce vulnerability.

Many well-tested policies are available to help reduce vulnerability to natural disasters. These range from building codes that can keep structures from collapsing in a storm, to land use regulations that limit construction in flood-prone areas, to environmental laws that preserve natural features, such as wetlands and forested slopes, that act as buffers against disasters. The rising toll of disasters demonstrates that nations are greatly underinvested in applying such policies (Guha-Sapir, et al. 2004), despite the certainty that they are effective, and that more disasters of greater magnitude will soon occur.

When advocates for the present climate policy regime connect natural disasters to global warming, presumably they hope to motivate policy makers to take action on emissions reductions as a way to help prevent future disasters. But if policy makers are not sufficiently motivated to apply relatively modest, well-proven interventions to help reduce vulnerability to the disasters happening today, why would they undertake extraordinarily complex, expensive, and long-term efforts to implement emissions reductions whose impacts on climate behavior won't be felt for 50 years or more, and whose benefits for disaster reduction are, at this point, completely unknown and are likely to be marginal compared to the common sense, near-term actions that policy makers refuse even now to take?

The idea that the UNFCCC climate regime, or the Kyoto Protocol specifically, amounts to a strategy to combat rising disaster vulnerability is both practically absurd and morally suspect. The problem isn't simply that Kyoto's marginal decreases in greenhouse gas emissions will have no effect on long-term climate behavior, although this is the case, but that emissions reduction programs don't reduce the entrenched social inequities, irresponsible development trends, and inadequate hazard reduction policies that led to the worst of Katrina's depredations and that are the cause of rising disaster vulnerability worldwide. In other words, the fact that science warns us that global warming may increase the severity of future weather events does not mean that reducing emissions is a reasonable or potentially effective way to prevent disasters. Reducing emissions is about as relevant to controlling the impacts of natural disasters as a nuclear nonproliferation treaty is to protecting public health.

The Limits of Science and the Logic of Technology

In 1997, the lead author of this chapter participated in a conference on science and the environment in Tulsa, Oklahoma, deep in the heart of America's hydrocarbon belt.[4] One of his roles at the conference was to moderate a group discussion on global climate change. Most of the participants in this discussion worked at reasonably high levels in the hydrocarbon industry, and all shared a reactive skepticism about global warming and climate change science. A few of the participants were scientists who offered well-argued technical critiques of the view that climate change was a threat to humanity; the rest were clear that they distrusted the motives of environmentalists, of academic scientists who study climate change, and of the administration of Democratic President Bill Clinton and, especially, Vice President Al Gore. Most were sympathetic to the view that global warming was a scientific "hoax" intended to impose an environmentalist political agenda on the nation.

Eventually, the subject of discussion turned from global warming to energy efficiency. Amazingly, the strong sense of the group that global warming was just a disguised political agenda was matched by an equally strong sentiment in favor of national policies to stimulate energy efficiency through technological innovation—as long as the policies were not highly prescriptive. It turned out that the group shared an explicit proefficiency ethic, and a sense that the role of technology was to help implement this ethic. They did not see increased energy efficiency as contrary to their own interests, but as a typically American path to greater profitability and energy independence via inventiveness, innovation, and competition.

Science provides the rational basis for Kyoto. Environmental groups use scientific findings to justify their advocacy of action on climate change; legislative bodies call on scientists to present their findings as a basis for political deliberation; and the UNFCCC process itself depends for technical support on the Intergovernmental Panel on Climate Change, which brings together hundreds of scientists

[4]Redefining Environmental Protection: The Case for Change, Tulsa, OK, August 11, 1997: www.nelpi.org/events/caseforchange/brochure.asp.

from around the world to continually assess and update the latest scientific information related to climate change, its possible future impacts, and the avenues available for reducing such impacts. The idea, simply, is that the emerging force of scientific knowledge about human effects on the climate system should impel, or compel, the innumerably diverse perspectives, interests, values, and experiences implicated in the world's hydrocarbon-intensive energy system to converge on a single course of action. Kyoto thus became the rational embodiment of right action, a scientific demand for a particular type of behavior from a highly diverse set of players.

The difficulty, however, is that science is always connected to action via the values and interests of those who want to act in a particular way. When a scientific fact—say, that the Earth's atmosphere is warming—becomes associated with a political agenda that supports a particular type of action—say mandated emissions reductions—the science must shoulder the values and interests of those who are pushing that agenda. Science, in other words, becomes a tool of political persuasion, a lens for focusing many values and interests on a single type of action. This difficult task is further compounded because, as science approaches the cutting-edge, it tends to raise as many questions as it resolves, so there is always room for debate about what the science is actually saying. When new scientific results are published, some will claim that the results support their particular version of what needs to be done, but others may not see it that way (Nelkin 1995; Sarewitz 2004). And even if scientists could confidently predict the societal consequences of global warming (which they can't), such knowledge would not dictate any particular path of action.

Technology is different. Technology is itself the embodiment of reliable action. Indeed, as the story of the Tulsa meeting shows, what's especially powerful about technologies is that often they can serve a variety of preferences simultaneously. A commuter who wants to reduce spending on gasoline, an environmental group that wants to reduce greenhouse gas emissions, and an automobile manufacturer that wants to develop new, profitable product lines all find their interests converging in, say, the development of hybrid vehicles. This is a trivially obvious example, but it says something profound about the relationships between technology and politics. People holding diverse and even strongly divergent values and interests may unite around a particular technology that can advance multiple interests. Technology, that is, can overcome political conflict not by compelling diverse interests and values to converge—the job assigned to climate science—but by allowing them to coexist in a shared sense of practical benefits. Values can also converge in opposition to the effects of a technology, of course, as illustrated by the conflicts around genetically modified foods.

The ability of technology to facilitate unlikely ideological bedfellows is well illustrated by a new project to develop a wind farm in the Gulf of Mexico off Galveston, Texas. Technological advances in oil platform design and wind turbines are converging to allow wind farms to be situated in deep water environments. The project is supported by a heterogeneous coalition of environmental and energy industry players motivated by opposite ideals: the former hoping to help phase out hydrocarbon energy by creating viable substitutes; the latter hoping to extend its life by adding new flexibility to the energy grid (Joyce 2006).

Few would disagree that global greenhouse gas emissions will only be substantially reduced over the long term through a global transition to energy technologies that no longer depend on hydrocarbon fuels—to a decarbonized global energy system. Technological advances might also allow the direct capture of greenhouse gases from the atmosphere. The Kyoto approach to creating the necessary technological transition depends on the idea that the science of climate change will create a global political convergence on the need for emissions reductions—and then, that this political convergence will cause the public and politicians to pursue policies that increase the costs of emissions, which in turn will stimulate the complex behavioral changes that lead to the necessary levels of conservation and innovation.

This is backwards. Whereas the demand for emissions reductions is a politically divisive and rarefied basis for global technological transformation, technological transformation itself is a politically unifying and inclusive principle for pursuing many beneficial objectives, including emissions reductions. There are many good reasons, in addition to global warming, to move the planet rapidly toward an energy system that is more efficient, more dependable, more technologically diverse, more equitable, less polluting, and less geopolitically destabilizing than the current, hydrocarbon-based system, and—consequently and crucially—there are many diverse, and sometimes even competing, values and interests that can be served by the pursuit of such a transition.

While the types of policy tools encouraged under the UNFCCC and Kyoto are meant, in part, to stimulate adoption of more efficient technology, their overall effect is to encourage states to work within the status quo and game the system to meet short-term targets, rather than to reward long-term and high-risk investment aimed at encouraging technological change over a highly uncertain timeline. The clearest sign that Kyoto policy is aimed at science-motivated behavioral and ideological convergence, rather than innovation-led technological pluralism, is the absence of any provisions in the treaty to stimulate research and development (R&D) on energy technologies and systems. As articulated by Hoffert, et al. (2002, 981): "a broad range of intensive research and development is urgently needed to produce technological options that can allow both climate stabilization and economic development." Yet, one of the most stunning contradictions in the global politics of global warming is this: since the late 1970s, as scientific understanding of climate change—and concern about its consequences—has continued to rise, the total public investment in energy R&D by industrialized nations has progressively fallen, perhaps by as much as 65% in constant dollars. Reliable data on private-sector energy R&D are not available, but the trends are probably similar (Runci 2005; also see Nemet and Kammen, in press, who document sharp declines in US private sector investment in energy R&D between 1994 and 2003). Kyoto does nothing to confront this contradiction, a failure that can only be explained if the problem of stimulating technical change is largely irrelevant to the politics surrounding Kyoto.

Indeed, the whole idea of mandated national emissions reductions reflects insensitivity to the highly decentralized, historically contingent, uneven manner in which new technologies emerge and diffuse. Similarly, the early absence of the private sector from negotiation processes leading to Kyoto demonstrates an unwillingness to engage the one constituency without which the necessary technological innovation cannot occur.

The current international climate regime asks nations to shoulder significant up-front costs in return for unknowable and inconsequential long-term benefits, while calcifying the terms of political discussion and the bureaucratic process for addressing climate change, encouraging cynicism and short-term action, radically circumscribing the range of relevant stakeholders who need to be involved, alienating key stakeholders who should be involved, and pushing the central component of any political and practical solution to the problem—technological change—to the back burner. The emerging symptoms of failure in the current climate regime painfully confirm that if Kyoto is only a first step—as its advocates continually insist—then it is a first step in the wrong direction.

Bush Saves Climate! . . . (Thousands Perish)

If the Bush Administration inadvertently saved the Kyoto Protocol, at least for a while, its intransigence on the global warming issue is having another, more salutary effect, particularly in the United States: stimulating new approaches to addressing climate change. While defibrillating Kyoto is still the name of the game for opposition politicians and mainstream environmental groups, a new environmental pragmatism is beginning to emerge from the margins. A widely distributed 2004 white paper by Shellenberger and Nordhaus, titled "The Death of Environmentalism," made the case for this pragmatism by highlighting an empirical embarrassment: "Over the last 15 years environmental foundations and organizations have invested hundreds of millions of dollars into combating global warming. . . . We have strikingly little to show for it" (2004, 6).

The white paper goes on to explain how the problem of reducing carbon emissions is inseparable from many other difficult political issues, such as the viability of the US auto manufacturing industry, and it is appropriately critical of policy fixes (such as tradable permit schemes, fuel efficiency standards, and carbon taxes) that are advanced without strategic consideration of their political context. Shellenberger and Nordhaus make the case for a climate change strategy "aimed at freeing the U.S. from oil and creating millions of good new jobs" (2004, 26) through the development of new energy technologies. In essence, they have recognized that technology is a much better organizing principle for mobilizing diverse interests than science-based policy prescriptions. The Apollo Alliance, a political coalition aimed at advancing the agenda laid out by Shellenberger and Nordhaus, makes no mention of global warming on its home page (www.apolloalliance.org) and talks instead of "rejuvenating our nation's economy by creating the next generation of American industrial jobs and treating clean energy as an economic and security mandate to rebuild America."

As a political strategy, this sort of oblique approach may be a bit too clever, yet the fact is that rising energy prices resulting from continuing instability in the Middle East has triggered a renewed interest in energy technology among mainstream politicians of both left and right in the United States. In addition, a recent National Academies report has managed to resuscitate some of the economic competitiveness hysteria of the 1980s, and in doing so singled out the need for a more robust investment in energy technology as a key element of a successful competitiveness strategy (Committee on Prospering in the Global Economy of the 21st Century, 2005). At a 2006 Congressional hearing stimulated by this report (Committee on Science, 2006), discussion of the need for energy-efficient technology was everywhere, while Kyoto and global warming were nowhere to be seen. Overall, the combination of high fuel prices, dependence on foreign oil, and concern about economic competitiveness are creating more attention for energy policy than has been seen in the United States since the early 1990s. Some of the proposals emerging from lawmakers even have a measure of pragmatic creativity, such as legislation (Senate bill S. 2045, "Health Care for Hybrids Act," introduced November 2005) offered by Democratic Senator Barack Obama to help struggling auto manufacturers cover the health-care costs of its retired workers in return for higher fuel-efficiency standards.

There is no simple recipe for catalyzing the technological transformation process necessary to reduce global greenhouse gas emissions significantly, but the history of technological change demonstrates that such pervasive transformations can occur in a matter of decades once the appropriate initial conditions exist. Furthermore, the history of energy technologies demonstrates a two-century trend toward decarbonization, while the more recent experience of agricultural, communication, and information technology revolutions demonstrates that conscious policy decisions can act to rapidly accelerate technological change (e.g., Nakicenovic 2002; Ruttan 2002). Of course knowing how to foster the appropriate initial conditions to nourish such change is precisely the problem, and the combination of relatively low hydrocarbon fuel prices, the need to replace existing energy technology infrastructure, and the rapid emergence of the energy-hungry Chinese and Indian economies compound the difficulties.

Nevertheless, the broad and diverse portfolio of policies and programs necessary to catalyze a long-term technological transformation to a low-carbon energy system is reasonably well understood, even if the path and timing of the transition cannot be precisely engineered. These measures include robust public funding for research spanning the gamut from exploratory to applied; pilot programs to test and demonstrate promising new technologies; public-private partnerships to incentivize private-sector participation in high-risk ventures (such as those now used to induce pharmaceutical companies to develop tropical-disease vaccines); training programs to expand the number of scientists and engineers working on a wide variety of energy R&D projects; government procurement programs that can provide a predictable market for promising new technologies; prizes for the achievement of important technological thresholds; multilateral funds for collaborative international research; international research centers to help build a global innovation capacity (such as the agricultural research institutes at the heart of the Green Revolution); as well as policy incentives to encourage adoption of existing and new energy-efficient technologies, which in turn fosters incremental learning and innovation that often leads to rapidly improving performance and declining costs.

In fact, significant aspects of such a portfolio were proposed and modestly funded during the Clinton Administration in the mid-1990s (Holdren and Baldwin 2002), but they were politically doomed from the outset because they were too narrowly promoted as climate change policies, rather than as advancing a broad set of national interests and public goals and goods. They did not survive into the Bush Administration; nor did they significantly find their way into the international climate regime. Indeed, the Kyoto approach is a disincentive to implementing many of the sorts of measures listed above because they will not contribute to a nation's ability to meet its short-term targets.

The Bush Administration has begun discussions with Australia, China, India, Japan, and Korea to develop a voluntary, technology-focused Asia-Pacific Partnership on Clean Development and Climate, the details of which are as yet unspecified, but which has already been condemned by Kyoto supporters as nothing more than a sop to the energy industry (e.g., Flannery 2006). There are many good reasons to be suspicious of the motives and commitment of these nations, starting with the Bush Administration and its clear preference for energy policies that focus on boosting domestic oil production. Yet there are also compelling reasons to consider the Partnership as a promising vehicle for a new, innovation-based approach to the greenhouse gas problem. For one thing, the participating nations will have a particular political ability to move away from Kyoto because they were never committed to the treaty's targets in the first place (with the exception of Japan—which isn't meeting them). In addition, together they represent about half of the world's current carbon emissions, are likely to account for the majority of future growth in energy demand, and are also likely to dominate the global technological innovation scene for the foreseeable future.

A particular international vehicle is probably less important, however, than sustained political attention on the need for more rapid innovation in the energy sector. In the United States, the question now is whether the political opportunity created by increasing competition for oil supplies, instability in the Middle East, the rising economic might of India and China, and (through 2009, at least) a highly unpopular president, can be translated into a strategic approach to transforming the world energy-technology system. Groups from outside the US environmental mainstream, such as the Apollo Alliance, the Clean Air Task Force and the Clean Energy Group, appear to have grasped the political potency of technological innovation as an organizing principle, but it is too early to say if these relatively modest seeds can flower into a significant and persistent political realignment.

The economic competitiveness angle, in particular, can be a powerful political organizing theme for accelerating investments in energy technologies worldwide. Consider that, when the United States decided, in the late 1990s, to begin to ramp up investments in R&D in the obscure field of nanotechnology, Japan and many European countries quickly followed suit to avoid being left behind. More broadly, as Steve Rayner (2004) has argued, the economic advantage that nations gain from more efficient energy systems could be a far more compelling inducement for other nations to pursue technological transformation than the cooperative approach that underlies Kyoto. One way for European nations to act on their rage over America's arrogant unilateralism would be to pursue aggressive energy innovation programs aimed at creating global leadership in energy technology—an approach that would have far more beneficial long-term environmental impacts than continued posturing over Kyoto. Such programs might also counter the inevitable loss of face that will occur when the difference between Europe's emissions reduction commitments and actual performance becomes obvious. British Prime Minister Tony Blair seems to have recognized the need to move beyond the Kyoto approach (Blair 2004), but it remains to be seen if fellow Kyoto-supporters in Europe are able to redirect their energies to this more pragmatic strategy.

And what of the victims of future Katrinas? Simply put, energy policies are an unconscionably inefficient and ineffectual approach to stemming the rising toll of natural disasters worldwide. In contrast to the international coalition of interest groups and governments that has organized around climate change and the Kyoto Protocol, no similarly potent political alliance has risen to advance the cause of reducing global disaster vulnerability. One reason for this neglect surely is that those who suffer most from natural disasters are the poor and politically disenfranchised. Another is that reducing disaster vulnerability can be construed as social adaptation to climate, and leading practitioners of climate change politics have consciously rejected positions that take adaptation seriously. As Al Gore (1993) explained, adaptation is "an important part of the underlying problem. Do we have so much faith in our own adaptability that we will risk destroying the integrity of the entire global ecological system? If we try to adapt to the changes we are causing rather than prevent them in the first place, have we made an appropriate choice?" (1993, 240).

We are well aware that the structure of the Kyoto Protocol, with its binding emissions targets for affluent nations, and programs for technology transfer to poorer nations, is meant to recognize the principal contribution of the industrialized world to global warming, and to signal the commitment by rich nations to accept their moral responsibility by taking the lead on emissions reductions. Kyoto has thus become a symbol of ethical global action that will reduce societal inequities related to the environment. For those who see Kyoto as an effective first step toward addressing global climate change, then, perhaps the most difficult aspect of our argument to swallow is that Kyoto does not, in fact, offer a viable path to greater environmental and socioeconomic equity. Indeed, the politics surrounding Kyoto has had the practical effect of distracting attention, resources, and action from policies that really could protect poor countries and poor people from climate impacts.

By bringing disasters into the global warming debate to score political points, yet refusing to embrace an agenda for actually reducing disaster vulnerability—indeed, by rejecting the basis for such an agenda—climate change politics has explicitly turned its back on the very people it exploits as rhetorical fodder to advance its cause. This is a political strategy that robs the global warming movement of its moral legitimacy and leaves climate change policy, after its faltering first step, facing backwards.

As this chapter goes to press, President Bush has just announced his support for a new round of international negotiations leading to a post-Kyoto plan for reduced greenhouse gas emissions. The President's announcement, although short on details, nevertheless signals that global warming has become a consensus issue in the U.S. Between the new Democratic majority in Congress, the continuing unpopularity of the president, the widespread public belief that Katrina was a climate-change phenomenon, and the publication of the latest Intergovernmental Panel on Climate Change assessment documenting, once again, the reality of human-caused warming, there is exceedingly little political space for denial. Yet the underlying dynamics of the problem have not changed; in some ways, the political consensus freezes the convenient common wisdoms into place, and thus keeps the essence of the climate challenge outside of most discussions of climate policy.

Acknowledgments

Thanks to Nicole Heppner and Rachel Smith for superb research support.

References

Austin, A. 2002. Advancing accumulation and managing its discontents: The U.S. antienvironmental countermovement. *Sociological Spectrum* 22:71–105.

Bell, R. 2006. The Kyoto placebo. *Issues in Science and Technology* 22(Winter):28–31.

Benedick, R. 2001. Striking a new deal on climate change. *Issues in Science and Technology* 18 (Fall):71–76.

Blair, T. 2004. Untitled speech to The Climate Group, April 27. http://www.theclimategroup.org/tcg_pmspeech.pdf.

Committee on Prospering in the Global Economy of the 21st Century. 2005. *Rising above the gathering storm: Energizing and employing America for a brighter economic future.* Washington, DC: The National Academies Press.

Committee on Science. 2006. Should Congress establish "ARPA-E," The Advanced Research Projects Agency—Energy?, March 9. http://www.house.gov/science/hearings/full06/March%209/index.htm.

Conference of Parties. 1998. Solidarity with Central America, Nov. 12. http://unfccc.int/cop5/l17-1.htm.

Flannery, T. 2006. The ominous new pact. *New York Review of Books* 53:24.

German Minister Stands Behind Criticism of Bush. 2005. *Spiegel Online*, August 31. http://service.spiegel.de/cache/international/0,1518,372405,00.html.

Gore, Al. 1993. *Earth in the balance: Ecology and the human spirit.* New York: Plume.

Gow, D. 2006. Power Tool, *Guardian Unlimited,* May 19. Available at: http://business.guardian.co.uk/economicdispatch/story/0,,1777038,00.html.

Guha-Sapir, D. et al. 2004. *Thirty years of natural disasters 1974–2003: The numbers.* Louvain-la Neuve, Belgium: Presses Universitaires de Louvain.

Hoffert, M. et al. Advanced Technology Paths to Global Climate Stability: Energy for a Greenhouse Planet. *Science* 298:981–987.

Holdren, J., and S. Baldwin. 2001. The PCAST energy studies: Toward a national consensus on energy research, development, demonstration, and deployment policy. *Annual Reviews of Energy and Environment* 26:391–434.

Jowit, J., and N. Temko. 2005. Prescott links global warming with Katrina. *The Observer* (United Kingdom), Sept. 11.

Joyce, C. 2006. Gulf energy purveyors get second wind. Report on National Public Radio "Morning Edition," May 9. http://www.npr.org/templates/story/story/php?storyld=5387574.

Kerr, R. 2005. Is Katrina a harbinger of still more powerful hurricanes? *Science* 309:1807.

Kerr, R. 2006. A tempestuous birth for hurricane climatology. *Science* 312:676–678.

Kolbert, E. 2005. Global warming. *New Yorker* 81:39–40.

Laird, F. 2000. Just say no to greenhouse gas emissions targets. *Issues in Science and Technology* 17:45–52.

Little, A. 2005. The school of Barack: Obama and a bipartisan crew of colleagues unveil eco-friendly bills on energy. *Grist,* Nov. 22. http://www.grist.org/news/muck/2005/11/22/obama/.

McCright, A., and R. Dunlap. 2003. Defeating Kyoto: The conservative movement's impact on U.S. climate change policy. *Social Problems* 50:348–373.

Nakicenovic, N. 2002. Technological change and diffusion as a learning process. In *Technological change and the environment,* ed. A. Grübler et al., 160–181. Washington, DC: Resources for the Future.

Nelkin, D. 1995. Science controversies: The dynamics of public disputes in the United States. In *Handbook of science and technology studies,* ed. S. Jasanoff et al., 444–456. Thousand Oaks, CA: Sage.

Nemet, G. and D. Kammen. 2007. U.S. energy research and development: Declining investment, increasing need, and the feasibility of expansion. *Energy Policy* 35:746–755.

Nordhaus, W. 2001. Global warming economics. *Science* 294:1283–1284.

Open Europe. 2006. *The high price of hot air: Why the EU emissions trading scheme is an environmental and economic failure.* http://www.openeurope.org.uk/research/ets.pdf.

Pielke, R. A., Jr. 2006. Disasters, death, and destruction: Making sense of recent calamities. *Oceanography* 19:138–147.

Pielke, R. A., Jr. et al. 2000. Turning the big knob: An evaluation of the use of energy policy to modulate future climate impacts. *Energy and Environment* 11:255–276.

Press Association. 2004. Tsunami highlights climate change risk, says scientist. *Guardian*, Dec. 31. http://education.guardian.co.uk/higher/sciences/story/0,12243,1381430,00.html.

Rayner, S. 2004. The international challenge of climate change: UK leadership in the G8 and EU. The Environmental Audit Committee, House of Commons, London. http://www.cspo.org/ourlibrary/documents/EACmemo.pdf.

Rayner, S. and E. Malone. 1997. Zen and the art of climate maintenance. *Nature* 390:332–334.

Retallack, S. 2001. We've saved Kyoto! (shame about the world's climate). *Ecologist* 31:18–22.

Runci, P. 2005. Energy trends: IEA, technical paper PNWD-3581. Richland, WA: Pacific Northwest National Laboratory/JGCRI. http://www.globalchange.umd.edu/?energytrends&page = iea.

Ruttan, V. 2002. Sources of technical change: Induced innovation, evolutionary theory, and path dependence. In *Technological change and the environment*, ed. A. Grübler et al., 9–39. Washington, DC: Resources for the Future.

Sarewitz, D. 2004. How science makes environmental controversies worse. *Environmental Science and Policy* 7:385–403.

Sarewitz, D., and R. A. Pielke Jr. 2000. Breaking the global-warming gridlock. *The Atlantic Monthly* 286:55–64.

Sarewitz, D. and R. A. Pielke, Jr. 2005. Rising tide: The tsunami's real cause. *The New Republic*, Jan. 17.

Schelling, T. C. 2002. What makes greenhouse sense? *Foreign Affairs* 81:2–9.

Schwartz, J. 2006. Army builders accept blame over flooding. *New York Times*, June 2.

Shellenberger, M., and T. Nordhaus. 2004. *The death of environmentalism: Global warming politics in a post-environmental world*. http://www.Thebreakthrough.org/images/Death_of_Environmentalism.pdf.

Travis, J. 2005. Scientists' fears come true as hurricane floods New Orleans. *Science* 309:1957–1959.

Victor, D. G. et al. 2005. A Madisonian approach to climate policy. *Science* 309:1820–1821.

Watson, R. 2003. Climate change: The political situation. *Science* 302:1925–1926.

18

The Interpretation of Air Temperature Trends and Their Impact on Public Policy

David R. Legates

I'm always fascinated by the way memory diffuses fact.

—Diane Sawyer

Humans tend to have short memories and for most purposes, this is beneficial. Many hurtful or unpleasant events fade into the past, allowing us to move on with our lives. But short memories are not beneficial when it comes to science. When our recollection of historical climatic events is distorted by this short memory, recent events can take on a greater historical importance than they should be given. A French historian, C. F. Volney, wrote in 1804:

> For some years it has been a general remark in the United States, that very perceptible partial changes in the climate took place . . . the state of things is now very different in [New England] . . . the seasons are totally altered; the weather is infinitely more changeable; the winter is grown shorter, and interrupted by great and sudden thaws. Spring now offers us a perpetual fluctuation from cold to hot, and from hot to cold . . . the heat of summer is less intense, but of longer continuance; autumn

> begins and ends later, and the harvest is not finished before the first week in November; in fine, winter does not display its severity before the end of December. . .consequently the same thing passes in America, as did formerly in Europe, and no doubt in Asia, and over the whole of the Old World, where history represents the climate as much colder formerly, than it is at present. . . . (1804, 266–273)

Indeed, current discussions about climate change tend to sound quite similar, proclaiming, for example, that recent events have never been so extreme. It is, of course, our perception through our short-term memory—as well as the advent of global and immediate news coverage—that often diffuses historical facts, making recent catastrophes and weather events seem more uncommon than they really are. Such was the case with Volney nearly two centuries ago and it is still the case today. And this becomes even more problematic when public policy—and the subsequent ramifications of that policy—become fueled by our biased and short-memory perception.

But the problem, of course, is that policy makers must rely on statements of fact before they can react—so that the more extreme statements elicit a more immediate response. Policy makers often have little interest in, or understanding of, qualified statements of possibilities or uncertainties. But the more extreme scenarios, often bolstered by exaggerated claims, are more likely to generate a response by policy makers. As Stephen H. Schneider has argued, this puts scientists who would be advocates into a "double ethical bind":

> On the one hand, as scientists we are ethically bound to the scientific method, in effect promising to tell the truth, the whole truth, and nothing but—which means that we must include all the doubts, the caveats, the ifs, ands, and buts. On the other hand, we are not just scientists but human beings as well. And like most people we'd like to see the world a better place, which in this context translates into our working to reduce the risk of potentially disastrous climatic change. To do that we need to get some broadbased support, to capture the public's imagination. That, of course, entails getting loads of media coverage. So we have to offer up scary scenarios, make simplified, dramatic statements, and make little mention of any doubts we might

> have. This "double ethical bind" we frequently find ourselves in cannot be solved by any formula. Each of us has to decide what the right balance is between being effective and being honest. I hope that means being both. (quoted in Schell 1989, 47)

Schneider insists that the "sound bite-communications process" is responsible for placing scientist-advocates into an ethical bind where the whole truth cannot be told in a manner which effectively conveys the message (1996). He further advocates for diligence on the part of scientists to distinguish between objective and subjective probabilities and explain the scientific basis for them to both the media and policy makers. But many scientist-advocates have failed at this ethical bind, as James Hansen explained: "Emphasis on extreme scenarios may have been appropriate at one time, when the public and decision-makers were relatively unaware of the global warming issue" but times now have changed (2003). However, extreme or blatantly false scenarios are not now nor were they ever appropriate to garner the attention of the public.

Unfortunately, times have not changed. For example, in discussing his new movie, *An Inconvenient Truth*, former Vice President Al Gore recently quipped:

> Nobody is interested in solutions if they don't think there's a problem. Given that starting point, I believe *it is appropriate to have an overrepresentation of factual presentations* on how dangerous it is, as a predicate for opening up the audience to listen to what the solutions are, and how hopeful it is that we are going to solve this crisis. Over time that mix will change. As the country comes to more accept the reality of the crisis, there's going to be much more receptivity to a full-blown discussion of the solutions. (quoted in Roberts 2006; emphasis added)

The idea that it is desirable to blatantly misrepresent a problem so that a solution can be adopted belittles the populace, demeans the scientific method, and undermines the policy-making process. But scientists are being pushed to overstate the facts regarding global warming either "for the good of humankind" or possibly to further their careers by garnering prestige or more research money. Indeed,

Linda Mearns, a climatologist at the National Center for Atmospheric Research, has warned that today, "There's a push on climatologists to say something about extremes, because they are so important. . . . but that can be very dangerous if we really don't know the answer" (Henson 2004). Unfortunately, many scientists and politicians have not heeded Mearns's advice.

Climate Change Data and its Uncertainties

"Global warming," by its strictest definition, is based on the premise that global air temperatures have been rising for at least the last 125 years. Much attention, therefore, is currently focused on monitoring the annually-averaged global atmospheric air temperature and the fact that ground-level air temperature measurements show that the Earth warmed approximately 0.6°C to 0.8°C over the last century (Figure 18.1).

Claims have been made that 1998 was the warmest year of the last century and the 1990s were the warmest decade. More recently, 2005 has been proclaimed as the warmest year or the second warmest year (depending upon which group is cited) in recorded history and in January 2007, it was speculated that 2007 might surpass 1998 as the warmest year. While the reduction of the climate of the earth to a single number—its air temperature, integrated both annually and globally—is virtually meaningless from a climatological standpoint, it plays well to the press and easily gets the attention of policy makers. This is largely because annually-averaged global air temperature appears very simple to understand; it succinctly shows, in a single, compact figure, the basic underlying premise of "global warming." While few climatologists dispute that global air temperatures have increased over the last century, it is the cause of such warming that has elicited such heated debate.

The underlying problem with the annually-averaged global atmospheric air temperature is that the uncertainty in obtaining such a number is extremely large and results from data quality issues that focus on both spatial and temporal biases. In particular, the ability of air temperature measurements to represent accurately global patterns is adversely affected by the proximity of air temperature measurements to major urban centers, the continuing inadequate coverage of many parts of the globe, and the fact that the number and loca-

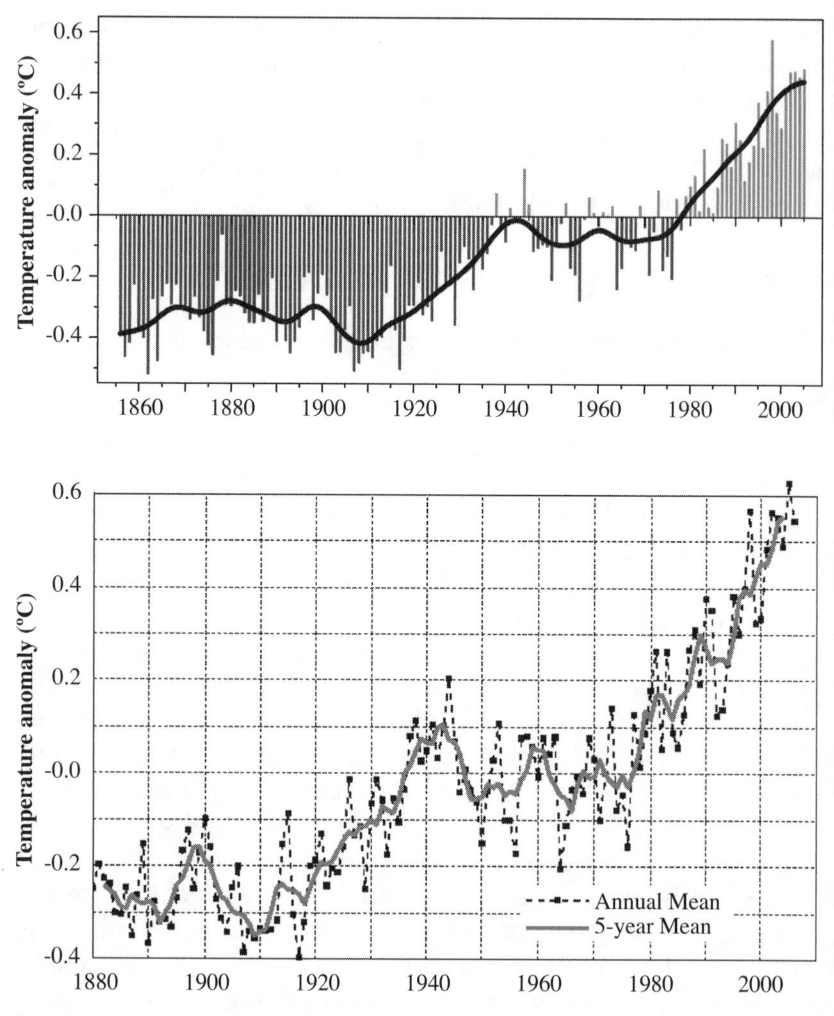

Figure 18.1 Time series of global surface air temperature (as available in February 2007) by *(top)* the University of East Anglia's Climatic Research Unit and the UK Meteorological Office's Hadley Centre (http://www.cru.uea.ac.uk/cru/info/warming/) (see also Jones, et al., 1999; Jones and Moberg 2003) and *(bottom)* the Goddard Institute for Space Studies (http://www.data.giss.nasa.gov/gistemp/graphs/Fig.A2_lrg.gif). In these graphs, air temperature is described as an anomaly from a common reference period.

tion of observation sites has varied by more than an order of magnitude over just the latter half of the 20th century. Ocean areas, which comprise nearly three-quarters of the Earth's surface, are virtually unrepresented in global air temperature and precipitation assessments. Moreover, high altitudes, high latitudes, and both the desert and rainforest regions of the tropics are considerably underrepresented, owing to the fact that the distribution of most meteorological stations tends to reflect the population distribution (Legates 1987, 84).

In addition, the temporal distribution of these stations has fluctuated considerably over the last 150 years. Few station data are available for the late 1800s and the early 1900s and while the number of stations rises considerably between the 1940s and 1970s, the number of available climate observations has dropped precipitously in recent decades (Figure 18.2).

Despite the widely recognized need to monitor climate change, this decrease is due largely to a reduction in the number of observing stations globally and a disturbing trend where many countries now sell their weather data as a source of income (Hulme 1994). Unfortunately, this has not resulted in an increase in data collection, as funds are not usually available for researchers to purchase these data.

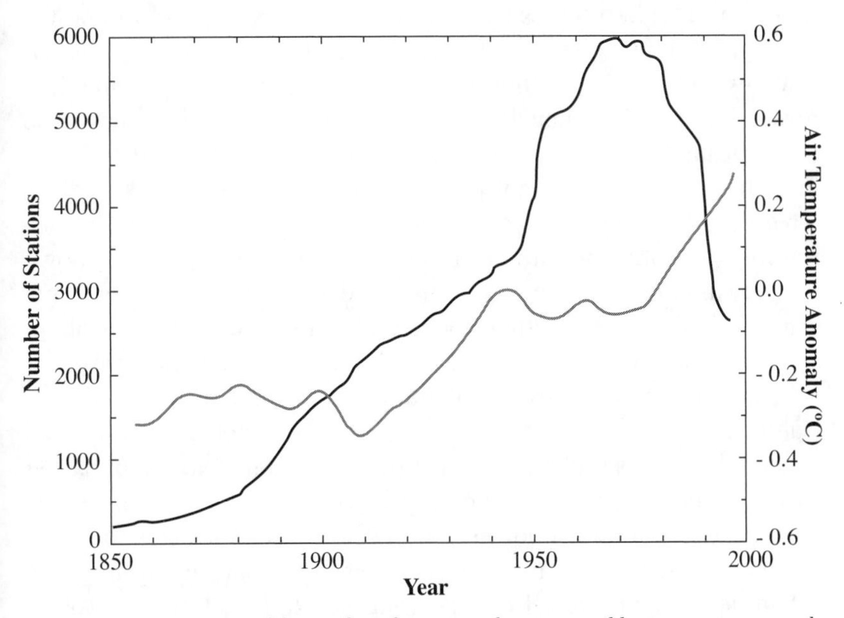

Figure 18.2 Time series of the number of stations with mean monthly air temperature available in the Global Historical Climatology Network Temperature Database by T. C. Peterson and R. S. Vose (solid black line, http://www.ncdc.noaa.gov/oa/climate/research/ghcn/ghcnoverview.html). Superimposed on this graph is the air temperature curve (gray line) from the University of East Anglia's Climatic Research Unit and the UK Meteorological Office's Hadley Centre (see Figure 1). Peterson and Vose note that the decrease in stations in recent years is due to extensive efforts to digitize historical data archives; by contrast, "only three data sources are available in near real time" which limits the ability to update current conditions.

In particular, the "global cooling" of the 1960s and 1970s matches an increase in the distribution of stations (see Figure 18.2). However, this pattern is not entirely spurious in that inadequate station networks have been shown to result in biased estimates of global averages of air temperature and that changing station densities can affect the calculation of mean air temperature (Willmott, et al. 1992). In particular, this bias can be significant as the differences resulting from variable station networks are often as large as 0.4°C —or more than half of the warming over the last century. This, of course, undermines the ability of existing spatial networks to represent adequately the true air temperature time series.

But even a station at which air temperature observations have been made for a long time period may be subject to a number of changes in location and site characteristics. Over time, many stations have either been moved or replaced, causing discontinuities in the measurements. After World War II, for example, many weather stations in the United States national network were moved from downtown locations to newly built airports on the urban fringe. Similar changes have affected other countries, most notably in precipitation measurement (Groisman 1991). Over the years, cities have grown so dramatically that many stations that showed little or no urban effect in 1940 are now significantly affected by the *urban heat island*—a term that describes the fact that a city will be warmer than its surrounding countryside. Studies in numerous urban areas have shown that city temperatures on warm summer days can be as much as 4.5°C warmer than the surrounding countryside and annually average about 2.5°C warmer. This is because of the increase in impervious surfaces (less heat is removed from concrete than soil via the evaporation of water), an overall decrease in winds (standing structures disrupt and reduce the wind's exchange of heat by convection), the use of darker surfaces, the canyon-like structure of cities (increasing solar energy absorption), and anthropogenic sources of heat. Although some researchers have taken significant measures to estimate and remove the urban heat island effect from the data record, it still poses a potential problem. Nevertheless, the effects of changing land use and land cover have been demonstrated to affect significantly climate model prognostications of warming, even to the extent to which some regional changes are of opposite sign than those posited in the Intergovernmental Panel on Climate Change Report (Feddema, et al. 2005).

Over time, observational differences and instrumentation changes have also occurred. Daily means computed from the daily minimum and maximum air temperature measurements, for example, often introduces a bias of as much as 1°C due to the asymmetry in the diurnal pattern of air temperature (Schaal and Dale 1977). In particular, variations in the time of observation for different countries can accentuate this problem (Baker 1975). For example, cooperative weather observers in the United States (who make daily observations for the National Weather Service) are more likely to take their observations in the morning rather than in the evening, which can decrease the mean daily air temperature below its true value (Jones, et al. 1986). Variable thermometer heights are also a potential source of error and the magnitude of the effect depends on the surface lapse rate in the planetary boundary layer.

Despite these biases, the biggest problems with estimating global air temperature change are those associated with using sparse spatial and temporal station distributions. Methods of spatial analysis have not been assessed adequately as these data networks are often spatially averaged to determine a terrestrial or global mean. In particular, air temperature anomalies from a long-term mean period of record (e.g., the 1961–1990 average in the University of East Anglia's representation—Figure 18.1, top) are considered because the unproven assertion is that anomalies exhibit less spatial variability than the actual observations. Although this reduces interpolation biases, valuable information (namely, the mean field) is lost. Nevertheless, research has shown that the error induced by the interpolation method may be as large as the estimates of air temperature trends over the last century (Robeson 1993).

Despite the significant scientific problems associated with obtaining annually averaged global air temperature estimates, these figures are presented regularly to the media and even small changes (out to the thousandths decimal place) are interpreted by some as important, despite the fact that the researchers themselves estimate the total error to be at least 0.1°C—or 12.5% to 16.6% of the estimated warming over the entire last century (Black 2005). But even more disturbing is the fact that these figures are regularly released to the public by three different research groups, often as early as mid-December. Although updated later, these "preliminary data" often overestimate the final result because data from late December has not yet occurred. More important, the preliminary statistic released in mid-December is compared with the updated figures from previous years that have included data from the entire year.

Recently, the annual estimates of globally-averaged air temperature are now "augmented" by including subjective adjustments for specific adverse effects. For example, beginning in 2005, James Hansen's group from the Goddard Institute for Space Studies in New York included an adjustment for the underrepresentation of the Arctic. This subjective adjustment was not included before 2005, but Hansen commented that "the inclusion of estimated Arctic temperatures is the primary reason for our rank of 2005 as the warmest year" (MSNBC 2005). Similarly, the National Climatic Data Center's own analysis suggested that 2005 was a close second to 1998, "in part because of how the Arctic was factored in" (MSNBC 2005). Amazingly, too, 1998 followed another change in methodology where a regression approach to sea surface temperature using empirical orthogonal functions was added to increase air temperatures during El Niño events (Hansen, et al. 1996). Such modifications to the way in which globally-averaged air temperatures are computed have not been retroactively applied to the earlier data record, which makes it impossible to compare the figures before and after each modification.

Climate Change Data Proxies and Further Uncertainties

If reconstructing globally-averaged air temperature from station observations is a daunting task (not to mention the relative uselessness of the resultant time-series), problems associated with using proxy records (e.g., tree rings, ice cores, ocean sediments, and historical data) to estimate air temperature over the last two centuries are myriad. Despite these problems, three recent research articles have been widely cited for their claim to depict air temperature trends dating back to AD 1000 (Mann, et al. 1998; Mann, et al. 1999) and, more recently, back to AD 200 (Mann and Jones 2003). Indeed, the United Nations' Intergovernmental Panel on Climate Change (IPCC) repeated the claims of the earlier of the two articles by citing these articles prominently.[1] The key ingredients of these "climate" recon-

[1]It should be noted that the section in the Third Assessment Report of the IPCC Report which touted highly the Mann et al. articles was written by M. E. Mann himself. Indeed, a co-author of the chapter relayed to me that Mann was admonished to include other contradictory reconstructions but despite assurances, none were ever included.

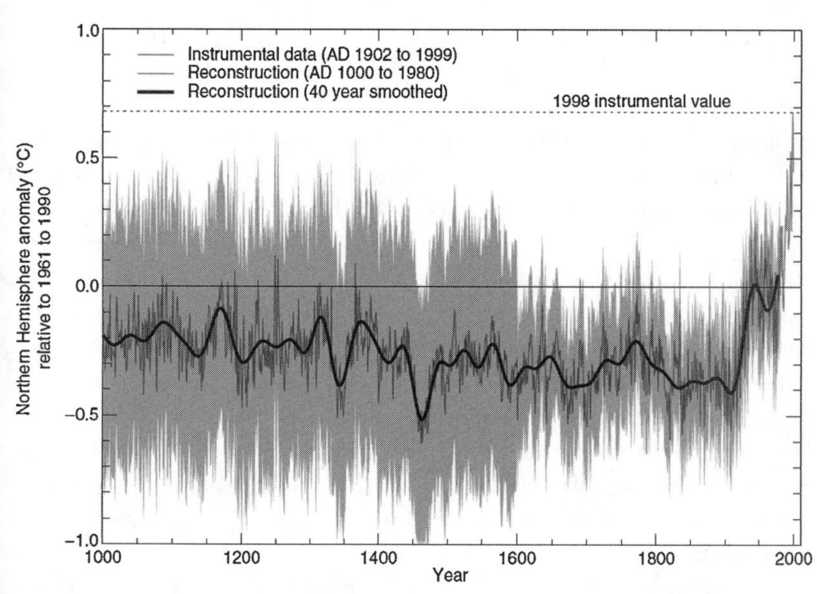

Figure 18.3 "Hockey Stick" reconstruction of air temperature for the Northern Hemisphere using proxy records (dark gray curve) and the instrumental record (light gray curve). The smoothed time-series is shown (black curve), as is the uncertainty represented by two standard errors (gray shading) (from the Technical Summary of the *Third IPCC Assessment Report*, Figure 5, page 29, 2001).

structions are the relatively flat "shaft" extending until about AD 1900, a rising "blade" that follows from AD 1900 to 2000, and a range of uncertainty that envelops the "shaft" (Figure 18.3).

Owing to these components, these curves are commonly called the "Hockey Stick." A number of independent research groups, however, have raised extensive concerns regarding the veracity of these claims by highlighting the problems with the underlying reconstructions that have persisted through most of these collaborative efforts, calling into question all three components of the "Hockey Stick."

Although the Hockey Stick suggests that globally- and hemispherically-averaged air temperatures remained relatively constant from AD 200 until about AD 1900 (i.e., the shaft), the widely-recognized Medieval Warm Period (from about AD 800 to 1400) and the Little Ice Age (from about AD 1600 to 1850) are absent. Although these terms do not directly refer to trends in air temperature, but rather, to periods of glacial extent—enlarged during the Little Ice Age and substantially reduced during the Medieval Warm Period—

they are correlated with the climate in that glacial mass balances certainly are affected by longer-term fluctuations in air temperature as well as precipitation (Grove 2001). More than 200 proxy records from around the globe, however, show that these climatic events were global in scale; that is, there is a significant decrease in air temperature during the period associated with the Little Ice Age and a significant increase in air temperature during the Medieval Warm Period (Soon and Baliunas 2003; Soon, et al. 2003). In particular, it has been noted from proxies of tree rings, ice cores, bore holes, and recorded observations, that in diverse regions such as southern South America, southern Africa, northern China, New Zealand and Australia, and western North America, marked warming occurred at the beginning of the last millennium, which was followed by significantly cooler conditions during the middle centuries. Indeed, as H. H. Lamb argued in his now famous book, the Little Ice Age appears to have been global in extent (1982).

The reason why the shaft of the Hockey Stick is flat and does not represent the true trends in the contributing data is errors in the collection and use of data from multiple sources, as well as errors in the calculation of the derived statistics (i.e., principal components). It was first demonstrated that the Hockey Stick calculations were 1. based on unjustifiable truncated or extrapolated trends from source data, 2. based on obsolete data, 3. based on incorrect calculations, and 4. associated with data sets using incorrect geographical locations (McIntyre and McKitrick 2003). More recent and extensive evaluation of the Hockey Stick methodology has revealed that the selection of the anomaly period for computing the long-term mean (i.e., AD 1902 to 1980 versus AD 1940 to 1980) adversely affects the first principal component, especially when the mean for these time periods varies significantly (McIntyre and McKitrick 2005). Specifically, since principal components analysis is a variance maximization procedure, the first principal component (which explains the most variance) becomes biased in favor of hockey stick–shaped patterns and its influence (explained variance) is exaggerated. In particular, the 15th-century period is described exclusively by the bristlecone pine species of North America, whose tree ring variance may be more related to precipitation variability than air temperature (Hughes and Graumlich 1996; Salzer and Kipfmueller 2005). Independent verification also shows that the fatal flaw with the Hockey Stick temperature reconstruction was its misrepresentation of longer-term trends (Esper, et al. 2004)—the hockey stick methodology inappropriately removes long-term trends. To construct the hockey stick climate trend, proxies were used with very limited data sets (often based on just one or two trees) for the beginning of the record and long-term cooling trends were removed by erroneously correlating air temperature trends with tree age (Esper, et al. 2004). Indeed, the Medieval Warm Period and the Little Ice Age are retained in other climate reconstructions:

> [The] Briffa (2000) and Esper et al. (2002) [air temperature reconstructions] display a pronounced [Medieval Warm Period] followed by a significant 200–300-year-long cooling trend associated with the Little Ice Age. Such a trend is broadly absent in [the work by Mann and his colleagues]. (Esper, et al. 2004)

Such methodological flaws were also highlighted by Pollack and Smerdon (2004), which eventually led to a published correction of the work (Rutherford and Mann 2004). Mann and colleagues admitted to underestimating the air temperature variations present in the proxy data after AD 1400 by more than 33% (Rutherford and Mann 2004), which explains why the hockey stick fails to incorporate the Little Ice Age cooling. Although they admitted to this error, Mann and his colleagues fail to comprehend the extent to which their historical reconstruction is undermined by it and the effect it has on air temperature trends in the latter portion of their reconstruction.

Recently, my colleagues and I more closely considered the "blade" of the hockey stick—how the observational record since AD 1900 has been "spliced" onto the proxy reconstruction (Soon, et al. 2004) (Figure 18.4).

Most scientists, including the IPCC reports, indicate that the Northern Hemisphere warmed only about 0.6°C (1°F) during the 20th century (Figure 18.1). This contrasts sharply, however, with the representation of the observational record by the Hockey Stick, where a warming over the last century of 0.95°C (1.5°F)—an air temperature rise more than 50% larger than our best estimates—is presented. Note that this warming estimate has grown substantially with subsequent published versions of the Hockey Stick, apparently to accommodate the continuing need to claim that the 1990s were the warmest decade of the last millennium, and subsequently of the last 1800 years when modifications were made to the uncertainty interval. We found

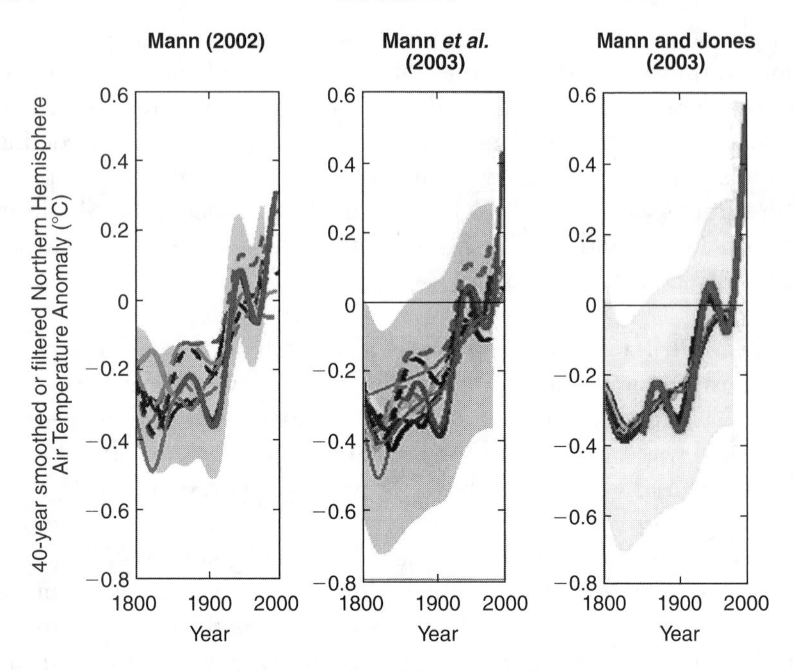

Figure 18.4 Three different presentations of the "Hockey Stick" reconstruction of air temperature for the Northern Hemisphere by Mann and colleagues (Mann 2002; Mann, et al. 2003; Mann and Jones 2003). Note that the end of the spliced observational record (thick curve)—the value in AD 2000—reaches to a higher value with each subsequent presentation of the Hockey Stick. Adapted from Figure 3 of Soon, et al. (2004).

that neither the representation of the observational record nor its subsequent "growth" could be attributed to the method that Mann and his colleagues claimed they used or to traditional statistical smoothing procedures (Esper, et al. 2004).

Using the Hockey Stick reconstruction of air temperature, its creators and proponents claim that the 1990s were the warmest decade since AD 200, with 1998 being the warmest year. However, the concomitant uncertainty assessment—the 95% confidence interval associated with the reconstructed estimates, shown by the shaded region in Figure 18.4—arises solely from a model "fit" characterization; that is, how well the reconstructed (proxy) air temperature time-series matched the observational data over the last century. Unfortunately, a model "lack-of-fit" is only one of a number of components of an uncertainty, or error, assessment; many of which include exacerbated extensions of the problems outlined earlier in estimating globally averaged air temperature from observational data. In addition to the statistical "lack of fit," uncertainty also arises from 1. the biases in obtaining global or hemispheric air temperature averages from the observational record (i.e., errors in the representation of the observational record), 2. temporal reconstructions that are based on a spatially-limited set of proxy records, 3. the inability of a proxy record to represent regional air temperatures, and 4. the observed variability in both the proxy record and the instrumental record. To provide a proper assessment of uncertainty for Hockey Stick reconstructions, these other potential sources of error must be incorporated into the uncertainty assessment. When these errors are considered, the exaggerated claim suggested by the Hockey Stick regarding the latter half of the 20th century becomes completely unfounded, as it is then impossible to argue that these conditions are statistically greater than any estimates from the last 1000 or more years (Lamb 1982; Legates 2004). The envelope of uncertainty, shown by the gray area in Figure 18.3, then includes the observations made in the 1990s. An assessment of the uncertainty in climate data—the importance of "The Little Figures That Are Not There," as discussed in the famous book *How to Lie with Statistics* (Huff 1954)—is critical to a proper interpretation of climate data.

Note that the Hockey Stick appears to be unresponsive to known climate singularities, such as large volcanic eruptions. In particular, the eruption of Tamboro in 1815 and weak solar magnetic activity (the height of the Dalton Minimum of sunspot activity) led to widespread cooling across the entire Northern Hemisphere in 1816, now known as the "Year Without A Summer" (Soon and Yaskell 2003). The latest 1800-year Hockey Stick reconstruction (Mann, et al. 1999), however, shows that of the period from 1811 to 1821, five years are warmer than 1816 and five years are colder. As decadal-scale fidelity is not preserved in the data set, claims regarding particular years or decades being the warmest are simply inappropriate.

The Hockey Stick and *Climate Research*

The importance of the global air temperature record—and, in particular, the Hockey Stick reconstruction—to the anthropogenic global warming argument cannot be overstated, as it featured prominently in an attempt to disrupt the climate change debate. In July of 2003, the Environment and Public Works Committee of the US Senate held a hearing[2] regarding climate temperature reconstructions. The stated goal of Chairman Inhofe was "to base our [policy] decisions . . . on sound science." Proponents of anthropogenic global warming had seized on the Hockey Stick to assert that air temperatures remained relatively constant for approximately 900 years, and only when industrial activity by humans began in the last century did air temperatures begin to rise precipitously. But at that time, the aforementioned flaws were beginning to be uncovered in the Hockey Stick representation of air temperature, and the first important rebuttals to the Hockey Stick had been published by Soon and his colleagues (Soon and Baliunas 2003; Soon, et al. 2003b), who concluded that both the Little Ice Age and the Medieval Warm Period could be detected in more than 200 individual proxy records from around the globe. But it was the first article by Soon and Baliunas, published in the journal *Climate Research,* that attracted the most attention—and elicited an invitation to Soon, Mann, and myself to testify before the US Senate Committee.

Central to the topic of this Senate hearing was an anonymous appeal to the publisher of the journal *Climate Research* requesting a retraction of the Soon and Baliunas paper.[3] Serving as judge, Dr. Otto Kinne (founder of *Climate Research* and the Director of *Inter-Research*, the publisher of *Climate Research*) made his own examination of the peer-review process that preceded publication of the Soon-Baliunas articles. In an e-mail to all editors and review editors of *Climate Research*, Kinne noted that the editor had done "a good and correct job," that the four reviewers solicited for each manuscript had "presented detailed, critical, and helpful evaluations," that the editor had "properly analyzed the evaluations and requested appropriate revisions," and that the authors had "revised their manuscripts accordingly."

Kinne then appointed Hans von Storch, long-time editor of *Climate Research,* as its new editor in chief, since none had existed to arbitrate such disputes. But von Storch's first act was an attempt to change the editorial policies that had governed *Climate Research* since its inception. He prepared an editorial that challenged the

[2] 108th Congress, Senate Hearing 108-359—Climate History and the Science Underlying Fate, Transport, and Health Effects of Mercury Emissions, July 29, 2003.

[3] I served as Review Editor of *Climate Research* during the event and was privy to the e-mail conversations that took place among the editors and Dr. Kinne.

methods used by Soon and Baliunas, asserting that an opinion piece published by the "Hockey Stick" proponents (Mann, et al. 2003) (and appearing in another journal) "convincingly demonstrated [to von Storch] that the manuscript . . . should not have passed the review process," even though Soon and his colleagues had provided a clear rebuttal (Soon, et al. 2003b) of the criticisms raised by the opinion piece. Von Storch then placed himself as the final arbiter, asserting that, in the future, the recommendation of an editor would be forwarded to von Storch, who would make the final decision of paper acceptance or rejection. He admitted that, while the formal review process was followed in the Soon/Baliunas case, the outcome was inadequate, such that his autonomous hand in controlling future publications was now required. But at the same time, von Storch further declared that Soon and Baliunas should be barred from future publications in *Climate Research*—despite the fact that Soon and Baliunas had not been charged with any offense—save criticism of the climate reconstruction suggested by the Hockey Stick.

With Soon set to testify within a week before the US Senate Committee, an immediate response to his proposed editorial was sought from the other eight editors. The response was mixed and Kinne eventually and appropriately weighed in that the von Storch proposal "would amount to a devaluation of editors and reviewers." He asked that the editorial not be rushed, but that a revised version, approved by all editors, be published in due course. Von Storch objected to the refusal to publish his editorial and summarily resigned as editor in chief, as did two additional editors. Several others, however, expressed their intention to resign had von Storch's original editorial appeared.

Subsequently, a number of researchers threatened a boycott of *Climate Research* as a result of the situation surrounding von Storch's resignation. In response, Kinne published an editorial (2003) that reviewed the peer-review process, supported von Storch's efforts, and condemned the Soon and Baliunas paper by concluding that, while their conclusions may have been valid, subsequent criticism indicated that appropriate revisions should have been requested prior to publication—despite the fact that four reviewers saw the manuscript and Kinne originally declared that both the editor and the authors had responded appropriately to the reviewers' comments. While we would like to separate science from politics and personal biases, such cannot be accomplished in these polarized times, especially in the realm of climate change.

While critical of Soon and Baliunas, von Storch has also been critical of the Hockey Stick itself. He and colleagues used a model representation of the climate/ocean system to simulate the variability in climate over the past millennium to examine the efficacy of air temperature reconstructions such as the Hockey Stick (von Storch, et al. 2004). Their findings indicated that methods such as those used in constructing the Hockey Stick underestimate the actual variability in air temperature and that "past variations may have been at least a factor of two larger than indicated by empirical reconstructions." In Congressional testimony, popular writings, and public presentations, von Storch has also been critical of the "climate alarmism" that has recently emerged within the scientific community, which he argues is an effort to gain media and political attention. For example, in an abstract for a public talk, von Storch says

> The "hockey stick" was elevated to an icon-status by the IPCC. While in the [Scientific Assessment], the reconstruction of the last millennium's temperature was presented with the proper caveats and uncertainties, in the publicly more visible [Summary for Policymakers] these caveats were less and less emphasized. The result is that in many quarters the hockey stick is considered to be an unquestionable indication of the detection and attribution of anthropogenic climate change. The problem was, and is, that the methodology behind the hockey stick has not been adequately tested. The methodology was not properly explained in the original *Nature* publication. Scientists still have difficulties [as to] what exactly is "in" the method. (2005)

After publication of his study and others that were critical of the Hockey Stick, he goes on to note:

> The response was surprising—almost no open response, a bit in the media, and many colleagues who indicated privately that such a publication would damage the good case of a climate protection policy. It would play into the hands of the "skeptics." It seems that exaggerated claims pass the internal quality checks of science relatively easily, whereas more reasoned and scientifically accurate claims find an unwelcome audience among scientists. The practice of scientists exaggerating threatening perspectives of anthropogenic climate change and its implications

> serves not only the purpose of supporting a policy perceived as "good" but also personal agendas of career and public visibility. The problem is, however, that the desired public attention can only be achieved if these perspectives are continuously topped by even more threatening perspectives. Thus, the credibility of climate science is endangered, and its important role of advising policy (in the naive sense of "knowledge speaks to power") becomes an unsustainable practice. We have a situation similar to the case of "tragedy of the commons" (von Storch 2005).

Having reached iconic status, the Hockey Stick appeared "untouchable" and was taken by many, both in scientific and public circles, as proven fact. However, the debate over the legitimacy of the Hockey Stick was not over.

The National Academy of Sciences and The Wegman Reports

In an attempt to provide a definitive answer to the raging debate over the validity of the Hockey Stick representation of global air temperatures over the last two millennia, The House Committee on Science charged the National Research Council of the National Academy of Science (NAS) with answering three specific questions relating to air temperature reconstructions. 1. What is the current scientific consensus on the temperature record of the last 1,000 to 2,000 years? 2. What is the current scientific consensus on the conclusions reached by Mann and his coauthors? 3. How important is the Hockey Stick debate to the overall scientific consensus on global climate change?

In its conclusions, the NAS report argues "there is less confidence in reconstructions of surface temperatures from 1600 back to AD 900, and very little confidence in findings on average temperatures before then" (2006). They found the Hockey Stick conclusion that late-20th century warming was unprecedented over the last thousand years to be plausible, but with less confidence that the warming was unprecedented prior to AD 1600. The period prior to 1600 AD is critical to the discussion as the existence and magnitude of warmth associated with the Medieval Warm Period is crucial to the debate. Simple assertions that we are in the warmest period of the last 400 years is not contested; indeed, such is true regardless of whether the new Hockey Stick model or the older Medieval Warm Period/Little

Ice Age model is correct. The NAS report, however, further notes that less confidence exists in the assertion that the 1990s is the warmest decade, with 1998 being the warmest year. Noting that the consensus of air temperature reconstructions are generally consistent, the NAS report asserts that the concept of a Medieval Warm Period and Little Ice Age are valid, at least for the Northern Hemisphere, but does state that data do not exist to suggest that the Medieval Warm Period was warmer than present conditions.

In July of 2006, the results of a report commissioned by the Committee on Energy and Commerce and the Subcommittee on Oversight and Investigations of the US House of Representatives were released, providing an independent evaluation of the Hockey Stick and its criticisms. The three-member panel, chaired by Edward J. Wegman, has come to be known as the Wegman Report, and featured an examination of the methodology used in the construction of the Hockey Stick by three prominent statisticians. Wegman is professor of statistics at George Mason University and chairs the National Academy of Sciences' (NAS) Committee on Applied and Theoretical Statistics. The Wegman Report (US House Committee on Energy and Commerce 2006) was commissioned after Chair Joe Barton (R-TX) wrote a letter to Mann and his colleagues requesting answers to specific questions about their publicly funded research. Although some viewed this as an act of intimidation on the part of the US House of Representatives toward specific scientists, the House Committee turned the matter over to Wegman, thereby allowing statisticians, and not politicians, to conduct a scientific evaluation of the Hockey Stick methodology.

Although much of the Wegman Report focuses on statistical and mathematical issues that are well beyond the scope of this essay, the overall report upheld most of the criticisms that were associated with the Hockey Stick. In particular, the report concluded that the publications by Mann and his coauthors which described their research methods were "somewhat obscure and incomplete" and the criticisms to be "valid and compelling" (U.S. House Committee on Energy and Commerce 2006, 4). The report also concluded that, while the methodology sounded reasonable, errors in the methodology "may be easily overlooked by someone not trained in statistical methodology" and noted that interactions between Mann and his coauthors and mainstream statisticians were limited.

But the Wegman Report went further, noting that at least 43 other researchers have coauthored papers with Mann and concluded that "our findings from this analysis suggest that authors in the area of paleoclimate studies are closely connected and thus 'independent studies' may not be as independent as they might appear on the surface" (p. 4). Noting the isolation of the paleoclimate scientists from the statistical community, despite the strong reliance on statistical methods, the report argued that too much reliance was given to peer review, which, in this case, was not necessarily independent. In terms of being able to reproduce the results, the report concluded "that the sharing of research materials, data, and results was haphazardly and grudgingly done" (p. 4).

In addition, the Wegman Report commented that "the Academy decided not to address the specific questions of the House Committee on Science and decided to focus the Academy study away from the specific questions and address broader issues" (p. 65). Thus, the Wegman Report went on to address the three questions that were specifically charged to the NAS panel.

With respect to the question "What is the current scientific consensus on the temperature record of the last 1,000 to 2,000 years?" the report concluded that strong evidence exists that air temperatures have been rising since 1850, but that the accuracy of climate reconstructions are a matter of debate and that no consensus exists. The panel also concluded that "the high level of variability in the proxy data as well as the lack of low frequency effects make the reconstructions more problematic than the advocates of these methods would have us believe" (p. 65).

The second question posed to the NAS was "What is the current scientific consensus on the conclusions reached by Mann and his coauthors?" The Wegman Report concluded that, although no real consensus on the Hockey Stick exists, Mann and his coauthors represent a "tightly knit group of individuals who passionately believe in their thesis" and, through their social network, tend to reinforce their ideas. The report adds, ". . . moreover, the work has been sufficiently politicized that they can hardly reassess their public positions without losing credibility" (p. 65).

Finally, with respect to the question "How important is the Hockey Stick debate to the overall scientific consensus on global climate change?" the Wegman Report notes that the Hockey Stick is "essentially irrelevant to the consensus on climate change" and reinforces the assertion that the globally-averaged air temperature has been increasing since 1850. The unfortunate issue, the report notes, is that the Hockey Stick "has been politicized by the IPCC and other public forums and has generated an unfortunate level of consensus in the public and political sectors and has been accepted to a large extent as truth."

Although asserting that the conclusions of Mann and his coauthors regarding the Hockey Stick are "plausible," it stops short of full endorsement, instead arguing that less confidence exists to suggest that the 1990s was the warmest decade and 1998 was the warmest year. The Wegman Report concluded rather harshly that "overall, our committee believes that Mann's assessments that the decade of the 1990s was the hottest decade of the millennium and that 1998 was the hottest year of the millennium cannot be supported by his analysis." Both reports allude to the widespread existence of both the Little Ice Age and, to a lesser extent, the Medieval Warm Period. As a final admission of its flaws, the Hockey Stick was conspicuously absent from the Fourth Assessment's *Summary for policy makers*, despite the fact that it had figured prominently in the IPCC's Third Assessment Report.[4]

The Keepers of the Data

Returning once again to the data, much of the dichotomy in this debate lies between climate modelers, who trust model prognostications of future climate change, and those who focus on the data and trends in the climate record. The largest organization of professionals who work directly with climate data at local, state, and regional levels is the American Association of State Climatologists (AASC). State Climatologists are appointed by their respective states to provide expertise on climate-related issues to state policy makers, governmental officials, legal professionals, and the general public. Since State Climatologists deal directly with the historical record and the observed trends, patterns, and periodicities, their collective view on the climate change issue provides an important perspective that must be considered. Indeed, many of those who reject the extreme anthropogenic global warming scenarios are State Climatologists.

In November 2001, the AASC released its policy statement on climate variability and change. It recognized the difficulty in recon-

[4]At the time of this writing, the Scientific Assessment of the Fourth Assessment Report was not finalized for quotation and/or citation.

ciling simulations of present-day conditions with observations and of specifying appropriate future scenarios of human-caused emissions. As a result, it concluded in its policy statement, "Climate predictions have not demonstrated skill in projecting future variability and changes in such important climate conditions as growing season, drought, flood-producing rainfall, heat waves, tropical cyclones, and winter storms." In short, it concluded that relying on simulations made by climate models to draw conclusions about the future is very risky, since they do not, and perhaps cannot, accurately simulate the present climate of the Earth.

Nevertheless, the conclusions of the AASC are quite moderate—that our environmental vulnerability is likely to change (not just from changes in our climate, but because of societal changes) and that policy responses should be "flexible and sensible," focusing on common-sense actions to select "policy alternatives that make sense for a wide range of plausible climate conditions regardless of future climate" and reduce both environmental and societal risk. Moreover, they agree that humans have indeed influenced their global and local climates, not just though increases in greenhouse gases but through changing land use (Willmott, et al. 1992) and sulfate emissions, which vastly complicate prediction scenarios. With respect to climate prediction, the AASC admits that it is a daunting task because it requires an assessment of an extremely varied and complicated system, filled with complex feedbacks and forcings, many of which are not well understood. The AASC also concludes that policy responses should be broad and not directed solely at policies aimed toward securing alternative sources of energy, as climate impacts many other sectors, including agriculture, insurance, water resources, ecosystem management, and disaster management and mitigation.

Such arguments are extremely constructive to the global change debate in that a well-reasoned and sensible approach is needed. But despite the changes in our future climate, we always will have severe weather events that require monitoring and a quick response from the public. Moreover, as the AASC has argued, an assessment of past climate conditions provides "a very effective analysis tool to assess societal and environmental vulnerability to future climate" change. It is, therefore, imperative that with respect to weather and climate observations, we must maintain a quality monitoring network by upgrading and modernizing our existing infrastructure and providing funding to secure our existing observational network. Unfortunately, in our current political climate, budgets for observing networks are being cut, while funding for climate change research has dramatically increased. This has had a negative effect—producing more studies that highlight the doom-and-gloom nature of future climatic (i.e., modeled) trends, while providing less information with which to assess the efficacy of model prognostications and proposed theories. But more important, this has left us ever less prepared for the immediate disasters that continue to come along and thus has inadvertently made us more vulnerable to environmental risks.

Conclusions

Ascertaining trends in globally-averaged air temperature, whether it be using observational data for the past 125 years or proxy data for the past two millennia, is a particularly difficult and politically charged task. It is, therefore, ironic that such trends are really useless, since we are more interested in how those events that create the greatest loss of life and have the greatest economic impact—floods, droughts, tropical and extra-tropical storms, severe local weather, sea-level rise, heat waves, and cold spells—and various ecosystem responses will change. Few of these events are directly impacted by changes in mean globally-averaged air temperature but rather, by regional and local changes in our climate. Nevertheless, it is imperative that scientific debate be supported and pursued. On one of his *Nightline* broadcasts, in which he chastised then Vice President Al Gore "for trying to use the media to discredit skeptical scientists," Ted Koppel concluded with:

> But the issues have to be debated and settled on scientific grounds, not politics. . . . The measure of good science is neither the politics of the scientist nor the people with whom the scientist associates—it is the immersion of hypotheses into the acid of truth. That's the hard way to do it, but it's the only way that works. (1994)

Despite the biased presentations, overstated claims, exaggerated views, and personal attacks that now seem to permeate the climate change debate, the American Meteorological Society, one of the largest professional organizations of climatologists in the country, recently adopted their statement on *Freedom of Scientific Expression*, which states, in part, (emphasis added):

> Advances in science and the benefits of science to policy, technological progress, and society as a whole depend upon the free exchange of scientific data and information as well as on open debate. The ability of scientists to present their findings to the scientific community, policy makers, the media, and the public *without censorship, intimidation, or political interference* is imperative. With the specific limited exception of proprietary information or constraints arising from national security, *scientists must be permitted unfettered communication of scientific results*. In return, it is incumbent upon scientists to communicate their findings in ways that *portray their results and the results of others, objectively, professionally, and without sensationalizing or politicizing the associated impacts*. . . . Thus, scientists, policy makers, and their supporting institutions share a special responsibility at this time for *guarding and promoting the freedom of responsible scientific expression*. (2006)

Following these guidelines will be essential if the science of global warming is to be examined in its proper context and if public policy that results from that science is to be appropriately applied and useful. What cannot be tolerated is a one-sided presentation that seeks to undermine and stifle valid criticisms and debate simply to force policy makers into actions that may be unwarranted and unwise.

References

American Association of State Climatologists. 2001. Policy statement on climate variability and change. http://www.ncdc.noaa.gov/oa/aasc/aascclimatepolicy.pdf.

American Meteorological Society. 2006. Freedom of scientific expression. *Bulletin of the American Meteorological Society* 87:515.

Baker, D. G. 1975. Effect of observation time on mean temperature estimation. *Journal of Applied Meteorology* 14:471–476.

Black, R. 2005. 2005 warmest on record in north. *BBC News*, Dec. 15.

Esper, J. et al. 2004. Climate reconstructions: Low-frequency ambition and high frequency ratification. *EOS, Transactions of the American Geophysical Union* 85:113–120.

Feddema, J. J. et al. 2005. The importance of land-cover change in simulating future climates. *Science* 310:1674–1678.

Groisman, P. 1991. Unbiased estimates of precipitation change in the Northern hemisphere extratropics. Fifth Conference on Climate Variations, American Meteorological Society, Denver, CO, October 14–18.

Grove, J. M. 2001. The initiation of the Little Ice Age in regions around the North Atlantic. *Climate Change* 48:53–82.

Hansen, J. E. 2003. Can we diffuse the global warming time bomb? *Natural Science*, August 1 http://naturalscience.com/ns.articles/01-16/ns-jeh.html.

Hansen, J. et al. 1996. Global surface air temperature in 1995: Return to pre-Pinatubo level. *Geophysical Research Letters* 23:1665–1668.

Henson, B. 2004. Going to extremes. *UCAR Quarterly* Winter. http://www.ucar.edu/communications/quarterly/winter04/extremes.html.

Huff, D. 1954. *How to Lie with Statistics*. New York: WW Norton & Company.

Hughes, M. K., and L. J. Graumlich. 1996. Climatic variations and forcing mechanisms of the last 2000 years. *Multi-millennial dendroclimatic studies from the western United States, NATO ASI Series* 141:109–124.

Hulme, M. 1994. The cost of climate data—a European experience. *Weather* 49:168–175.

Jones, P. D., and A. Moberg. 2003. Hemispheric and large-scale surface air temperature variations: An extensive revision and an update to 2001. *Journal of Climate* 16:206–223.

Jones, P. D. et al. 1986. Northern hemisphere surface air temperature variations: 1851–1984. *Journal of Climate and Applied Meteorology*. 25:161–179.

Jones, P. D. et al. 1999. Surface air temperature and its changes over the past 150 years. *Reviews of Geophysics* 37:173–199.

Kinne, O. 2003. Climate research: An article unleashed worldwide storms. *Climate Research* 24:197–198.

Koppel, T. 1994. Is environmental science for sale? *Nightline*, Feb. 24.

Lamb, H. H. 1982. *Climate history and the modern world*. London: Methuen.

Legates, D. R. 1987. A climatology of global precipitation. *Publications in Climatology* 40:84.

Legates, D. R. 2004. Breaking the "Hockey Stick." Brief analysis no. 478. National Center for Policy Analysis. http://www.ncpa.org/pub/ba/ba478.

Mann, M. E. 2002. The value of multiple proxies. *Science* 297:1481–1482.

Mann, M. E., and P. D. Jones. 2003. Global surface temperatures over the past two millennia. *Geophysical Research Letters* 30:1820.

Mann, M. E. et al. 1998. Global-scale temperature patterns and climate forcing over the past six centuries. *Nature* 392:779–787.

Mann, M. E. et al. 1999. Northern hemisphere temperatures during the past millennium: Inferences, uncertainties, and limitations. *Geophysical Research Letters* 26:759–762.

Mann, M. E. et al. 2003. On past temperatures and anomalous late-20th century warmth. *EOS* 84:256.

McIntyre, S., and R. McKitrick. 2003. Corrections to the Mann et al. (1998) proxy data base and Northern hemisphere average temperature series. *Energy and Environment* 14:751–771.

McIntyre, S., and R. McKitrick. 2005. Hockey sticks, principal components, and spurious significance. *Geophysical Research Letters* 32, 12 February: DOI:10.1029/2004GL021750.

MSNBC. 2005. 2005 warmest year on record, data indicates. http://www.msnbc.msn.com/id/11009001/.

National Academy of Sciences. 2006. Surface temperature reconstructions for the last 2,000 years. Washington, DC: National Academies Press.

Pollack, H. N., and J. E. Smerdon. 2004. Borehole climate reconstructions: Spatial structure and hemispheric averages. *Journal of Geophysical Research* 109: DOI: 10.1029/2003JD004163

Roberts, D. 2006. Al Revere. *Grist Magazine*, May 9. http://www.grist.org/news/maindish/2006/05/09/roberts/.

Robeson, S. M. 1993. Spatial interpolation, network bias, and terrestrial air temperature variability. *Publications in Climatology* 46:1–51.

Rutherford, S., and M. E. Mann. 2004. Correction to "Optimal surface temperature reconstructions using terrestrial borehole data." *Journal of Geophysical Research* 109: DOI:10.1029.

Salzer, M. W., and K. F. Kipfmueller. 2005. Reconstructed temperature and precipitation on a millennial timescale from tree rings in the Southern Colorado Plateau, USA. *Climate Change* 70:465–487.

Schaal, L. A., and R. F. Dale. 1977. Time of observation temperature bias and climate change. *Journal of Applied Meteorology* 16:215–222.

Schell, J. 1989. Our fragile Earth. *Discover* 10:45–48.

Schneider, S. 1996. Don't bet all environmental changes will be beneficial. *American Physical Society News* 5:5.

Soon, W-H., and S. L. Baliunas. 2003. Proxy climate and environmental changes of the past 1,000 years. *Climate Research* 23:89–110.

Soon, W-H. and S.H. Yaskell. 2003. A year without summer. *Mercury* May–June, 13–22.

Soon, W-H. et al. 2003a. Comment on "On past temperatures and anomalous late-20th century warmth." *EOS* 84:473–476.

Soon, W-H. et al. 2003b. Reconstructing climatic and environmental changes of the past 1,000 years: A reappraisal. *Energy and Environment* 14:233–296.

Soon, W-H. et al. 2004. Estimation and representation of long-term (>40 year) trends of Northern-hemisphere-gridded surface temperature: A note of caution. *Geophysical Research Letters* 31: DOI:10.1029/2003GRL019141.

U.S. House Committee on Energy and Commerce. 2006. Ad hoc committee report on the "Hockey Stick" global climate reconstruction. http://www.energycommerce.house.gov/108/home/07142006_Wegman_Report.pdf.

Volney, C. F. 1804. *The view of the climate and soil of the United States of America.* London: C. Mercier and Co.

von Storch, H. 2005. Hockey sticks, the tragedy of the commons and sustainability of climate science. Abstracts of Center for Ocean-Land-Atmosphere Studies, Calverton, MD, Oct. 20.

von Storch, H. et al. 2004. Reconstructing past climate from noisy data. *Science* 306:679–682.

Willmott, C. J. et al. 1992. Influence of spatially variable instrument networks on climate averages. *Geophysical Research Letters* 18:2249–2251.

19

Rising Atmospheric Carbon Dioxide and Plant Biology: The Overlooked Paradigm

Lewis H. Ziska

Introduction

As human populations continue to expand, concurrent increases in energy and food will be required. Consequently, fossil-fuel burning and deforestation will continue to be human-derived sources of atmospheric carbon dioxide (CO_2). The current annual rate of CO_2 increase (~0.5%) from these sources is expected to continue with concentrations exceeding 600 parts per million (ppm) by the end of the current century (Schimel, et al. 1996).

Because carbon dioxide absorbs heat leaving the earth's atmosphere, there is widespread agreement that increasing CO_2 will result in increasing global temperatures. The extent to which temperatures increase, and the potential biological consequences—from sea-level rise, to the spread of malaria—have been discussed and debated extensively in both the scientific and popular literature.

Unfortunately, given the focus on global warming, it is seldom acknowledged that no matter what the end effect of rising temperatures, the ongoing increase in atmospheric carbon dioxide, *of and by itself*, has affected, and will continue to affect, all life on the planet.

The fact that carbon dioxide will continue to impact life directly is not derived from new science. Plants supply food, energy, and carbon to all living things. Life for us, and *for every ecosystem on the planet*, depends on the ability of plants to generate complex carbohydrates and chemical energy from just four basic resources: sunlight, nutrients (e.g., nitrogen, phosphorous), water and . . . carbon dioxide.

Imagine that, since 1960, the amount of sunlight reaching the earth had increased by 20%. Would plant biology be affected? Of course. Light is one of the four resources required by plants, and changing the amount of sunlight selects those plant species that respond positively or negatively to that resource (e.g., sun- or shade-adapted plants). Similar differential responses among plant species would be evident if water or nutrient availability had increased 20% over this same period.

Since 1960, the amount of carbon dioxide in the atmosphere has risen from 315 to 378 ppm, an increase of approximately 20%. To put this in perspective, plants evolved at a time of high atmospheric carbon dioxide (4-5 times present values), but concentrations appear to have declined to relatively low values during the last 25–30 million years (Bowes 1996). The values have been low enough, long enough, that evolution has selected for a small percentage of plants, principally tropical grasses that have maximum photosynthetic rates even at the current low CO_2 concentrations. However, these grasses (termed "C_4" plants) only comprise about 3–4% of all known plant species, the bulk (95%) of the 250,000+ plant species (termed "C_3" plants), lack optimal levels of carbon dioxide. For these plants, the recent rise and projected increase in atmospheric carbon dioxide represents an upsurge of an essential resource. There are hundreds of studies showing that both recent and projected increases in atmospheric carbon dioxide can significantly stimulate growth, development, and reproduction in a wide variety of C_3 plants (see Kimball 1983; Kimball, et al. 1993; Poorter and Navas 2003 for reviews examining the response to future CO_2 concentrations, and Sage 1995 for a review of the response to recent CO_2 increases).

So, a basic question remains: if plants are necessary for life to exist, and rising carbon dioxide is affecting how 95% of them grow and function, why aren't we discussing the global impact of rising

CO_2 on plant biology/ecology, in addition to the greenhouse effect?

CO_2 and Agricultural Systems

One obvious reason may be related to the "green is good" concept. In other words, given that carbon dioxide makes plants grow more, and plants are beneficial, isn't this is a constructive result? Why are we complaining? The impact of rising carbon dioxide on plants, particularly agronomic crops, has even been lauded by some as "a wonderful and unexpected benefit from the industrial revolution" (Robinson and Robinson 1997).

Are there really benefits to be gained in agriculture from rising carbon dioxide? Let us consider for a moment rice and wheat, two global cereals whose production supplies the calories for almost four billion of the world's six billion people. A number of studies have, in fact, demonstrated that both rice and wheat can show a positive response to increasing atmospheric carbon dioxide (Mandersheid and Weigel 1997; Horie, et al. 2000). Indeed, variation among different lines of rice and wheat in response to carbon dioxide is such that we could begin a large-scale effort to select for the most CO_2 sensitive cultivars (e.g. Ziska, et al. 1996) (Figure 19.1).

Such an effort holds significant promise, as we could, potentially, increase overall yields of these cereals at a time when the earth's population is rapidly expanding. Of course, there are challenges in such an approach; in particular, there is ubiquitous evidence indicating a decline in protein concentration for cereals as CO_2 increases (Jablonski, et al. 2002; Ziska, et al. 2004b) (Figure 19.2). Alternatively, sufficient variation may exist to make it possible to select for increased yields without sacrificing grain quality (Hall and Ziska 2000).

However, before we give in to the temptation of viewing an increase in a needed resource as being uniformly beneficial, it is important to remember that species respond differently to the same resource. Here is a simple example: in the 1950s when fertilizer (e.g., nutrients, a resource) was cheap, adding supraoptimal amounts of nitrogen was tested as a means to reduce weed competition with crops, the logic being that if there was a surplus amount of nitrogen, then competition for that resource would be reduced. Instead, scientists found that competition from weeds *increased* and crop yields were

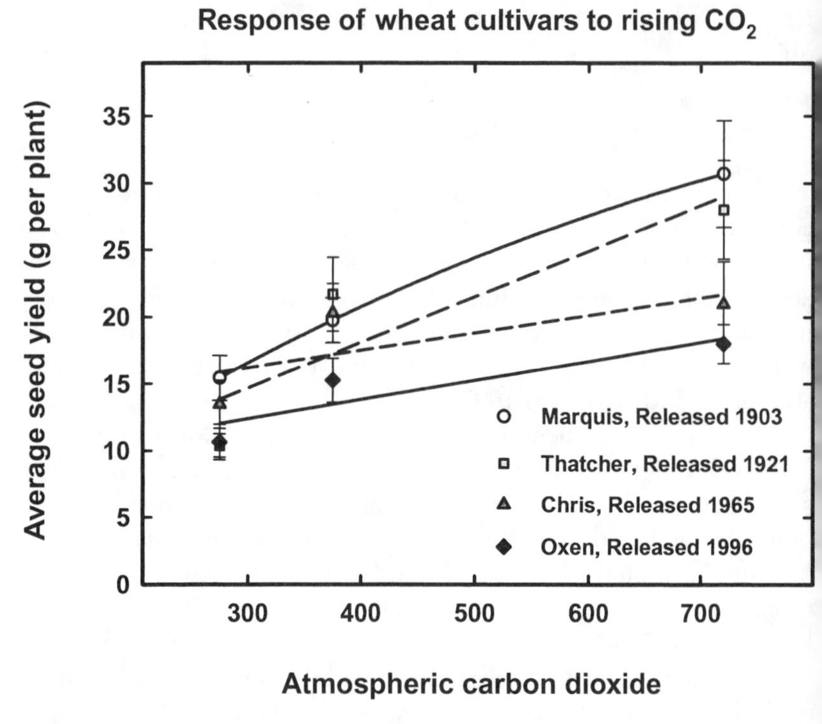

Figure 19.1 Average seed yield ± standard error of four different lines of Spring wheat in response to carbon dioxide concentrations at the beginning of the 20th century (ca. 300 ppm), the current level of carbon dioxide (ca. 400 ppm) and that projected for the end of the 21st century (ca. 720 ppm). These data suggest that sufficient variability exists among wheat lines to begin selecting for increasing seed yield in response to rising carbon dioxide.

reduced further (e.g., Vengris, et al. 1955). Why? Because weeds were able to utilize the additional resource (N) much more efficiently and less was available to the crop.

So, if we change a resource, we not only change the growth of an individual plant species, we differentially affect the growth of *all* the plant species within that community. Agriculture, in its simplest sense, represents a managed community that consists of the crop (desired plant species), and weeds (undesired plant species). Weeds, since the inception of agriculture, have limited the ability of human society to maximize crop productivity and maintain food security. In rice alone, the direct loss in production as a result of weed competition is estimated at approximately 20%, with losses climbing to 100% if weeds are not controlled (IRRI 2002).

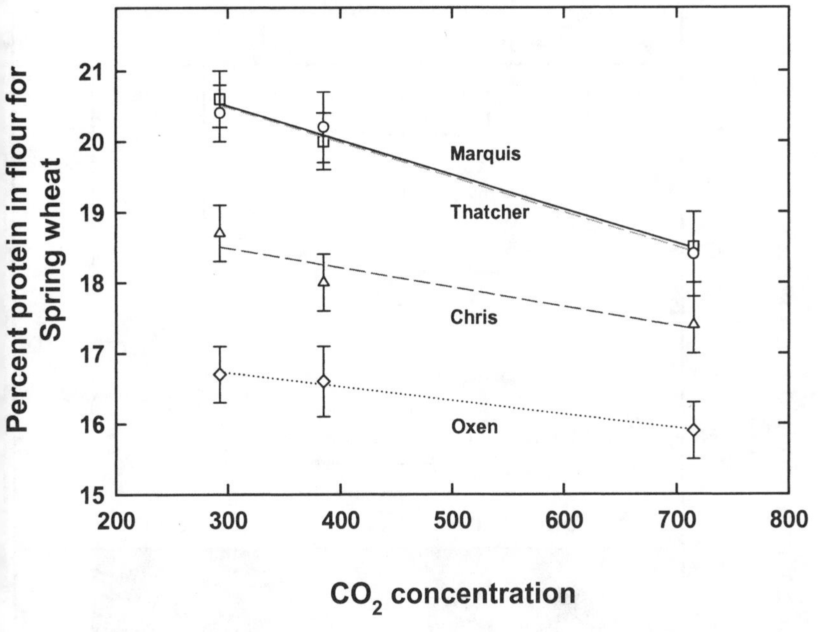

Figure 19.2 Percent protein in flour as a function of wheat variety and rising carbon dioxide concentration. Wheat varieties and carbon dioxide treatments are as described in figure 19.1. The overall trend for wheat is a decline in percent protein with increasing carbon dioxide.

Unfortunately, a number of studies indicate that among plant species in an agricultural system, weeds, rather than crops, are likely to show the strongest relative response to rising carbon dioxide (Ziska 2000; Ziska 2003b). That is, even though individual plants of rice or wheat can respond to carbon dioxide, the greater response of weedy species to CO_2 may result in increased competition and exacerbated losses in crop production (Figure 19.3). This is analogous to the supraoptimal nitrogen experiments where weeds responded more to excess nitrogen than the crop. Assuming that the studies conducted so far represent general trends, then rising carbon dioxide could actually reduce yield within agricultural systems on a global scale.

The basis for the greater response among weedy species to a resource increase is not entirely clear. In some instances, whether a crop or a weed is a C_3 or C_4 plant will determine its relative response to carbon dioxide and its competitive abilities (see Table 19.1, Ziska and Runion 2006). Many of the worst weeds for a given crop, however, are wild (uncultivated) plants of the same genus or species (e.g., rice and wild rice, oat and wild oat, sorghum and shattercane) and

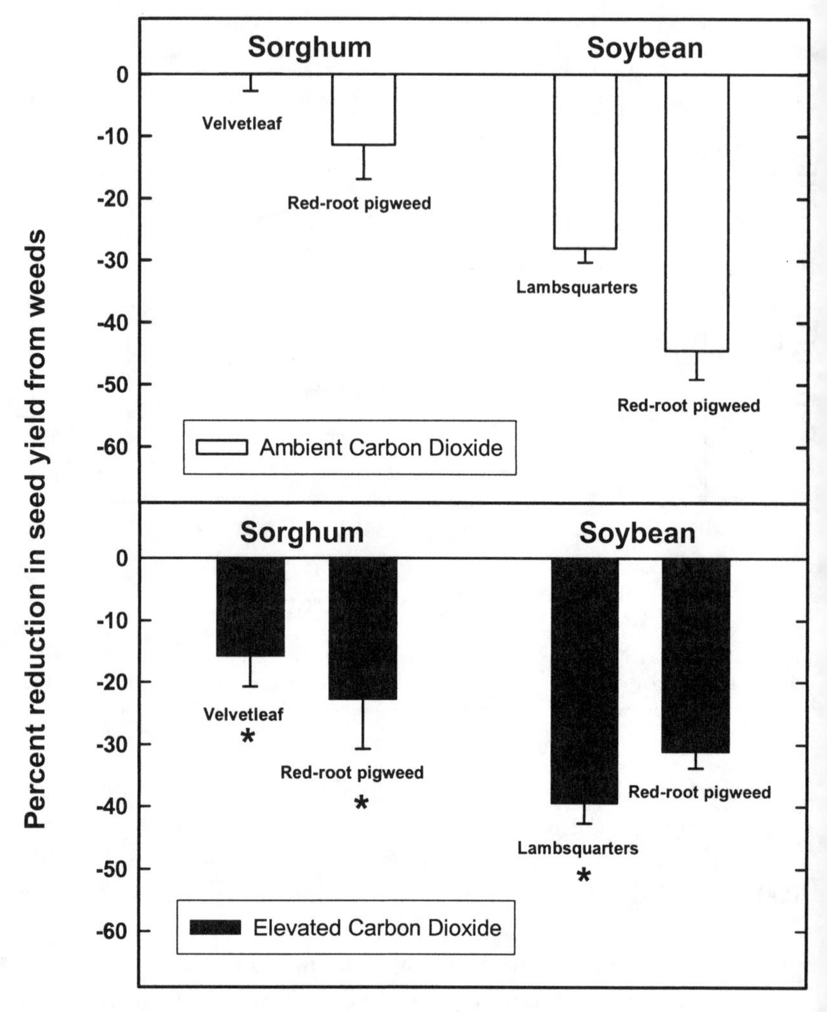

Figure 19.3 Percent reduction in seed yield for sorghum and soybean (a C4 and C3 crop respectively) as a function of competition from a C3 (velvetleaf, lambsquarters) and C4 weed (red-root pigweed) at ambient carbon dioxide and at 200 ppm above ambient. Weed spacing was two plants per meter of crop row in all cases. Increasing carbon dioxide resulted in a greater loss in crop seed yield from weedy competition (indicated by the asterisk) in all cases, except red-root pigweed in soybean. See Ziska 2000 and Ziska 2003a for additional details.

are likely to have the same photosynthetic pathway. Alternatively, it has been suggested (Treharne 1989) that the greater range of responses observed for weeds with increasing atmospheric CO_2 is related to their greater genetic diversity relative to crops; in other words, the greater the gene pool, the more likely it is for a species to respond to a resource change. In the ongoing struggle to maintain ever-increasing yields to feed a hungry world, we are relying on a narrow genetic base and increasing crop uniformity. While this has been a successful strategy in the past, we may have—inadvertently—conferred a disadvantage on current crop lines as they face increasing competition from their wild relatives for additional carbon dioxide. Whether we can recognize our strategic shortcomings and change tactics—i.e., select for genetically diverse, CO_2-responsive crop lines that can increase agricultural productivity—remains to be seen.

CO_2 and Native Systems

Unlike agricultural systems, most plant communities are complex and consist of many different plant species. Here again, it is tempting to think that rising carbon dioxide will result in increased growth of such communities. And so it may (see Rasse, et al. 2005 for marsh systems; DeLucia, et al. 1999 for loblolly pine), but, as with agricultural systems, the stimulation of growth will almost certainly be uneven (see Poorter and Navas 2003) with resulting "winners" and "losers" among plant species. Needless to say, given the complexity of such systems (e.g., forests, wetlands, prairie, desert, etc.), understanding the full implications of an increase in carbon dioxide will be difficult. Nevertheless, we can begin assessing some of the potential roles rising carbon dioxide might play by examining a simple example of invasive plant species and ecosystem function.

Invasive plants (i.e., exotic or alien species) are those plant species, nonnative to a given geographical area, whose geographic introduction within a community results in extensive economic or environmental damage. Millions of acres of productive rangelands, forests, and riparian areas have been overrun by such invaders, with a subsequent loss of native flora. E. O. Wilson, the noted ecologist, has observed that, "On a global basis, the two great destroyers of biodiversity are, first, habitat destruction and, second, invasion by exotic species" (1999). It has been estimated that more than 200 million acres of natural habitats (primarily in the western United States) have already been lost to invasive, noxious weeds, with an ongoing loss of 3000 acres a day (Westbrooks 1998). The invasive plant species that are most harmful to native biodiversity are those that significantly change ecosystem processes, to the detriment of native species.

One example of environmental damage related to invasive plants is the frequency and spread of natural or anthropogenic fires within native communities. For much of the western United States, cheatgrass (*Bromus tectorum*) originally introduced from central Asia, grows quickly in dry environments, colonizing open spaces between perennial, native shrubs with a fine flammable material that increases the frequency of fire events (Billings 1990, 1994). As fire events increase, native species diminish, while cheatgrass, a fire-adapted species, becomes dominant within the community. At present, cheatgrass dominates almost 17 million acres, or 17% of federally-owned lands in the western United States, with an additional 62 million acres at high risk (Jayne Belnap, US Geological Survey, personal communication). A study of three cheatgrass populations collected from different elevations within the Sierra Nevada range revealed that even small, recent changes in atmospheric carbon dioxide (approximately 50 ppm) could increase growth rate and combustibility of cheatgrass, while reducing digestibility (Ziska, et al. 2005b), all factors linked to increasing the amount of above-ground biomass, with potential effects on fire frequency and species diversity within native plant communities.

The cheatgrass observations are consistent with other data on carbon dioxide sensitivity from a range of invasive plants, suggesting that, on average, invasive species may show a stronger response to both recent and projected changes in atmospheric carbon dioxide than other plant species (Ziska and George 2004). It can be argued that such a strong response of invasive plants will be dependent on other resources, such as water and nutrients, and may not mimic experimental evaluations in situ. This is a fair criticism. Yet, it is also worthwhile to note that for invasives in managed, agricultural systems, water and nutrients may be optimal. Even in native systems such as those for cheatgrass, optimal water and nutrients may be periodically available depending on fluctuations in weather (Young, et al. 1987).

Still, actual field-based data from plant communities remains the best proof of whether rising carbon dioxide is altering the success of invasive plants (and subsequent ecosystem function). Unfortunately, field data are rare, and only a handful of studies have addressed this question with respect to projected carbon dioxide increases. A comparison of the impact of increasing CO_2 concentration on an invasive, noxious weed, yellow star thistle (*Centaurea solstitialis*), demonstrated a significant increase in biomass in monoculture, but a nonsignificant impact when yellow star thistle was grown within a grassland community (Dukes 2002), suggesting that rising carbon dioxide did not stimulate the growth of this plant species preferentially. In contrast, work with the invasive honey mesquite (*Prosopis glandulosa*) and the native species little bluestem (*Schizachyrium scoparium*) suggests that the woody invader, honey mesquite, *is* preferentially stimulated by CO_2 concentration (Polley, et al. 1994). Research on Japanese honeysuckle in a forest understory also demonstrated a strong CO_2 concentration–linked growth response and subsequent increase in percent cover (Belote, et al. 2003). Experiments with a woody invasive species in Switzerland (*Prunus laurocerasus*) showed a stronger CO_2 concentration–response relative to native trees (Hattenschwiler and Korner 2003), also suggesting preferential growth of a nonnative species. Finally, elevated CO_2 concentration increased the productivity and invasive success of a noxious invasive rangeland weed associated with fire outbreaks (*Bromus madritensis*, spp. rubens) in an arid ecosystem (Smith, et al. 2000). Overall, four of the five seminal studies suggest that rising levels of CO_2 can preferentially increase the growth of invasive plant species within a plant community. As rising CO_2 preferentially selects for a few species within a plant community, the implications with respect to basic ecological function, particularly species diversity, are troubling.

CO_2 and Human Systems

So far we have provided evidence that the rise in carbon dioxide per se, has, or will have, significant impacts on both managed (e.g., agricultural weeds) and unmanaged (e.g., invasive species) plant communities. Clearly, these impacts will have indirect effects on human society by altering the goods and services provided by these systems. Yet it is likely that there will be direct effects as well.

As with ecosystem function, the impact of rising CO_2 on human systems has not been completely elucidated. However, we are beginning to understand how rising CO_2 could interact with plant biology to affect one aspect—public health. At first glance, such an effect seems implausible. After all, plants do not carry disease. Yet, there are a number of ways in which plants directly affect human health, including allergenic reactions, contact dermatitis, and pharmacology/toxicology. All of these means, in turn, are likely to be affected by the rise in atmospheric carbon dioxide.

One of the most common plant-induced health effects is related to allergies, which are experienced by approximately 30 million people in the United States (Gergen, et al. 1987). Symptoms include sneezing, inflammation of nose and eye membranes, and wheezing. Complications such as nasal polyps or secondary infections of the ears, nose, and throat may also be common. Severe complications, such as asthma, permanent bronchial obstructions, and damage to the lungs and heart can occur in extreme cases.

Although there are over four dozen plant species that produce airborne allergens, common ragweed (*Ambrosia artemisiifolia*) causes more problems than all other plants combined (Wodehouse 1971). Initial indoor studies examining the response of ragweed to recent and projected changes in carbon dioxide demonstrated an increase in both ragweed growth and pollen production, with qualitative increases in ambient a 1, the principle protein that triggers allergic reactions (Ziska and Caulfield 2000; Wayne, et al. 2002; Singer, et al. 2005). Additional outdoor experiments that exploited an urban-rural transect that differed in carbon dioxide concentration, also showed the sensitivity of ragweed pollen production to CO_2 in situ (Ziska, et al. 2003). Overall, these data indicate a probable link between rising CO_2 (both at the local urban level, and that projected globally) and ragweed pollen production, with subsequent effects on allergic rhinitis.

Over a hundred different plant species are associated with contact dermatitis, an immune-mediated skin inflammation. Chemical irritants can be present on all plant parts, including leaves, flowers, and roots, or can appear on the plant surface when injury occurs. One well-known chemical is urushiol, a mixture of catechol derivatives. This is the compound that induces contact dermatitis in the poison ivy group (*Toxicodendron*/*Rhus* spp.). Currently, sensitivity to urushiol occurs in about two of every three people, and amounts as small as one nanogram (i.e., one billionth of a gram) are sufficient to induce a rash. Over two million people in the United States suffer annually from contact with members of the poison ivy group (i.e., poison ivy, oak, or sumac). The amount and concentration of these chemicals varies with a range of factors including maturity, weather, soil, and ecotype. However, recent research from the Duke Forest Free-Air CO_2 Enrichment facility also indicated that poison ivy growth and urushiol toxicity is highly sensitive to rising CO_2 levels (Mohan, et al. 2006). Overall, these data suggest a probable link between rising carbon dioxide levels and increased contact dermatitis.

Plants display a wide range of chemical diversity. Many of these chemicals are historically acknowledged as having pharmaceutical value (Table 19.1). Even in developed countries, where synthetic drugs dominate, 25% of all prescriptions dispensed from community pharmacies from 1959 through 1980 contained plant extracts or active principles prepared from higher plants (e.g., codeine, Farnsworth, et al. 1985). For developing countries, however, the World Health Organization (WHO) reported that more than 3.5 billion people rely on plants as components of their primary health care (WHO 2002). Furthermore, for both developed and developing countries, there are a number of economically important pharmaceuticals derived solely from plants (e.g., tobacco), whose economic value is considerable (see Table 2 in Raskin, et al. 2002).

Table 19.1 Plant-Derived Pharmaceutical Drugs and Their Clinical Usage Although many of these drugs are synthesized in developing countries, the World Health Organization estimates that as many as 3.5 billion people still rely on botanical sources for medicines (WHO 2002). Recent work on atropine and scopolamine indicates that increasing carbon dioxide and/or temperature will alter the concentration and or production of these plant-derived compounds (Ziska, et al. 2005b).

Drug	Action/Clinical Use	Species
Acetyldigoxin	Cardiotonic	*Digitalis lanata*
Allyl isothiocyanate	Rubefacient	*Brassica nigra*
Atropine	Anticholinergic	*Atropa belladonna*
Berberine	Bacillary dysentery	*Berberis vulgaris*
Codeine	Analgesic, antitussive	*Papaver somniferum*
Danthron	Laxative	*Cassia spp.*
L-Dopa	Anti-Parkinson's	*Mucuna spp.*
Digitoxin	Cardiotonic	*Digitalis purpurea*
Ephedrine	Antihistamine	*Ephedra sinica*
Galanthamine	Cholinesterase inhibitor	*Lycoris squamigera*
Kawain	Tranquilizer	*Piper methysticum*
Lapachol	Anticancer, antitumor	*Tabebuia spp.*
Ouabain	Cardiotonic	*Strophantus gratus*
Quinine	Antimalarial	*Cinchona ledgeriana*
Salicin	Analgesic	*Salix alba*
Taxol	Antitumor	*Podophyllum peltatum*
Vasicine	Cerebral stimulant	*Vinca minor*
Vincristine	Antileukemic agent	*Catharanthus roseus*

There are an increasing number of studies that are beginning to address how rising carbon dioxide affects the production or concentration of these pharmaceuticals. For example, growth of wooly foxglove (*Digitalis lanata* Ehrh.) and production of digoxin (a cardiotonic used in heart surgery) were increased at 1,000 ppm CO_2 relative to ambient conditions (Stuhlfauth and Fock 1990). Similarly, production and concentration of atropine and scopolamine (strong anticholinergics—chemicals that block the transmission of nerve impulses) was stimulated with both recent and projected carbon dioxide increases (Ziska, et al. 2005a).

The effect of CO_2 on such chemicals, however, is a two-edged sword. Given the subtle distinctions that exist between toxicity and pharmacology, CO_2-induced changes in their production or ratio will almost certainly influence their efficacy and use in human systems, particularly in developing countries (Uzun, et al. 2004). In addition, there are chemicals that are known poisons, with no pharmaceutical benefit. Poison hemlock (*Conium maculatum*), oleander (*Nerium aleander*), and castor bean (*Ricinus communis*) are so poisonous that tiny amounts can be fatal if ingested (e.g., ricin in castor beans has a greater potency than cyanide). Unfortunately, the impact of carbon dioxide on the concentration or production of such poisons is almost completely unknown. This is unfortunate, given that in 2001 alone, 73,000 cases of accidental plant ingestion were reported for children under the age of six in the United States (Dr. Rose Anne Soloway, American Association of Poison Control Centers, personal communication).

Rising CO_2 and Human Control Efforts

But even if rising carbon dioxide stimulates the growth of undesirable plants, won't we still be able to limit where and when such species grow? Can't we control the establishment and success of unwanted plant species either by mechanical, chemical, or biological means? Wouldn't this limit or negate any adverse effects (direct or indirect) associated with rising carbon dioxide?

Control of undesirable plant species is a difficult task. One only need consider the spread of an invasive weed like kudzu (*Pueraria lobata*) in the southeastern United States to appreciate the fact that no matter the amount of human effort, kudzu is here to stay. In fact, human control of selected species within an ecosystem requires a tremendous amount of effort and financial commitment, and is often unsuccessful.

But what about highly managed systems, such as agriculture? For the United States and many developed countries, chemical methodologies allow for cheap, effective control in agronomic production. Actually, a single herbicide, glyphosate (commercially sold as "Round-Up"), is so effective in controlling weeds that more than three-quarters of the US soybean crop, and over a third of the US corn crop have been genetically modified to be glyphosate resistant (e.g., Gaskell et al. 1999).

So is the solution to simply spray to control any undesirable plant pests? Unfortunately, there is an increasing number of studies (Ziska, et al. 1999; Ziska and Teasdale 2000) that demonstrate a decline in pesticide efficacy with rising CO_2 levels (Figure 19.4). The basis for the observed decline in efficacy is unclear. In theory, rising CO_2 levels could hamper absorption of pesticides into leaves by reducing the number or aperture of stomata (pores in the leaf that control exchange of gases and liquids) or by altering leaf thickness or size. In addition, CO_2-induced changes in transpiration could limit uptake of soil-applied pesticides. For weed control, timing of application may need to be adjusted if elevated CO_2 decreases the time the weed spends in the seedling stage (i.e., the time of greatest chemical susceptibility). Overall, it is likely that weeds could still be controlled chemically, either through additional sprayings, or increased herbicide concentrations, but this would almost certainly alter the environmental and economic costs of pesticide usage.

What about other means of control? In the developing world, mechanical control is still the predominant means to prevent weed-induced crop losses. Tillage, a common form of weed control, cuts and discs roots. But one response to rising CO_2 levels is an increase in below-ground root growth relative to above-ground shoot growth (Ziska 2003). For some invasive weeds, asexual propagation from small root segments is commonly observed. For Canada thistle, an invasive North American weed, rising CO_2 can double root growth relative to shoot growth in the field (Ziska, et al. 2004). As a consequence, increasing tillage as a control measure would lead to additional plant propagation for this species in a higher CO_2 environment.

Biological control of pest plants by natural or manipulated means is also likely to be affected by increasing atmospheric carbon dioxide

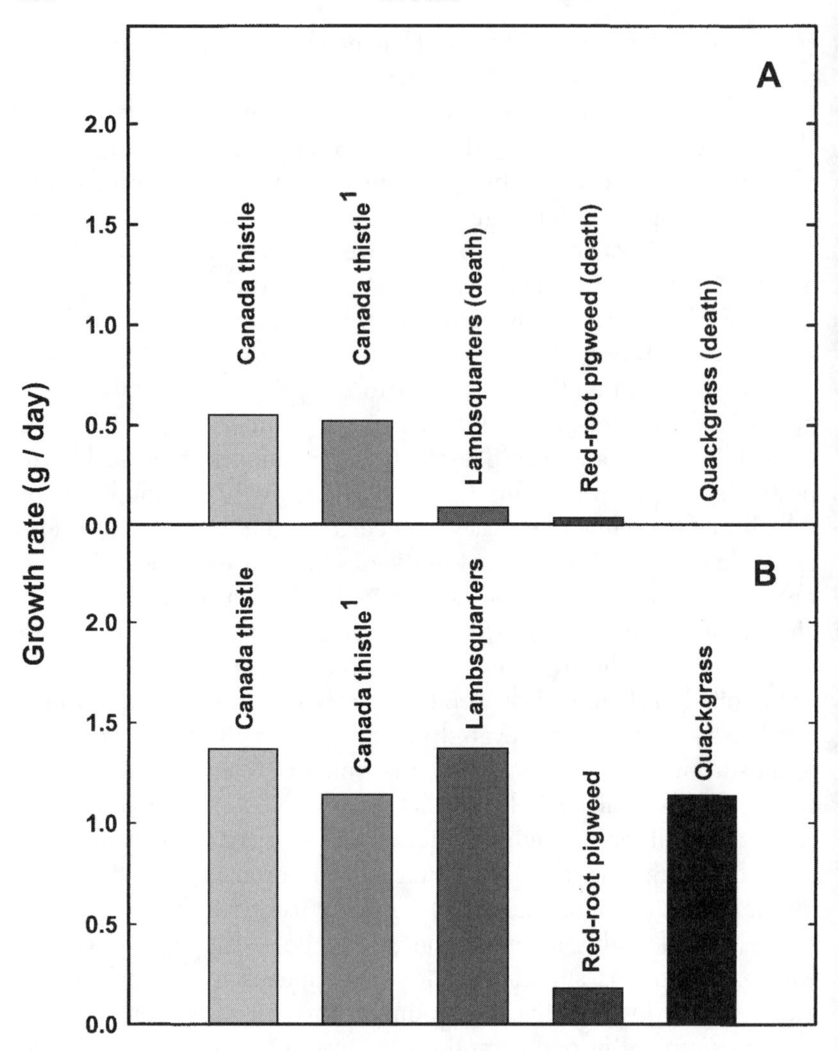

Figure 19.4 Change in growth rate (gram dry matter per day) for agronomic and invasive weeds when sprayed at recommended rates of herbicide at either: (A) current CO_2 or (B) future CO_2 levels (600-800 ppm). All growth rates less than 0.1 resulted in plant death. Herbicide was glyphosate in all cases except Canada thistle[1], which was sprayed with glufosinate. Overall, these data indicate a greater resistance to chemical control in weedy species as a function of rising atmospheric carbon dioxide.

(Norris 1982; Froud-Williams 1996). Since rising carbon dioxide can affect the development, growth, and reproduction of any given plant, such changes would alter any pest-plant synchrony with subsequent changes in control. In addition, rising CO_2 levels are likely to reduce the concentration of leaf nitrogen (Jablonski, et al. 2002). If insects are used as a biocontrol agent, this is likely to affect feeding patterns and the degree of induced damage. Overall, however, direct experiments to determine CO_2-induced changes in biocontrol of unwanted species have not been explicitly conducted.

CO_2 and Plant Biology, Revisited

Although the title of this book is *Controversies in science and technology*, there is no question that rising carbon dioxide levels will differentially stimulate the growth and function of plant species on a global basis, thereby affecting the flow of energy and carbon through ecosystems. Indeed, it seems fair to anticipate that, as carbon dioxide increases, ecosystem composition itself will change (e.g., cheatgrass and fires). In any event, increasing atmospheric carbon dioxide per se will have a number of consequences for both managed and unmanaged plant communities, and, hence, for all living things. It is imperative then, that we begin, in earnest, to assess the positive and negative aspects of these consequences in regard to both the natural environment and human society. It is regrettable that in the debate regarding rising CO_2 levels and global warming, that the direct impact of increased carbon dioxide on plant biology, and the role of plants in sustaining life, remains underappreciated by all sides.

Ultimately, one of the fundamental challenges we face in the 21st century is the unprecedented level of human-induced change. In addition to the rapid rise in atmospheric carbon dioxide, humans are significantly altering rates of nitrogen deposition (Wedin and Tilman 1996), the extent of tropospheric ozone (Krupa and Manning 1988), and land-use patterns (Pielke, et al. 2002). These induced changes will also, in time, transform life on a panoptic scale. It isn't just about warming anymore.

References

Belote, R. T. et al. 2003. Response of an understory plant community to elevated [CO_2] depends on differential responses of dominant invasive species and is mediated by soil water availability. *New Phytologist* 161:827–835.

Billings, W. D. 1990. Bromus tectorum, a biotic cause of ecosystem impoverishment in the Great Basin. In *The Earth in transition: Patterns and processes of biotic impoverishment*, ed. G. M. Woodwell, 310–322. Cambridge, UK: Cambridge University Press.

Billings, W. D. 1994. Ecological impacts of cheatgrass and resultant fire on ecosystems in the western Great Basin. In *Ecology and management of annual rangelands: US Forest Service general technical report*, ed. S. B. Monsen and S. G. Kitchen, 22–30. Ogden, UT: Intermountain Research Station.

Bowes, G. 1996. Photosynthetic responses to changing atmospheric carbon dioxide concentration. In *Photosynthesis and the environment*, ed. N. R. Baker, 387–407. Dordrecht, Netherlands: Kluwer Publishing.

DeLucia, E. H. et al. 1999. Net primary production of a forest ecosystem with experimental CO_2 enrichment. *Science* 284:1177–1179.

Dukes, J. S. 2002. Comparison of the effect of elevated CO_2 on an invasive species (*Centaurea solstitialis*) in monoculture and community settings. *Plant Ecology* 160:225–234.

Farnsworth, N. R. et al. 1985. Medicinal plants in therapy. *Bulletin of the WHO* 63:965–981.

Froud-Williams, R. J. 1996. Weeds and climate change: Implications for their ecology and control. *Aspects of Applied Biology* 45:187–196.

Gaskell, G. et al. 1999. Worlds apart? The reception of genetically modified foods in Europe and the U.S. *Science* 285:384–387.

Gergen, P. J. et al. 1987. The prevalence of allergic skin-test reactivity to eight common aeroallergens in the US population: Results from the second National Health and Nutrition Examination survey. *Journal of Allergy and Clinical Immunology* 80:669–679.

Hall, A. E., and L. H. Ziska. 2000. Crop breeding strategies for the 21st century. In *Climate change and global crop productivity*, ed. K. R. Reddy and H. F. Hodges, 407–419. New York: CAB International Press.

Hattenschwiler, S., and C. Korner. 2003. Does elevated CO_2 facilitate naturalization of the nonindigenous *Prunus laurocerasus* in Swiss temperate forests? *Functional Ecology* 17:778–785.

Horie, T. et al. 2000. Crop ecosystem responses to climate change: rice. In *Climate change and global crop productivity*, ed. K. R. Reddy and H. F. Hodges, 81–101. New York: CABI Publishing.

IRRI. 2002. Pests, diseases and weeds of rice. http://www.riceweb.org/research.

Jablonski, L. M. et al. 2002. Plant reproduction under elevated CO_2 conditions: A meta-analysis of reports on 79 crop and wild species. *New Phytologist* 156:9–26.

Kimball, B. A. 1983. Carbon dioxide and agricultural yield: An assemblage and analysis of 430 prior observations. *Agronomy Journal* 75:779–788.

Kimball, B. A. et al. 1993. Effects of increasing atmospheric CO_2 on vegetation. *Vegetation* 104/105:65–75.

Krupa, S.V., and W.J. Manning. 1988. Atmospheric ozone: Formation and effects on vegetation. *Environ. Pollut.* 50:101–137.

Mandersheid, R., and H. J. Weigel. 1997. Photosynthetic and growth responses of old and modern spring wheat cultivars to atmospheric CO_2 enrichment. *Agriculture, Ecosystems and the Environment* 64:65–71.

Mohan, J. E. et al. 2006. Biomass and toxicity responses of poison ivy (Toxicodendron radicans) to elevated atmospheric CO_2. *Proceedings of the National Academy of Sciences* 103:9086–9089.

Norris, R. F. 1982. Interactions between weeds and other pests in the agroecosystem. In *Biometeorology in integrated pest management*, ed. J. L. Hatfield and I. J. Thomason, 343–406. New York: Academic Press.

Pielke, R.A. et al. 2002. The influence of land-use change and landscape dynamics on the climate system: Relevance to climate-change policy beyond the radiative effect of greenhouse gases. *Philos. Trans. R. Soc. Lond.* 360:1705–1719.

Polley, H. W. et al. 1994. Increasing CO_2: Comparative responses of the C_4 grass *Schizachyrium* and grassland invader *Prosopis*. *Ecology* 75:976–988.

Poorter, H., and M-L. Navas. 2003. Plant growth and competition at elevated CO_2: On winners, losers and functional groups. *New Phytologist* 157:175–198.

Raskin, I. et al. 2002. Plants and human health in the twenty-first century. *TRENDS in Biotechnology* 20:522–531.

Rasse, D. P. et al. 2005. Seventeen years of elevated CO_2 exposure in a Chesapeake Bay Wetland: Sustained but contrasting responses of plant growth and CO_2 uptake. *Global Change Biology* 11:369–377.

Robinson, A., and Z. Robinson. 1997. Science has spoken, global warming is a myth. *The Wall Street Journal*, December 4.

Sage, R. F. 1995. Was low atmospheric CO_2 during the Pleistocene a limiting factor for the origin of agriculture? *Global Change Biology* 1:93–106.

Schimel, D. et al. 1996. Radiative forcing of climate change. In *Climate change 1995: The science of climate change*, ed. J. T. Houghton et al., 22–46. Cambridge, UK: Cambridge University Press.

Singer, B. D. et al. 2005. Increasing Amba1 content in common ragweed (*Ambrosia artemisiifolia*) pollen as a function of rising atmospheric CO_2 concentration. *Functional Plant Biology* 32:667–670.

Smith, S. D. et al. 2000. Elevated CO_2 increases productivity and invasive species success in an arid ecosystem. *Nature* 408:79–82.

Stuhlfauth, T., and H. P. Fock. 1990. Effect of whole season CO_2 enrichment on the cultivation of a medicinal plant, *Digitalis lanata*. *Journal of Agronomy and Crop Science* 164:168–173.

Treharne, K. 1989. The implications of the 'greenhouse effect' for fertilizers and agrochemicals. In *The greenhouse effect and UK agriculture*, ed. R. C. Bennet, 67–78. Ministry of Agriculture, Fisheries and Food, UK.

Uzun, E. et al. 2004. Traditional medicine in Sakarya province (Turkey) and antimicrobial activities of selected species. *Journal of Ethno-Pharmacology* 95:287–296.

Vengris, J. et al. 1955. Plant nutrient competition between weeds and corn. *Agronomy Journal* 47:213–216.

Wayne, P. et al. 2002. Production of allergenic pollen by ragweed (*Ambrosia artemisiifolia* L.) is increased in CO_2-enriched atmospheres. *Annals of Allergy, Asthma and Immunology* 80:669–679.

Wedin, D.A., and D. Tilman. 1996. Influence of nitrogen loading and species composition on the carbon balance of grasslands. *Science* 274:1720–1723.

Westbrooks, R. 1998. *Invasive plants, changing the landscape of America: Fact book*. Washington, DC: Federal Interagency Committee for the Management of Noxious and Exotic Weeds (FICMNEW).

Wilson, E. O. 1999. *The diversity of life*. New York: W. W. Norton Press.

Wodehouse, R. P. 1971. *Hayfever plants*, 2nd Edition. New York: Hafner Publishing.

World Health Organization. 2002. Traditional medicine: Growing needs and potential. *WHO Policy Perspectives on Medicines* 2:1–6.

Young, J. A. et al. 1987. Cheatgrass. *Rangelands* 9:266–270.

Ziska, L. H. 2000. The impact of elevated CO_2 on yield loss from a C_3 and C_4 weed in field-grown soybean. *Global Change Biology* 6:899–905.

Ziska, L. H. 2003a. Evaluation of the growth response of six invasive species to past, present, and future carbon dioxide concentrations. *Journal of Experimental Botany* 54:395–404.

Ziska, L. H. 2003b. Evaluation of yield loss in field sorghum from a C_3 and C_4 weed with increasing CO_2. *Weed Science* 51:914–918.

Ziska, L. H., and F. A. Caulfield. 2000. Rising carbon dioxide and pollen production of common ragweed, a known allergy-inducing species: Implications for public health. *Australian Journal of Plant Physiology* 27:893–898.

Ziska, L. H., and G. B. Runion. 2006. Rising atmospheric carbon dioxide and global climate change: Assessing the potential impact on agroecosystems by weeds, insects, and diseases. In *Agroecosystems in a changing climate,* ed. P. C. D. Newton et al., Boca Raton, FL: CRC Press, Taylor & Francis Group, pages 261–290.

Ziska, L. H., and J. R. Teasdale. 2000. Sustained growth and increased tolerance to glyphosate observed in a C_3 perennial weed, quackgrass (*Elytrigia repens*), grown at elevated carbon dioxide. *Australian Journal of Plant Physiology* 27:159–164.

Ziska, L. H., and K. George. 2004. Rising carbon dioxide and invasive, noxious plants: Potential threats and consequences. *World Resource Review* 16:427–447.

Ziska, L. H. et al. 1996. Intraspecific variation in the response of rice (*Oryza sativa* L.) to increased CO_2 and temperature: growth and yield response of 17 cultivars. *Journal of Experimental Botany* 47:1353–1359.

Ziska, L. H. et al. 1999. Future atmospheric carbon dioxide may increase tolerance to glyphosate. *Weed Science* 47:608–615.

Ziska, L. H. et al. 2003. Cities as harbingers of climate change: Common ragweed, urbanization and public health. *Journal of Allergy and Clinical Immunology* 111:290–295.

Ziska, L. H. et al. 2004a. Changes in biomass and root:shoot ratio of field-grown Canada thistle (*Cirsium arvense*), a noxious, invasive weed, with elevated CO_2: Implications for control with glyphosate. *Weed Science* 52:584–588.

Ziska, L. H. et al. 2004b. Quantitative and qualitative evaluation of selected wheat varieties released since 1903 to increasing atmospheric carbon dioxide: Can yield sensitivity to carbon dioxide be a factor in wheat performance? *Global Change Biology* 10:1810–1819.

Ziska, L. H. et al. 2005a. Alterations in the production and concentration of selected alkaloids as a function of rising atmospheric carbon dioxide and air temperature: Implications for ethnopharmacology. *Global Change Biology* 11:1798–1807.

Ziska, L. H et al. 2005b. The impact of recent increases in atmospheric CO_2 on biomass production and vegetative retention of Cheatgrass (Bromus tectorum): Implications for fire disturbance. *Global Change Biology* 11:1325–1332.

PART 5

Biology and Gender: Scientific Careers and Scientific Theories

20

Perception Bias, Social Inclusion, and Sexual Selection: Power Dynamics in Science and Nature

Patricia Adair Gowaty

Power striving is everywhere. Some have power; some wish for it. Women wish for jobs in science; women scientists with jobs in science wish for opportunity equity and pay equity. Sexist bias affects the likelihood that people think women scientists competent (Valian 1999; Wennerås and Wold 1997) and whether others read, understand, remember, and cite their work (Valian 1999). Sexism reaches into the hypotheses we cast and test (Allen, this volume), the experiments and observational protocols we construct, the conclusions we draw (Roughgarden, this volume) and the things we do to others in our professional roles as physicians (Matta, this volume; Dreger, this volume) and teachers (Dreger, this volume). Sexism at its most basic is about the power to exclude and include. And no one seems exempt from gender schemas (Valian 1999) that produce gendered perception biases.

The five articles in this section are about power dynamics in the science of sex and gender. Dreger, a medical humanist, describes the collusion of nature and nurture in the production of sex and gender. Allen, a plant pathologist, argues that sexist perceptions have molded and limited scientific hypotheses about sex determination. Matta, an historian of science, reviews the deadly history of medical hubris and its horrific aftermath, which resulted in mutilation of hermaphrodites and persons with ambiguous genitalia. Barres, a developmental neurobiologist, writes about the origins of the institutionalized sexism of science in the common perception bias that women have limited scientific competence compared to men, and Roughgarden, a mathematical ecologist, describes her effort to replace what she calls the "sexual-selection system" with what she calls "social selection."

My impression is that, taken together, these articles contribute to another conversation: one about social inclusion. This metapoint is explicit in Dreger, but obvious in all. The dialectics—the dynamic interactions between what individuals perceive about their sex and gender and what others perceive about them—affect opportunities to join and thrive in the institutions of science (Barres, this volume; Valian 1999). Perhaps even more disheartening, perception bias and empathy affect our very abilities to do science. It is this dialectic between our perceptions and others' perceptions that influences the power dynamics of sexism in society (what we do and what others do to us because of our sex and genders) and in science (what hypotheses we articulate and pursue, and how we test them, and interpret them). This dynamic creates opportunities for sexism to unfold in any question we might ask—scientific and otherwise—about sex, gender, and the social behavior of sexual organisms. When political activists detect sexism in society, their usual response is to argue for social inclusion: for example, the extension of the rights of straight men to women, gays, bisexuals, and transsexuals. When feminist scientists detect sexism in science, their usual response is to argue for inclusion of alternative, more inclusive, perspectives in hypotheses and methods.

"Doing Science"

Because most of what caught my attention in these articles was about "doing science," in this brief section I review some of my views about science to create the backdrop for the rest of my remarks.

I am convinced by Valian (1999) that few scientists are exempt from biased perceptions about the competence of women scientists compared to men and by Hrdy (1986) that empathy is a primary source of favored hypotheses about sex, gender, and social behavior. Both of these ideas make it clear that the eye of the beholder matters.

I wish to emphasize that some knowledge is closer to the truth than other "knowledge" and that I have never been convinced that women "do science," i.e., engage the methodology of science, differently from men (Schiebinger 1989; Schiebinger 2001). I am most interested in knowledge that comes from scientifically robust methods. For me, this means that the most important question that scientists must ask is how do we test (possibly biased) hypotheses? How we attempt to test our hypotheses should derive from accepted and emerging scientifically valid methods. While I think women and men engage the methodologies of science in the same ways, I often observe that women are interested in some hypotheses and conclusions more than men are and vice versa. The things we do with our knowledge are often political and sometimes very personal.

To be clear, in this chapter when I say "science" I mean a system of methods for eliminating and controlling for human deceit and self-deception (common sources of bias) that get in the way of discerning the truth. The things scientists do to get to the truth include testing alternative hypotheses with crucial predictions (Platt 1964) when we can, or stand-alone hypotheses when we cannot, testing the assumptions of hypotheses and their alternatives and their predictions using experiments or systematic observations controlled against perceived or apparent biases. One of the signs of a scientist "doing science" is apparent in our behavior when we confront hypotheses with data: do we abandon hypotheses when assumptions do not fit or predictions fail? I hope it is clear that the origins of hypotheses do not necessarily require the absence of bias, but that the testing of hypotheses absolutely requires controls against bias.

Scientists "doing science" are often contentious, e.g., fighting over priority or the robustness of observations for a given conclusion. Because contention in doing science is often the origin of new and better ways to do science (discovery of new statistics, new distributions, better controls, etc.), and the mother of new and better hypotheses, I think of science, even the most seemingly competitive parts of doing science, as a profoundly cooperative endeavor. In addition, even the most creative and original, who manage to see further than others, are standing on the shoulders of giants—a collaborative intellectual program.

Two of our traits of social inclusion in science, i.e., signs of our cooperation in "doing science" are honesty and giving credit where credit is due, which includes keeping track of and recognizing in our papers the history of previous works that inform and impinge on our insights, experiments, and syntheses. Making up data—the worst kind of lying in science—is, when discovered, lethal to individual careers. A failure to give credit for previous work breaks the contract among scholars and is, after all, a form of plagiarism, easier to discover than data fabrication in these days with so many search engines literally at our fingertips, but no less a defection on the cooperative science contract.

Perception Bias and Nepotism: Social Inclusion Traits in the Administration of Science

Take the case of former Harvard president Larry Summers, who in 2005 said that the failure of women to advance in scientific careers was due to differences in aptitude between women and men rather than to discrimination. Summers suffered opprobrium for his disingenuous remarks, which were seemingly uninformed and woefully ignorant, as Ben Barres's reprinted *Nature* commentary (this volume) discusses. Among the studies Barres discusses is the landmark one reported by Winnerås and Wold (1997), remarkable because it identified, not just the statistical trends showing women's lower achievement and status in science, but a *mechanism* of sexism in science. The Winnerås and Wold study used a matched pairs design and showed that women applicants had to publish approximately three or four more papers in *Science* or *Nature* or 20 more in journals with impacts of three or below than the male applicants to be acknowledged to have equivalent scientific status and productivity. The problem was not in the women applicants, but in the minds of their peer reviewers.

Perception bias against women is a consistent undervaluing of women and their work relative to men, and it is associated with a "societal assumption that women are innately less able than men" (Barres, citing Valian 1999)—what Valian calls a "gender schema." One reason Summers's remarks sickened me was that I experienced the "Larry Summers Hypothesis" as a defection on the cooperative contracts of honesty and citation in science.

Nepotism, i.e., "friendship bonuses," are the favorable judgments of friends or friends of friends that are also a biasing factor in peer review. In Winnerås and Wold's study, women postdoctoral applicants were able to make up for negative evaluations due to perception bias based on gender through nepotistic "friendship bonuses" if they were somehow associated with the reviewers (say, if their reviewer was a friend of their major advisor). To receive a score similar to her competency-matched pair male who had a friendship bonus, a female applicant without a friendship bonus needed about 44 papers in journals with an impact factor of three or 131 papers in journals with an impact factor of one. Nepotism is the result of recognition of relationship through an ol' boys club, perhaps, or a particular training laboratory. Nepotism is everywhere in the administrative institutions of science, but it should not be part of the cooperative acts of "doing science" as I defined science above. Nepotism is part of the affiliative networks that determine, through review and vote, who will be allowed to do science or be admitted to honoraries like the National Academy of Sciences, and although most of us say we are not biased, it seems we are (see citations in Barres, this volume; Valian 1999).

What the Winnerås and Wold study did was reject the hypothesis that competence equal to that of men applicants is adequate for admission of women applicants to the social institutions of science—at least at the postdoctoral level in Sweden. We do not yet know whether similar bias exists in the United States, United Kingdom, or in the rest of Europe because few or none of the funding agencies in those countries allow the level of open access to records as the Swedish did. Obviously, to be included in the club of science, the bar is higher for women (Valian 1999). It is impossible now to claim that the premier trait of social inclusion in the institutions of science is scientific competence.

What club do sexism and nepotism—two demonstrated mechanisms of social *exclusion* from the institutions of science—support?

I do not know whose interests sexism and nepotism serve. When I am in the mood for conspiracy theories, I think it is part of a global antiwoman initiative to which ol' boy misogynists consciously contribute. When I am calmer, I remember that the vast majority of men I know are feminists, in truth, as committed to the advancement of women as the contributors to this volume. Even when they lapse and forget to have gender-balanced committees and workshops, when they are reminded of their gender skew in attention, they readily correct, just as I do. In another mood, I have hypothesized, as have many others before me, that exclusion of women from positions of power, privilege and influence—jobs, status, and honoraries—are part of the forces that control and manipulate women's sexuality and reproduction for the benefit of men (Gowaty 2003a; Gowaty 2003b).

I suspect that most contention in reviews (both ad hoc and published) of science is selfishness: defensiveness against hypotheses that are alternatives to our own or hypotheses that make us nervous, that challenge our worldviews, data that fail to support our pet ideas or ways of doing things, or interpretations with which we do not agree. I think we have a responsibility in review to evaluate hypotheses, methods, data, and interpretations in as "disinterested" a way as possible. We must learn to control our evaluations against the possibilities of prejudices that arise from nepotism (an affiliative force that excludes the unrelated), sexism (an antagonistic force that excludes women), homophobia (an antagonistic force that excludes gays, lesbians, bisexuals, and transsexuals) and nationalism (an antagonistic force that excludes nonwestern or non-English–speaking scientists). I hope that antisexist, antihomophobic, antinepotistic, antinationalistic memes will soon be as much a part of the social administration of power and privilege in science as our memes for doing science. It is not entirely clear how to eliminate these prejudices from science administration, but Virginia Valian's (1999) "Remedies" chapter has some good ideas, which I recommend. In the meantime, I take personal action. In the hopes of remaining "indifferent" in reviews, I began signing all of my reviews from the very first ones I did in 1974, negative and positive. This small meme (a cultural ritual) makes me self-consciously own and own up to all I might say in a review. Sometimes I think it helps; more controlled studies like Winneras and Wold's might tell us if I am right. Signing my reviews, however, has sometimes made me "the bad guy," even when I was trying very hard to be helpful (Roughgarden, this volume).

Nepotism, sexism, homophobia, and nationalism affect the administrative institutions of science. Perception bias similarly affects what physicians and surgeons have done in the name of science.

Perception Bias and the Line of Demarcation: Ambiguous Genitalia, Hermaphrodism, and Homosexuality

Perception bias made it hard for physicians and surgeons "to draw the line of demarcation between physical and moral perversion" (Matta, this volume, quoting Lydston 1889) in the nineteenth and twentieth centuries. Christina Matta's paper (this volume) reviews an incredibly painful history of social exclusion and mutilation suffered by persons with ambiguous genitalia and hermaphrodism. These reports horrified me. I actually shuddered, because it is still true today that many physicians do procedures without benefit of randomized, replicated, and controlled experiments and observations to evaluate what is "wrong," who might be suffering, and what the effects of "procedures" might be. In the nineteenth and twentieth centuries, hermaphrodites and those with ambiguous genitalia were socially excluded and mutilated by physicians and surgeons who projected sex categories and moral judgments onto them, and assigned roles to them. Physicians responsible for this sorry history justified their actions in the absence of anything that I would call "data" and in relation to their own biased view of "morality." What struck me as I read was how little science guided the actions of these physicians and surgeons. One can hope that contemporary professional ethics require physicians and surgeons to base their decisions on random, replicated, and controlled studies that evaluate the effects of mutilation (a.k.a. "procedures") on those who exhibit ambiguous genitalia and hermaphrodism.

Social Inclusion, Developmental Gender-Bending, and Reproductive Capacity

The wonderfully witty Alice Dreger (this volume) made me giggle, because the complexities that exercise her tickle my heart and befuddle the pigeonholers who struggle so hard to categorize us all. I found her words and her approach deeply instructional; I think her words and her heart disarm the potentially controversial by illuminating what might be feared because it is misunderstood. For example, on the first day of class, she asks her students at the Northwestern University School of Medicine to answer two crucial and surprising questions: 1. Do you think you are a man or a woman? 2. What makes you think that? By focusing on how individuals answer these questions, usually in total ignorance of what their gametes say about their sex, what their chromosomes say about their sex, or even whether or not they have ovaries or testicles, she makes the point that our perceptions about our own sex and gender are developmentally constructed (an interplay of nature and nurture). Some of us exclaim, "Thank development that gender expression is so rich and variable!" Her student who responded, "I know I'm a guy, because my lover is gay" made Dreger's day and mine, too.

Dreger draws five sensible conclusions from her review of the many ways she knows that sex and gender are confounded and sometimes perhaps confused during development. I repeat them here because I think they are very important. 1. In day-to-day interactions (e.g., getting a driver's license or getting married), it is an individual's social presentation that matters. 2. Both intrinsic biological factors and the environment in which they are expressed contribute to "every individual's experience of sex and gender identity." 3. Looking at a person almost never provides enough information to predict what their karyotype (a description of the variation in their chromosomes) will show, what organs an individual has inside, or what gender identity that individual will end up with. These are complex developmental phenomena, meaningful to individuals even when they fall wide of simplistic lines of gender demarcation. 4. Because development results in variation, any attempt to neatly divide people into only two sex types is ultimately a social action, so far not a scientific one. 5. Knowledge about the nature of sex provides no formula for social justice, but it might help us begin to think about social justice in relation to gender.

It seems that we make a slew of assumptions when we declare our sexes and gender identities. Like Dreger, I think our declarations interesting (neither good nor bad). What might be most important is how the declarer feels. Does it feel right to them? My emphasis is on "feels right" to the individual who experiences this or that sex-gender identity. What I mean is that if a person has XY chromosomes (usual in men), but lacks hormone receptors for a masculinizing hormone, testosterone, that person will have the secondary sexual characteristics of a woman and often feel like a woman, and be comfortable about her feminine gender presentation, a real life example that Dreger reviews. Just because she happens to have an X and a Y chromosome, rather than the XX, which is more typical of women, is no reason she should cast herself, or anyone else should cast her, for life in a role she would never have thought about assuming before she was karyotyped.

Dreger's review is about proximate mechanisms of sex and gender. She writes about *how* we are what we are. She is concerned with the mechanisms, for example, the hormonal cascades that fuel sex and gender differences, and the modifications of cellular communications that interrupt typical developmental pathways. She is also concerned primarily with the proximate mechanisms of internal thoughtful justifications we have for what we are. Her perspective is an important complement to evolutionary studies of sex and gender.

The complexes of proximate mechanisms associated with sex and gender include sexual orientation, which Dreger and Matta mention briefly. I think about sexual orientation as what individuals *think* about, not who they have sex with. This is because some modern studies (Hamer, et al. 1993) that used Kinsey indicators to assign subjects to "gay" and "straight" showed that while individuals are easily binned into two categories based on indicators of which sex of individual they think about when they are thinking about having sex, subjects' behavior is much more variable. Gay men frequently admit to sex with women; lesbians often have sex with men: their behavior is bisexual. Thus, the behavior of self-reported homosexual men and women is ambiguous relative to their self-reports about sexual orientation. As far as I know, there are three categories of "which sex an individual thinks about when thinking about having sex": men, women, or men *and* women. Thinking about sexual orientation from the perspective of the five papers in this volume has made me wonder if there are more than the three sexual orientations (gay, straight, and bisexual) implied by Kinsey studies. Do some individuals have more specific orientations than "I think about men," "I think about women," or "I think about women and men"? Do some have "sexual orientations" toward particular gender expressions? How many categories of sexual orientation are there? It is very interesting that our discussions of gender diversity seem so narrowly focused on "gay," "straight," and "bisexual."

These questions reinforce to this evolutionary biologist that sexual orientation might be about conceptive and nonconceptive advantages, i.e., about reproduction *and* the fitness-enhancing advantages of within- and between-sex coalition-building for the achievement of common goals (de Waal 1997; de Waal 1995; Parish 1996; Wrangham 1993). Consider bonobos. Sexual "orientation" (in quotes, because no one can ask them who they think about when they think about having sex) and social inclusion in bonobos, our other closest relative (Horn 1979) is different from common chimpanzees and humans. Females are usually dominant in interactions over food and sex, and they use their dominance to control reproduction and social behavior (Parish 1994; Parish and De Waal 2000). Even though, like the other great apes, males are philopatric and females disperse into nonnatal groups, adult females (nonrelatives) bond with each other using same-sex sex as the proximate bonding mechanism (Manson, et al. 1997; Parish 1996). Female same-sex sexuality is so important that their clitorises are well forward of placement in humans or in common chimpanzees, an adaptation that facilitates their exuberant G-G (genito-genital) rubbing (Parish 1994). In bonobos, "communication sex" (Wrangham 1993) is associated with female "hypersexuality." Wrangham hypothesized that female-female coaliations mediated by their nonconceptive sex with one another, reduced the effects of male coercion of female reproductive decisions. Communication sex that appeases socially tense situations is more common in female bonobos than in males (Manson, et al. 1997; Parish 1996). This patterns suggests that it functions to build bonds between unrelated females, a hypothesis that has support in comparison to wild white-faced capuchin (*Cebus capucinus*), in which unrelated males, which disperse into their breeding groups, engage in more "communication sex." Frans de Waal, a frequent observer of great apes, including bonobos, hypothesized (de Waal 1997; de Waal 1995; de Waal 2003) that primate communication, including same-sex sex, which is not uncommon in other primates (Vasey 1995), is a form of negotiation that transcends the usual categories used for signals. He further suggested some of the objectives that bonobos and other great apes might be trying to achieve by these exchanges (Vervaecke, et al. 2000), including access to a breeding group, an idea that has fascinated me for a long time (Gowaty 1977). The point that considering bonobos makes is that same-sex sexual orientation might mediate the formation of same-sex coalitions to achieve fitness benefits, and that the benefits may be sexually selected or naturally selected.

To my mind, as a Darwinian evolutionary biologist, it is important to understand why some do and do not reproduce, why some have more offspring than others, and whether these common gender variants enhance reproductive success or survival, as a consideration of bonobos suggests they do. Because I am a biologist, I am also interested in the steps in development that result in all this variation and how developmental variation feeds evolution (West-Eberhard 2003). Variation is variation, neither good nor bad, but it is a major source of variation in fitness and evolution. What I want to know is how particular variants serve or do not serve individual survival and reproductive success, given the social and ecological environments individuals live within. I would welcome studies that take into account the development of individual social presentations and how they correlate with individuals' views of what sex and gender they are and with typical morphological, physiological, and behavioral syndromes that we associate with sex and gender categories, and sexual orientations. Such information might be useful in stopping mutilating surgeries of infants and others with genitalia that do not look like some observer "thinks they should" (Matta, this volume). There are many other questions one might ask. I would like to know how many sets of infant or adult genitalia one needs to see to have an expert opinion of what is "normal," and if "normal" varies with ecological and developmental circumstances, such as whether the mother was well nourished or not and whether this affected her resources-to-young allocations. Does this or that gender variant help individuals to survive or reproduce better than their neighbors? And, if so, how? And, are there ecological forcers that favor these variants? Never mind that secondary sexual characteristics vary or that genitals are "ambiguous"; the question that interests me is: "do the variants serve the fitness interests of their bearers?" Never mind that the secondary sexual characteristics or the genitals are "nonnormal"; the individual, not physicians or surgeons, should assign their sex and gender identities.

Put into the context of gender variation among nonhuman animals, these questions are not as odd as they might at first appear. Their answers might add to our understanding of variation in the way that detailed observation of spotted hyenas, *Crocuta crocuta* (East, et al. 1993; Frank and Glickman 1994; McFadden, et al. 2006; Muller and Wrangham 2002; Place and Glickman 2004; Szykman, et al. 2003; Van Horn, et al. 2004) has been doing for 25 years or, more recently, as have ruffs, *Philomachus pgnax*, shorebirds that breed in the high arctic (Jukema and Piersma 2006). In general, it is not uncommon for immature male birds to look like adult females, but it is remarkably uncommon in sexually dimorphic species for any adult males to look like females. Yet, with the ruff, some males are permanently outfitted in female plumage. Ruffs are highly sexually dimorphic (typical males are larger than females) and, until recently, renowned for having two distinctly male morphs in two distinct plumage types. The most elaborately plumaged males defend territories on a lek, to which females come to mate. Researchers call these ornamented males, independents. Another male morph, tolerated by the independents and called satellites, hangs around the peripheries of the independents' territories and copulates with females when the independent is distracted, perhaps when he is copulating with other females. The third male morph also copulates with females, accounting for some of the previous incorrect observations of "female-female" copulations in ruffs. They are intermediate in size between the other males and females. In the spring they develop femalelike feathers, but their testes are 2.5 times larger than the independent and satellite males. They are called *faeders*, a middle Dutch word meaning "father." Their discovery has increased the variation in known sexual phenotypes in this species, and researchers are busily looking for the likely genetic underpinnings of this third male morphology and documenting the behavioral variation of these female-plumaged males, who sometimes also copulate with independents (Susan McRae and David B. Lank, personal communication). To an unwary observer these copulations look like male-female copulations. Sexual presentation in ruffs is complicated and interesting, as is sexual presentation in hyenas (just read the titles of the paper cited above to get the flavor or, better yet, read some of them) and humans.

Some gender warriors (see Dreger, this volume for description) talk about getting rid of gender altogether. Dreger thinks it will never happen because most people feel "some kind of gender" that arises through the complexities of development and that individuals often enjoy their genders. Indeed. I wonder why anyone would seek such homogenization in the first place. Why are people uncomfortable with those who are different from them? What social-psychological complex in gender warriors does variation excite? If we were all the same, would it enforce social inclusiveness? I doubt it.

Empathy: Nonscientific Source of Hypotheses

Hypotheses sometimes seem to come out of the air. Hypotheses actually come out of us and very often reflect our empathetic responses to the behavior of nonhuman animals and our projections toward other people, a point Sarah Hrdy (1986) made 20 years ago. Empathy is an important source of testable hypotheses to explain many types of behavior. Scientists who study social behavior—similar to so many preachers, or recently, US Congressmen, who project their own sins on their constituents—empathetically project. It really is hard to "be in someone else's shoes," which made Judy Stamps (1997) argue that the most important reason to have individuals representative of all ages, both sexes, multiple genders, and ethnicities doing science was that such social inclusion would decrease the time to truth—especially when our empathetic projections result in competing hypotheses. And, once an idea is in the literature and "in the air," one no longer has to identify with the idea—even a formerly idiosyncratic idea—in order to confront the idea with data (i.e, to test it).

Thus, I think that worldviews seldom come out of science, contra what Judith Butler says. Roughgarden, paraphrasing Butler says: ". . . which evolutionary system of sex, gender, and sexuality ultimately prevails scientifically determines our worldview of nature itself" (this volume). I think this view of the directionality of influence from science to worldview, by and large, has the force of influence going in the wrong direction. Worldviews, particularly those about sex, gender, sexuality, and reproduction, reliably come out of lived experiences, our empathetic projections, and our desires. Science has the power to test worldviews, to confront them with appropriately controlled and replicated observations and experiments. Thus, science does indeed have the power to change what we reliably know, but it does not change what we wish for, what we project onto others, or what sorts of perception biases we carry (think of Larry Summers, for whom all the data in the world made no difference).

Empathetic projections, a type of "perception bias," need do no harm to doing science, especially if we take the Judy Stamps approach of social inclusion of diverse scientists. One can use one's social concerns to shape an hypothesis. There is no requirement to control against bias in an hypothesis, but it is absolutely required to control against known, suspected, or possible bias in tests of any hypothesis. Our responsibilities when communicating the idea as an hypothesis are to identify as many underlying assumptions and predictions as clearly as possible, and to test or urge tests of these hypotheses using controls against perceived or apparent bias and alternative hypotheses. It is useful to lead the way when presenting new hypotheses to suggest how they might be tested and what important controls against our salient social assumptions are necessary.

The fact that scientists who comfortably reflect status quo political positions are often ignorant of their biases and therefore unable to control against them, puts feminist and other socially aware scientists into a privileged position when doing science of sex, gender, and social behavior. Knowing our biases allows us to control for some of them when making systematic observations and doing experiments.

Here's an example. Years ago, when I began a series of studies with collaborators on the fitness effects of mate choice, my collaborator on flies, Wyatt Anderson said, "We have to study male choice too." I resisted this at first, because the hypothesis that interested me and fueled my drive to do the experiments was about socially imposed constraints on females' reproductive decisions (Gowaty 1996; Gowaty 1997a; Gowaty 1997b; Gowaty 2003a; Gowaty 2003b; Gowaty and Buschhaus 1998). Wyatt insisted. I gave in, saying to myself that testing for the fitness effects of male mate preferences, if they existed, would be a great control on our experiment. Our experiments showed that not just females, but males too, had fitness-enhancing preferences for alternative potential mates. Experimentally induced constraints on the expression of reproductive decisions decreased male fitness, just as it decreased female fitness. Our data reject the anisogamy hypothesis (Parker, et al. 1972) that says that the sex with the largest gametes will be the choosy sex, because *Drosophila pseudoobscura*, our study species, has huge sexual asymmetries in gamete size because males have very tiny sperm. Wyatt saved this scientist from making the common error of focusing on only females or only males; instead of assuming the theory was right, we tested it. My point is that it is hard to keep up with our own prejudices (in my case a favored hypothesis about sexual conflict, female behavior, and fitness). Most of us, not just the guys advantaged by the status quo, need all the help we can get to keep our minds clear and our eyes on the prize of discovering the nature of things.

Perception Bias Obscures Hypotheses about Females

Although my perception bias obscured an hypothesis about males, Caitilyn Allen (this volume) argues that perception bias more likely obscures hypotheses about females. Her chapter begins with Aristotle's incredible quote, "Femaleness is deformity," making it easy for us moderns to wonder how we ever thought he said anything clever. Allen confronts the difficult issue of what determines which questions we ask in the first place, what determines which hypotheses get our attention. She rightly worries that the absence of a hypothesis is usually invisible. Thus, this type of bias in science is insidious. It was thinking about such absences that inspired Judy Stamps to argue, as I mentioned already, that we need everyone's perspective actively added to the cacophony of hypotheses about this or that. Sometimes taking seriously a perspective-not-our-own may be controversial, but it ought to be considered a source of alternative hypotheses, and, all the better, if the predictions differ and are testable.

Allen argues persuasively that "social gender expectations have confounded the study of biological sex determination," so that we have a distorted view, much like Aristotle's, of the processes of sex determination. She recognizes two forms of social bias. The first bias arose from the misclassification as males or females of intersex individuals depending on the perception bias of practitioners about whether a penis was present or not. The second bias was associated with the Victorian view that females are passive and males active, a bias that affected Darwin's characterization of sex differences.

There seems little doubt that the mask of theory can hide the face of nature (Lawton, et al. 1997) an argument that Roughgarden (this volume) also makes.

Social Inclusion: Another Mechanism of Sexual Selection

Roughgarden (this volume) says that in the evolutionary study of nonhuman social behavior there is a rigidity of gender categories, such that sex and gender are synonymous. According to Roughgarden, commitment to a gender binary that she thinks is inherent to the "sexual-selection *system,*" facilitates failure of students of social behavior to attend to alternative explanations for the interesting patterns in nonhuman animals of multiple male and female morphs, the existence of males that look like females and females that look like males, and homosexuality. She also argues that the propositions buttressing the "sexual-selection *system*" are mostly false. Roughgarden's views are controversial, I think, because her claims do not ring true to other evolutionary biologists, even those like me, Hrdy (2004), and Allison Jolly (2004) with great sympathy for her efforts to promote (more) attention to gender variation in nonhuman animals.

For example, Roughgarden says that sexual selection is a *system*; most investigators think sexual selection is *an hypothesis* to explain the evolution of bizarre or showy traits. Many are confused (Day, et al. 2006; Ghiselin 2006; Hurd 2006; Lessells, et al. 2006; Miller 2006; Pizzari, et al. 2006; Shuker and Tregenza 2006; Stewart 2006) when Roughgarden (2006; this volume) describes, as an "alternative to sexual selection," a "new" hypothesis that fits the operational definition of sexual selection. She confuses her within-discipline readers when she renames the process by which "social inclusionary traits" mediate within-sex differential reproduction as "social selection" (Jolly 2004), a term that has been in use to describe a similar theory since the 1970s (West-Eberhard 1979; West-Eberhard 1983; West-Eberhard 1984; West-Eberhard 2003). And some of her criticisms seem more appropriate for the 1970s and 1980s (Hrdy 2004). Most germane to this essay is how different the views about females from 2007 are from the 1970s, when many of us thought the Victorian characterization of coy, passive females was wrong (Gowaty 1996; Hrdy 1974; Hrdy and Hrdy 1976; Sherfey 1972). Since then, many mainstream evolutionary biologists feel the field has moved on from the old ideas (1) that sex roles are fixed (Gowaty, et al. 2003; Langmore and Davies 1997), (2) that females do not behaviorally compete and are coy and passive, (3) that males do not have fitness-enhancing mate preferences (Bonduriansky 2001), or (4) that there are only two rigidly defined genders (think of the male ruffs that I described above).

As I have tried to emphasize throughout this chapter, evolutionary biologists have made many discoveries of unexpectedly complex gender variation in nonhuman animals that has led many of these passionate naturalists to novel explanations for the evolution of these traits. I hope that these discoveries will continue to illuminate an enthusiastic and respectful discussion of the exuberant variation in social presentation and behavior of humans, something that Roughgarden also seeks.

Inclusion of Subjects in Experimental and Observational Protocols

The biases that Allen describes and that Roughgarden fears—the biases embedded in our hypotheses—are particularly hard to deal with. As Allen notes, we cannot see the invisible, but now that we know the overwhelming effects of sex and gender on our biases, it is rational, efficient, and relatively easy to do all of our experiments and observations on both males and females, all the time, no exceptions. Remember when we had to argue for inclusion of women in studies of epidemiology and health at NIH? Remember Bernadine Healy? Here's a rule we might use regularly: always include both sexes and their known genders in tests of our hypotheses that have anything to do with sex, gender, and social behavior. If we leveled the observational field so that we observed females as often as males, at the very least we would know more about female animals, including ourselves. I can imagine how hard this might be for individual scientists with more status quo expectations than mine. When I was wrapped up in a favored hypothesis, I was saved from carrying out a limited and limiting experiment. Because of my generous collaborator, I can now say, in view of our extraordinary results (Anderson, et al. 2006), that we might all just routinely test females when we ask questions about males and routinely include males when we ask questions about females. If you have to, think of the inclusion of opposite sex individuals as a control.

I want to end with a prediction about new observations that investigators will make because we have new genetic tools that allow us to recognize more gender variation than ever before. When I spoke with the ruff observers, who are studying the copulation behavior of males that look like females, I asked if genotyping allowed for the discovery of *faeders*. The answer in that case was a hearty "no." Careful, observant natural historians, who were remarkably familiar with size variations of females and males, made that discovery. Nevertheless, now that we can easily assign chromosomal sex to the individual animals we observe in the field, I believe that more of us who follow the lives of individuals over days, months, and years will discover that there is more to being a ruff, a bonobo, or a guppy, just as Dreger describes for humans, than being in the catchall categories of "male" and "female."

References

Anderson, W. W. et al. 2006. Experimental constraints on mate preferences in *Drosophila pseudoobscura* decrease offspring viability and fitness of mated pairs. *Proceedings of the National Academy of Sciences*. Vol. 104, 4484–4488.

Bonduriansky, R. 2001. The evolution of male mate choice in insects: A synthesis of ideas and evidence. *Biological Reviews* 76:305–339.

Day, T. et al. 2006. Debating sexual selection and mating strategies. *Science* 312:691.

de Waal, F. 1997. Bonobo dialogues. *Natural History* 106:22–25.

de Waal, F. B. M. 1995. Bonobo sex and society. *Scientific American* 272:82–88.

de Waal, F. B. M. 2003. Darwin's legacy and the study of primate visual communication. In *Emotions inside out:130 years after Darwin's The Expression of Emotions in Man and Animals,* ed. P. Eckman et al., 7–31. *Annals of The New York Academy of Sciences,* vol. 1000.

East, M. L. et al. 1993. The erect penis is a flag of submission in a female-dominated society: Greetings in Serengeti Spotted Hyenas. *Behavioral Ecology and Sociobiology* 33:355–370.

Eberhard, W. G. 1996. *Female control: Sexual selection by cryptic female choice*. Princeton: Princeton University Press.

Frank, L. G., and S. E. Glickman. 1994. Giving birth through a penile clitoris-parturition and dystocia in the spotted hyena (*Crocuta crocuta*). *Journal of Zoology* 234:659–665.

Ghiselin, M. T. 2006. Debating sexual selection and mating strategies. *Science* 312:691–692.

Gowaty, P. A. 1977. Female choice in social animals. *Abstracts of Animal Behavior Society Annual Meeting*, 40.

Gowaty, P. A. 1996. Battles of the sexes and origins of monogamy. In *Black partnerships in birds*, ed. J. L. Black, 21–52. *Oxford series in ecology and evolution*. Oxford: Oxford University Press.

Gowaty, P.A. 1997a. Principles of females' perspectives in avian behavioral ecology. *Journal of Avian Biology* 28:95–102.

Gowaty, P. A. 1997b. Sexual dialectics, sexual selection, and variation in mating behavior. In *Feminism and evolutionary biology: Boundaries, intersections, and frontiers*, ed. P. A. Gowaty, 351–384. New York: Chapman Hall.

Gowaty, P. A. 2003a. Power asymmetries between the sexes, mate preferences, and components of fitness. In *Women, evolution, and rape*, ed. C. Travis, 61–86. Cambridge, MA: MIT Press.

Gowaty, P. A. 2003b. Sex roles, contests for the control of reproduction, and sexual selection. In *Sexual selection in primates: New and comparative perspectives*, ed. P. M. Kappeler, and C. P. van Schaik, 163–221. Cambridge: Cambridge University Press.

Gowaty, P. A., and N. Buschhaus. 1998. Ultimate causation of aggressive and forced copulation in birds: Female resistance, the CODE hypothesis, and social monogamy. *American Zoologist* 38:207–225.

Gowaty, P. A. et al. 2003. Indiscriminate females and choosy males: Within- and between-species variation. *Drosophila Evolution* 57:2037–2045.

Hamer, D.H., S. Hu, and V.L. Magnuson. 1993. A linkage between dna markers on the x-chromosome and male sexual orientation. *Science* 261 (5119):321–327.

Horn, A. D. 1979. Taxonomic status of the Bonobo Chimpanzee. *American Journal of Physical Anthropology* 51:273–281.

Hrdy, S. B. 1974. Male-male competition and infanticide among langurs (*Presbytis-Entellus*) of Abu, Rajastan. *Folia Primatologica* 22:19–58.

Hrdy, S. B. 1986. Empathy, polyandry, and the myth of the coy female. In *Feminist approaches to science*, ed. R. Bleirer, 119–146. New York: Pergamon Press.

Hrdy, S. B. 2004. Evolution's rainbow: Diversity, gender and sexuality in nature and people. *Nature* 429:19.

Hrdy, S. B. and D. B. Hrdy. 1976. Hierarchical relations among female hanuman langurs (Primates-*Colobinae*, *Presbytis-Entellus*). *Science* 193:913–915.

Hurd, P. L. 2006. Debating sexual selection and mating strategies. *Science* 312:692–693.

Jolly, A. 2004. Evolution's rainbow–Diversity, gender, and sexuality in nature and people. *Science* 304:965–966.

Jukema, J., and T. Piersma. 2006. Permanent female mimics in a lekking shorebird. *Biology Letters* 2:161–164.

Langmore, N. E., and N. B. Davies. 1997. Female dunnocks use vocalizations to compete for males. *Animal Behaviour* 53:881–890.

Lawton, M. F. et al. 1997. The mask of theory and the face of nature. In *Feminism and evolutionary biology*, ed. P. A. Gowaty, 63–115. New York: Chapman and Hall.

Lessells, C. M. et al. 2006. Debating sexual selection and mating strategies. *Science* 312:689–690.

Manson, J. H. et al. 1997. Nonconceptive sexual behavior in bonobos and capuchins. *International Journal of Primatology* 18:767–786.

McFadden, D. et al. 2006. Masculinized otoacoustic emissions in female spotted hyenas (*Crocuta crocuta*). *Hormones and Behavior* 50:285–292.

Miller, G. 2006. Debating sexual selection and mating strategies. *Science* 312:693.

Muller, M. N., and R. Wrangham. 2002. Sexual mimicry in hyenas. *Quarterly Review of Biology* 77:3–16.

Parish, A. R. 1994. Sex and food control in the uncommon Chimpanzee: How bonobo females overcome a phylogenetic legacy of male dominance. *Ethology and Sociobiology* 15:157–179.

Parish, A. R. 1996. Female relationships in bonobos (*Pan paniscus*): Evidence for bonding, cooperation, and female dominance in a male-philopatric species. *Human Nature: An Interdisciplinary Biosocial Perspective* 7:61–96.

Parish, A. R., and F. B. M. De Waal. 2000. The other "closest living relative": How bonobos (*Pan paniscus*) challenge traditional assumptions about females, dominance, intra- and intersexual interactions, and hominid evolution. In *Evolutionary Perspectives on Human Reproductive Behavior*, ed. D. LeCroy and P. Moller, 97–113. *Annals of The New York Academy of Sciences*, vol. 907.

Parker, G. A. et al. 1972. The origin and evolution of gamete dimorphism and the male-female phenomenon. *Journal of Theoretical Biology* 36:529–553.

Pizzari, T. et al. 2006. Debating sexual selection and mating strategies. *Science* 312:690.

Place, N. J., and S. E. Glickman. 2004. Masculinization of female mammals:

Lessons from nature. *Advances in experimental medicine and biology* 545:243–253.

Platt, J. R. 1964. Strong inference. *Science* 146:347–353.

Roughgarden, J. et al. 2006. Reproductive social behavior: Cooperative games to replace sexual selection. *Science* 311:965–969.

Schiebinger, L. 1989. *The mind has no sex? Women in the origins of modern science.* Cambridge, MA: Harvard University Press.

Schiebinger, L. 2001. *Has feminism changed science?* Cambridge, MA: Harvard University Press.

Sherfey, M. J. 1972. *The nature and evolution of female sexuality.* New York: Random House.

Shuker, D. M., and T. Tregenza. 2006. Debating sexual selection and mating strategies. *Science* 312:693–694.

Stamps, J. A. 1997. The role of females in extra pair copulations in socially monogamous territorial animals. In *Feminism and evolutionary biology,* ed. P. A. Gowaty, 294–319. New York: Chapman and Hall.

Stewart, J. 2006. Debating sexual selection and mating strategies. *Science* 312:694.

Szykman, M. et al. 2003. Rare male aggression directed toward females in a female-dominated society: Baiting behavior in the spotted hyena. *Aggressive Behavior* 29:457–474.

Valian, V. 1999. *Why so slow?* Cambridge, MA: The MIT Press.

Van Horn, R. C. et al. 2004. Role-reversed nepotism among cubs and sires in the spotted hyena (*Crocuta crocuta*). *Ethology* 110:413–426.

Vasey, P. L. 1995. Homosexual behavior in primates: A review of evidence and theory. *International Journal of Primatology* 16:173–204.

Vervaecke, H. et al. 2000. The pivotal role of rank in grooming and support behavior in a captive group of bonobos (*Pan paniscus*). *Behaviour* 137:1463–1485.

Wennerås, C., and A. Wold. 1997. Nepotism and sexism in peer review. *Nature* 387:341–343.

West-Eberhard, M. J. 1979. Sexual selection, social competition, and evolution. In *Proceedings of the American Philosophical Society* 51:222–234.

West-Eberhard, M. J. 1983. Sexual selection, social competition, and speciation. *The Quarterly Review of Biology* 58:155–183.

West-Eberhard, M. J. 1984. Sexual selection, competitive communication and species-specific signals in insects. In *Insect communication: Proceedings of the 12th Symposium of the Royal Entomological Society of London,* ed. L. Trevor, 283–324. London, Academic Press.

West-Eberhard, M. J. 2003. *Developmental plasticity and evolution.* New York: Oxford University Press.

Wrangham, R. W. 1993. The evolution of sexuality in Chimpanzees and Bonobos. *Human Nature* 4:47–79.

21

Social Selection versus Sexual Selection: Comparison of Hypotheses

Joan Roughgarden

Introduction

Toward the end of his career, in 1871, Charles Darwin initiated what has become a universal biological theory of sex roles. These writings reflect concern for traits like the peacock's tail and the deer's antlers, called male ornaments, which seemed to have no importance for improving survival. Instead, Darwin turned to the possible role such traits might have in mating. Darwin theorized that male ornaments evolved because of a certain social dynamic between the sexes: females prefer males on the basis of the ornaments they exhibit, and males compete with each other for access to females. Darwin wrote that females had an innate "aesthetic" for beautiful features and by choosing mates with these features, the females bred the males to be beautiful. He wrote, "Many female progenitors of the peacock must . . . have . . . by the continued preference of the most beautiful males, rendered the peacock the most splendid of living birds." Some ornaments might be nonfunctional, like the peacock's tail was assumed to be, whereas others might be weapons such as antlers that males were thought to use in combat, with the victor winning access to females. Although male-male combat might conceivably circumvent female choice by limiting it only to the victors, females were assumed to prefer the winners anyway because the winners were, ipso facto, the best of the males. Thus, female preference for victorious males caused males to become "vigorous and well-armed . . . just as man can improve the breed of his game-cocks by the selection of those birds which are victorious in the cock-pit." All in all, males evolve to be beautiful *and* well-armed because of female mate choice.

Although Darwin's story of male-male combat combined with female choice might conceivably be true in some species, in Darwin's thinking the peacock's tail and the deer's antlers were emblematic of male-female social relations generally. Darwin wrote, "Males of almost all animals have stronger passions than females," and "the female . . . with the rarest of exceptions is less eager than the male . . . she is coy." The phrases "almost all" and "with rarest of exceptions" show that Darwin was enunciating a theory not solely for the gender roles in peacocks and deer, but for all of nature, excepting a few rarities we need not worry about.

Darwin's theory of sex roles is called "sexual selection." This theory has grown into a "system" of propositions concerning gender expression, sex roles, and sexuality. With the word *system*, I mean first, a set of propositions that are logically interconnected and have independent truth status. At first glance, some of the propositions within the system, say, A and B, appear to express facts, and other propositions, say, T, offer theoretical explanations for those facts. But in a system like sexual selection, if it emerges that A or B is not explained by T, then T is still reaffirmed as true, but as simply not applying to the "special" case of A. A separate proposition, T′ is then added to deal with A. In this way, the system grows because no observation can challenge T, and T is amended as needed. Thus, T has a truth status independent of the facts it was originally intended to explain. The male and female gender roles that Darwin first enunciated to account for male ornaments have attained a truth status that cannot be challenged, and many researchers cannot even imagine the possibility that sexual-selection theory is false as a whole.

For example, in a letter responding to a recent critique of sexual selection in which we asserted that the entire theory was incorrect (Roughgarden, Oishi, and Akçay 2006), evolutionary biologists David M. Shuker of the University of Edinburgh and Tom Tregenza of the University of Exeter declared that claiming sexual selection to be critically flawed was "a quite extraordinary statement, comparable with dismissing Einstein's general theory of relativity." I leave it to the reader to decide which is more "extraordinary"—equating sexual selection with the general theory of relativity, or asserting that the entire theory of sexual selection is incorrect.

By referring to sexual selection as a "system," I also allude to the phrase "sex-gender system" coined in 1975 by the anthropologist and gender theorist Gayle Ruben. Ruben refers to how gender categories emerge from cultural processes that build on the "raw material" of biology. Although biological and anatomical markers are used to assign people to genders, how the genders are characterized and what performance and appearance are expected from them, are human demands forced onto the biological substrate. Furthermore, the expectations flowing from however gender is defined in various cultures wind up being "the part of social life which is the locus of the oppression of women, of sexual minorities" (Ruben 1975). Although gender categories may not be originally constructed for the deliberate purpose of oppressing others, they end up authorizing such oppression. They create big difference where little exists: "[G]ender can be seen as a socially imposed division of the sexes . . . which exaggerates the . . . differences between the sexes. . . . Male and female it creates them, and it creates them heterosexual."

We have much to glean from the thirty years of discourse in gender theory that has ensued since Ruben's writings. For example, let us return to the logical structure of sexual-selection theory, and ask, why would a discipline insist that some theoretical propositions remain accepted as true regardless of whether they continue to explain the facts they were originally intended for? It's because the theoretical propositions themselves invest power, they establish the "norms" against which "exceptions" may then be defined, and thereby become the locus of oppression against those left unprivileged by the theory. This politicizing of science undercuts its credibility, and many humanists steeped in gender-theoretic discourse doubt that science can achieve anything approaching the objectivity it claims for itself.

Unlike most gender theorists, I do think that science works, and is an innovative development of Western culture that yields true facts and explanations of nature. But it only works if we really follow our own guidelines—that is, state alternative hypotheses and test them against the data. When we encounter a theoretical proposition being sustained despite contrary evidence, then we are professionally obligated to develop and entertain alternative hypotheses. In that spirit, I offer an alternative to sexual-selection theory that is equally extensive, but, point-by-point, different from it.

I've termed the alternative to sexual selection that I propose "social selection" (Roughgarden 2004). This phrase has been previously used for a different concept by West-Eberhard. She writes, "The special characteristics of sexual selection discussed by Darwin apply as well for social competition for resources other than mates" (West-Eberhard 1983). So-called social competition is "competition in which an individual must win in interactions or comparisons with conspecific rivals in order to gain access to some resource. The contested resources might include food, hibernation space, nesting material, mates, or places to spend the night. Seen in this broader perspective, *sexual selection* refers to the subset of social competition in which the resource at stake is mates. And *social selection* is differential success . . . in social competition."(italics in the original) Thus, social selection according to West-Eberhard is a generalization of the central competitive narrative of Darwin's sexual selection.

The concept of social selection that I propose is completely unrelated to the West-Eberhard usage. Social selection is selection for, and in the context of, the social infrastructure in a species within which offspring are produced. The social relationships in the social infrastructure would generally include cooperation as much or more than competition, and would revolve about negotiation more than "winning." Social selection in my formulation does not extend sexual selection; it replaces sexual selection. Although it's conceivable that the Darwinian picture of females selecting the winners at male-male combat might develop as a social system under my formulation of social selection, I think this outcome rarely, or never, actually exists except perhaps as a coincidence when other considerations have been the primary causal forces.

Finally, which evolutionary system of sex, gender, and sexuality ultimately prevails scientifically determines our world view of nature itself. The sexual-selection system proposes a worldview of nature that emphasizes conflict, deceit, and dirty gene pools. If this Darwinian picture of nature is true, so be it. But is it true? We'll never know unless it is tested against a full-fledged alternative, not picking on small points, but challenging the entire structure.

A dispute about sexual selection theory has ramifications beyond evolutionary biology. Sexual-selection theory is taken as axiomatic in evolutionary psychology, the discipline that purports to explain human desire and beauty using sexual-selection theory. A collapse of sexual-selection theory threatens the future of evolutionary psychology as a discipline. Moreover, the sexual-selection narrative has been widely assimilated into popular culture. Many people have their careers and even self-perception heavily invested in saying that science affirms the sexual-selection narrative as the biological norm. The controversy over sexual selection has become personal and vicious, with back-and-forth charges of "agendas," accompanied by homophobic slurs slipped into reviews and leaked to reporters. At a biology master's student seminar in Amsterdam recently, I was heckled and nearly prevented from speaking. This unprofessional conduct only plays into the hands of those who doubt that science can ever be the objective enterprise it claims to be.

This essay presents a point-by-point exposition of how sexual selection compares with social selection. I hope you will then be in a position to see not only how these systems offer alternative explanations for particular behaviors, but also how they add up to different worldviews of nature.

Starting the Table

A table with all the issues appears in Table 21.1, with related points grouped together. Sexual selection and social selection approach the subject of sex, gender, and sexuality from different starting positions. The fourth entry on the right consists of sexual selection in its original sense, which I term its "central narrative." Sexual selection starts with a male-female binary and attributes universal properties to both sexes—males as passionate and females coy. Sexual selection then proceeds up to why sexual reproduction and the male-female binary exist to begin with, and also down to how sexual reproduction is manifested.

In contrast, the upper left of the table is where social selection starts out—why sexual reproduction exists at all. Not even sexual reproduction, much less a male-female binary, are assumed to be given. Social selection then proceeds down to whether a male-female binary exists, and then on to how sexual reproduction is manifested. This sequence insures that explanations for how sexual reproduction takes place are logically consistent with why sexual reproduction exists to begin with, whereas the sequencing in sexual selection has led to hypotheses pertaining to how sexual reproduction is manifested that contradict any rationale for why sexual reproduction should have evolved in the first place.

So let's start at the top, realizing that this direction visits the points in the sexual-selection system somewhat out of sequence from its point of view.

Origin of Sex and Male-Female Binary

Sexual Reproduction

According to sexual selection, sexual reproduction evolved from asexual reproduction as a mechanism to cleanse the gene pool of deleterious mutations. The idea postulates that family lines inevitably accumulate deleterious mutations. A mating between different lines generates offspring who don't have these mutations, rejuvenating the stock (Haigh 1978; Chao 1990). The problem with this idea is accounting for why deleterious mutations are present in the lines to begin with, because if mutations are indeed deleterious, they are eliminated by natural selection when they first appear, and so don't accumulate.

According to social selection, the function of sexual reproduction compared with asexual reproduction is to maintain a diverse gene pool needed for long-term survival of populations in an ever-fluctuating environment. The gene pool of an asexual population becomes overcommitted to particular genotypes after a run of environmental circumstances favoring those genotypes, and when the environment changes, the population suffers a crash that lowers its geometric mean fitness below that of a corresponding sexual population (Roughgarden 1991). (The geometric mean through time of the instantaneous arithmetic mean fitness is higher in a random-mating diploid sexual population than in a corresponding asexual population. Long-term survival of a population depends on a high geometric mean fitness.) The proximate cause of environmental fluctuation might include the waxing and waning of parasites (Hamilton, Axelrod, and Tanese 1990).

Both sexual and social selection agree that sexual reproduction is a cooperative act—the objective is to produce offspring with combined genomes. They differ on the advantage that this cooperation provides. However, the existence of any cooperative rationale to sexual reproduction, regardless of which is correct, contradicts sexual selection's later postulate that conflict underlies all male-female relationships. If serious conflict were ever-present in sexual reproduction, species wouldn't bother to reproduce sexually and would reproduce asexually instead.

Table 21.1 Evolutionary Systems of Sex, Gender and Sexuality

Social Selection	Sexual Selection
Sex increases geometric mean fitness	Sex eliminates deleterious mutations
Gametic binary to maximize number of viable gametic encounters	Gametic binary from egg-sperm competiton, quality-quantity trade-off
Hermaphroditism primitive, dioecy derived specialization	Dioecy primitive, hermaphroditism specialized, size-threshold hypothesis
Sex roles local, none necessarily universal	Males universally passionate, females coy
Social behavior as reproductive system: selection arising from differences in number reared, nongendered bargaining/ payments for offspring control	Social behavior as mating system: selection arising from difference in mating success, females as limiting resource, male competition for access to females
Females choose for compatibility and capability of providing direct benefits	Females choose for "good genes"
Males equivalent in genetic quality	Males comprise hierarchy of genetic quality
Bateman's principle an artifact, fitness definition identical for males and females number of offspring successfully reared	Bateman's principle: male fitness ↑ number of mates, female fitness independent of mates> 1 and = number of eggs, variance of mating success indicates strength of sexual selection
Behavior as developmental strategy, social system as evolutionary strategy. Two tiers—evolution as ESS, behavior as NCE or NBS	Behavior as evolutionary strategy, single tier, behavior as ESS
Parental investment, F=M, egg~ejaculate	Parental investment, F>M, "cheap sperm"
Sexual cooperation primitive, conflict derived	Sexual conflict primitive, cooperation derived
Male promiscuity derived, reflects exclusion from control of offspring rearing	Male promiscuity primitive, results from Bateman's principle reflecting cheap sperm
Monogamy: economic efficiency	Monogamy: males entrapped by females, and/or males have no alternative prospects

(*continued*)

Table 21.1 Evolutionary Systems of Sex, Gender and Sexuality (cont'd)

Social Selection	Sexual Selection
EPP/EPM—System of genetic side payments	EPP/EPM—Cheating
Secondary sexual traits—admission tickets to power-holding cliques, social inclusionary traits, condition indicators of capacity to supply direct benefits	Secondary sexual traits—favored by females to produce sexy and/or victorious sons Containing good genes, condition indicators of genetic quality
Sexual monomorphism—absence of same-sex power-holding cliques	Sexual monomorphism—absence of any female "aesthetic" for male ornaments
Sex-role reversal—possible outcome of male-female ecological circumstance	Sex-role reversal—restated as reversal of operational sex ratio (OSR) bargaining in local, contradicts Darwinian universal sex roles
Multiple male/female morphs—prezygotic and postzygotic helpers, social niches	"Alternative" mating strategies—pejorative descriptions: "sexual parasites," "sneakers"
Feminine males—males communicate with female "body english," marriage brokers	Feminine males—"female mimics" deceive masculine males to "steal" their matings
Masculine females—females communicate with male "body english," territoriality	Masculine females—"female ornaments"—imperfect suppression of male traits
Homosexuality—physical intimacy to coordinate action and sense team welfare	Homosexuality—accident, or deceit, or practice for later heterosexuality
Human attractiveness—beauty in eye of beholder, compatibility, health indicates ability to deliver direct benefits	Human attractiveness—handsomeness in males signals good genes, beauty in females signals fecundity
Human brain—social inclusionary trait, needed to raise young in human society, fast evolution—runaway social selection	Human brain—in males to attract females, in females to appreciate males, a male ornament imperfectly expressed in females

Gametic Sexual Binary

The size difference between sperm and egg is the basis for defining male and female in a sexually reproducing species. Many sexually reproducing algae and fungi have only one size of gamete, and for these species the concepts of male and female are not defined, even though the reproduction is sexual in the sense that offspring are produced by combining the genomes of two parents. However, most sexually reproducing species do have two gamete sizes, one tiny—sperm—and one large—egg—and for these species the male-female binary is well defined insofar as gametes are concerned. Because most sexually reproducing species do have two gamete sizes, in this sense the existence of a male-female binary can be considered a very broad, almost universal, empirical generalization for sexually reproducing species. Evolutionary theory for the origin of this gametic dimorphism in size therefore amounts to an evolutionary theory for the origin of male and female as biological categories.

According to sexual-selection theory, the protosperm and protoegg are playing a game against each other (Parker, Baker, and Smith 1972; Charlesworth 1970; Bulmer and Parker 2002; Matsuda and Abrams 1999). Imagine that initially both the protosperm and protoegg are the same size (a condition called "isogamy"). Then the protosperm tries, so to speak, to cheat a little, becoming a little smaller than the protoegg, so that more sperm can be produced with the leftover energy. A numerical advantage in sperm production then allows that size of sperm to outcompete the less numerous sperm with the original size. But a slightly smaller zygote now results and it is less viable than the full-size zygote. Therefore, the protoegg responds by increasing its size, restoring zygote viability to its original level. The egg's compensating in this way is better than becoming smaller itself to match the smaller sperm, because then the zygote would suffer a very deleterious double loss of investment. This analysis of the sperm and egg's "best response" to each other culminates in one gamete becoming nearly as large as the zygote, and the other gamete becoming as tiny as possible (a condition called "anisogamy"). In sexual-selection theory, the distinction between male and female arises from a battle—the sexes are created *ab initio* as combatants.

According to social selection theory, a parent is simply trying, so to speak, to divide up the material it can place into eggs and sperm to maximize the *number of gametic contacts* that produce viable zygotes (Kalmus 1932; Scudo 1967; Cox and Sethian 1984; Randerson and Hurst 2001; Iyer and Roughgarden 2007). If both protosperm and protoegg are the same size, then only so many of them will bump into each other in the ocean's water where life began. The number of gametic contacts increases as gametes become more numerous, conceptually forming a larger and more dense cloud of gametes. In principle, if both gametes could be made tiny, then when a cloud of sperm mixes with a cloud of eggs, the highest number of contacts will occur. But egg and sperm can't both be tiny and still produce zygotes big enough to survive. So, the maximum number of contacts producing viable zygotes then occurs when one of the gametes is nearly the desired zygote size while the other is as small as possible. Thus, the evolution of the gametic dimorphism in size is simply the mechanism to maximize the number of contacts between gametes, subject to the constraint that the resulting zygotes must be big enough to survive. Social selection views the sexual-selection narrative about the sperm-egg game and sperm-sperm competition as gratuitous and irrelevant to the evolution of the sperm-egg size dimorphism.

Whole-Organism Sexual Binary

If a species does have a size binary among its gametes, the next issue is whether the whole organism obeys a similar binary—that is, if gametes can be classified as male and female, can whole organisms be classified as male and female too? Generally, no. It seems the most common body plan among multicellular organisms is for an individual to make both male and female gametes at the same time, or at different times during its life. Species in which each organism makes only one size of gamete during its life are called "dioecious," and in these species, the individuals can be classified as male or female according to the gamete size produced. Species in which some organisms make both gamete sizes during their lives are called "hermaphroditic." If an individual makes both sizes at the same time, it is called "simultaneously hermaphroditic," and if it makes them at different times, it is called "sequentially hermaphroditic." In sequential hermaphroditic species, individuals "change sex," going from male to female (switching from making small gametes to making large gametes) or vice versa. Simultaneous and sequential hermaphroditism are not the only possibilities. Individuals in some species are "crisscrossing hermaphrodites" who go from male to female and then back to male, or vice versa, etc.

According to sexual selection, the whole-organism binary is taken as a starting point, and hermaphroditism is viewed as a special case arising in peculiar circumstances. For example, the size-threshold hypothesis for sequential hermaphroditism argues that a male must attain a certain large size to successfully control and defend a harem of females (Bullough 1947; Warner 1982; Charnov 1982; Sadovy and Shapiro 1987; Ross 1990). If sex change is possible in such a species, as it is in many coral-reef fish, then individuals should start out as female to produce a batch of eggs. Then, once a large enough body size has been attained, it's advantageous for the female to switch to male so that it can fertilize more than one batch of eggs. Stories have also been advanced for other social situations in which it might seem best to be female when large, and thus switch from male to female after growing through a threshold size. The problem with all the size-threshold models is that they don't explain the sex crisscrossing species. In crisscrossing coral-reef fish, the individuals don't grow past a threshold size, then shrink below it, only to regrow past the threshold again (Kuamura, Nakashima, and Yogo 1994; Munday, Caley, and Jones 1998).

According to social selection, the whole-organism binary is taken as a specialization, and hermaphroditism is viewed as the original baseline condition. This stance agrees with the overall pattern of hermaphroditic body plans being widespread in the oldest animal phyla and classes, such as the marine invertebrates, as well as throughout the plant kingdom, whereas dioecy—separate sexes in separate individuals—occurs primarily in the more recent phyla and classes of animals such as terrestrial arthropods (spiders and insects) and in terrestrial vertebrates. Indeed, research on gender in plants also views dioecy as a specialization (Doust and Doust 1983; Policansky 1981; Geber, Dawson, and Delph 1999). I've conjectured that the evolutionary rationale in animals driving the specialization of one sex per body plan is mobility, especially in terrestrial habitats where the cost of mobility is high compared with marine and freshwater habitats. Male individuals are specialized to provide, so to speak, the "home delivery of sperm" (Roughgarden 2004). The importance of mobility is evidenced by the amount of energy terrestrial plants spend to pay animals to transport their pollen and seeds. Pollen and seed wastage is greatly reduced by paying animals to carry pollen and seeds. Similarly, outfitting the small gametes with motile tails transforms a cloud of sperm from simply drifting by diffusion to the eggs but allows directed movement toward them, which would be much more efficient, and is presumably why sperm have tails. Thus, according to social selection theory, the male-female binary at the whole-organism level exists as a response to ecological conditions in which it's efficient for one body type to be devoted to transporting gametes while the other body type remains relatively stationary. In lineages where hermaphroditism is derived from dioecious ancestors (whose own ancestors are invariably hermaphroditic) the hermaphroditism is merely a reversion to the ancestral state expected if the special ecological conditions propelling the evolution of dioecy disappear. According to social selection, the whole-organism male-female binary has no deep and universal biological significance, unlike sexual reproduction itself and the gametic male-female binary.

Central Narrative

The central narrative of male-female relationships is where sexual selection begins. Sexual selection takes the existence of a whole-body male-female dichotomy as unproblematic and sets out to explain how the males and females interact. In contrast, social selection gets to this point only after considering the circumstances in which the male-female dichotomy can evolve to begin with.

Universal Sex Roles

According to sexual selection, males and females conform to near-universal templates, which, as Darwin wrote, are that males are "passionate" and females "coy." Today's jargon substitutes the words "promiscuous" for "passionate" and "constrained" for "coy," as this quotation from University of Chicago geneticist Jerry Coyne, confidently asserts: "We now understand. . . . Males, who can produce many offspring with only minimal investment, spread their genes most effectively by mating promiscuously. . . . Female reproductive output is far more constrained by the metabolic costs of producing eggs or offspring, and thus a female's interests are served more by mate quality than by mate quantity" (Coyne 2004). Also, genetic "quality" now substitutes for Darwin's female "aesthetic." Even with these shifts in terminology, the central narrative of Darwinian sexual selection remains the same—promiscuous males fight each other and display to females who then choose the most handsome victors as their mates. Coyne's confidence in this sexual-selection narrative seems widely shared among biologists today.

Nonetheless, in 1996, Gowaty introduced an interpretation of "constrained" that emphasizes variation in social and ecological options among females, for a given presence of males. In a review of this article, Gowaty argues that Coyne is incorrect in interpreting "constrained" as referring to metabolic costs that females incur in egg production and offspring rearing. She writes, "The Constrained Female Hypothesis, which introduced the idea of ecological and social constraint on females is an idea I introduced with a paper in 1996. The constrained female hypothesis is about variation in females and their abilities to feed themselves and their offspring without male help. Today, I'm happy to realize that when most people use the term 'constrained female,' they are referring to my concept and not to coy females, which was the way Jerry Coyne used it. I hope you will not perpetuate Coyne's misuse. He really just meant that females are limited by the number of offspring they can produce. His misuse of the term is about relative reproductive rates of males and females; my idea is about among-female variation in female capabilities relative to other females." Well, I agree that females may vary among themselves in the options available to them socially and ecologically, and this could lead to intrasexual selection among females, for a given presence of males. I think, however, that Coyne is correct in representing the consensus understanding of how the word "constrained" is interpreted among biologists today—it refers to the physiological costs faced by females compared to males in the production of offspring, and is the basis of the Darwinian narrative of intersexual selection leading to different sex roles for males and females. As I see it, the issue is not whether Coyne correctly uses the *word* "constrained," but whether what he's claiming with that word is correct—are females more "constrained" by their reproductive physiology than males, thereby producing the gender roles that Darwin claims are nearly universal?

In a nutshell, the Darwinian sex roles are nowhere close to being universal. In thousands of species too numerous to name, females are as passionate and promiscuous as males, and males as constrained as females, or more so, because of tending or guarding eggs.

There are no general surveys of reproductive habits across all dioecious animal species, and the universal templates for males and females that Darwin claimed are at best unsubstantiated and apparently false as generalizations. In fish, for example, surveys show that, of those species in which one or more parents care for the eggs, the male is more likely than the female to be the care provider (Reynolds, Goodwin, and Freckleton 2002). In birds, there is often biparental care (Lack 1968; Cockburn 2006), whereas, in mammals, the female usually provides the care (Clutton-Brock 1989). In insects, male choice is often as choosy as female choice (Bonduriansky 2001). It is hard, moreover, to distinguish "care" from "control" and the parent who is caring for the eggs or young might actually be more concerned with the control of the young than in the provision of care (Roughgarden 2004), a point also made by Gowaty (1996). All in all, no general pattern has actually been demonstrated about male/female sex roles throughout the animal kingdom, although the stereotypes that Darwin enunciated are widely believed.

In contrast, according to social selection, no necessary and universal sex roles exist anyway. What each sex does is subject to negotiation in local circumstances. Any statistical regularities in sex roles that may emerge reflect a statistical commonness of circumstance together with what constitutes a best bargain in such circumstances. Social behavior emerges from local ecology, and does not express any biological universal. If local ecology shows statistical regularities, so will the sex roles that emerge in those ecologies.

Purpose of Reproductive Social Behavior

According to sexual-selection theory, reproductive social behavior comprises a "mating system" (Emlen and Oring 1977). Within a mating system, natural selection arises from differences in "mating success" and particular behaviors are understood by how they contribute to attaining mating success. In a mating system, the females are regarded as a "limiting resource" for males, and males compete for access and control of mating opportunities with females to acquire paternity. The sexual-selection narrative explains what happens within a reproductive social system primarily in terms of "mating."

The problem with sexual selection here is that evolution does not depend on mating, *per se*, but on number of offspring successfully reared. The next generation's gene pool is influenced only by successfully reared offspring. Sexual selection elevates one component of reproduction, namely mating, into an end in itself. This is a fundamental error.

According to social selection theory, reproductive social behavior comprises an "offspring-rearing system." Within an offspring-rearing system, natural selection arises from differences in number of offspring successfully reared, and particular behaviors are understood by how they contribute to building or maintaining the social infrastructure within which offspring are reared. The principal male-female social dynamic is to determine bargains and exchange side payments that establish control over offspring and stabilize the offspring-rearing social infrastructure.

Objective of Female Mate Choice

Female choice plays a pivotal role in sexual-selection theory, tracing to Darwin's claim that the peahen's aesthetic caused peacocks to evolve the most beautiful of tails. Today's narrative largely avoids the term "aesthetic" for female choice, but instead envisions that females select males for their genetic quality. Whatever sense of "aesthetic" a female bird might have is understood to be a finely honed ability to distinguish which males have the best genes. The rationale for a female to choose males with "good genes" is to endow their own sons with the traits they found attractive. By acting on their preference for good genes, females insure that their own sons are sexy and destined to succeed in the mating game. This rationale is called the "sexy son hypothesis."

The problem is that data are scanty for female choice being motivated by indirect future genetic benefits compared with direct present-day ecological benefits. Females choose males who provide food and/or protection, with the importance of genes being rather moot. Few sexual-selection advocates today will argue that selection for indirect genetic benefits is common, but often special cases are cited in which the male is supposed to provide only his genes to the offspring and is supposedly not involved in parental care or protection in any way. However, such suppositions invariably are not demonstrated—they are simply asserted. Just because the male doesn't remain at the nest doesn't mean he doesn't contribute to the female's direct interests in some other way. Frequently, a male interacts with a female in the nonbreeding season, or is part of a predator alarm network, etc. The claim that "all the male contributes is his genes" is impossible to verify. A male doesn't have to contribute much immediate ecological benefit to outweigh the present value of an uncertain future genetic benefit.

In her review of this article, Gowaty notes that the phrase "good genes" is used in two ways, one referring to the best genes in a hierarchy of genetic quality, say, of perfection in a showy trait, and the other referring to compatible genes, say, of complementary genes at immune coding loci. She claims that data support the second usage, if not the first. But this second usage of "good genes" distorts the original Darwinian intent of female choice. Female choice for compatibility (best for her) is different in principle from female choice for the best overall. The Darwinian narrative stipulates that females are selecting according to the same metric, and to shift the theory's focus away from genetic quality to genetic compatibility is changing the theory in a way I find disingenuous. In politics, this would be called "gut and amend," where legislation has its wording gutted and replaced by something else, and then supplemented with amendments, all to get votes. Gutting and amending Darwinian female choice to refer to genetic compatibility instead of genetic quality yields a theory more likely to pass muster with reality, but misleads the public into thinking that the theory of sexual selection has been successfully tested.

Furthermore, data do not in fact support female choice either for best-overall genes or best-for-her genes. A recent review of 95 studies covering 50 species of birds finds limited to no support on behalf of female choice for any genetic benefits of any type (Akçay and Roughgarden 2007). In her review of this article, Gowaty reports "an emerging consensus is that direct effects predominate in birds; while in insects indirect effects do." But then, the peacock is a bird, after all, and so this consensus implies that the sexual-selection narrative has now abandoned the peacock and settled for beetles. Others, too, find little support, either theoretically or empirically, for indirect benefits in comparison to direct benefits (Kirkpatrick and Ryan 1991).

In contrast, according to social selection, females choose mates based on maximizing the number of young she can produce that are successfully reared by her own efforts, as well as in conjunction with help from the mate, plus assistance from the social infrastructure. According to social selection, female choice is indifferent to male genes, but is focused on whether the male is likely to deliver on any promised direct benefits (Forsgren 1997). She should use an expectation of direct benefits from a male discounted by the probability that the male will renege on, or somehow be prevented from, delivering on his promise, as the criterion for choice. Thus, a premium will be placed on the compatibility and health of the prospective partner. Health is important not as an indicator of "good genes," but as a sign of competency to provide direct benefits. According to social selection, both males and females must choose each other, because the offspring represent a common investment in which both have a shared interest.

Interest in the "aesthetic" nature of showy characteristics has not disappeared, however, but has been refocused on explaining why the traits involved in sexual selection seem so arbitrary. A predisposition arising from neuroanatomy may account for why the peacock's tail, specifically, rather than some other trait, is socially significant (Burley and Symanski 1998). Any aesthetic, in this sense of explaining the arbitrary nature of a social marker trait, is consistent with both sexual and social selection.

Male Genetic Quality

According to sexual-selection theory, males can be ranked in a hierarchy of genetic quality. In addition to the good genes that females are seeking in their mates, there are bad genes they aim to avoid.

The problem with the supposition that male variation can be ranked in a hierarchy of genetic quality is that it is impossible to maintain a stable polymorphism of good and bad genes in the face of continued directional selection against the bad genes. If, generation after generation, female choice aims to weed out the males with bad genes, then eventually no bad genes should remain. Therefore, sexual-selection theorists have to concoct genetic schemes, typically involving high mutation rates spanning polygenic loci, to replenish the supply of bad genes that are being continually eliminated by female choice. The genetic dilemma of maintaining a hierarchy of genetic quality in the face of sustained directional selection against the bad genes is called the "paradox of the lek" (Rowe and Houle 1996; Tomkins, Radwan, Kotiaho, and Tregenza 2004). This paradox appears to be an insurmountable and fatal flaw in sexual-selection theory.

The role of female choice in eliminating bad genes lends a eugenic tone to discussions of sexual selection, which is seen as a process that contributes to "conservation value" by pruning the gene pool to enhance the prospects of species survival (Radwan 2004).

Genetic variation may indeed exist among males, and some males may indeed be especially compatible genetically with certain females, because of say, histocompatibility and immune complementarity, and females might therefore conceivably select their mates on the basis of genetic compatibility, not quality. Again, one must be clear that the Darwinian picture of sexual selection hypothesizes that all the females share the same perception of hierarchy. The narrative specifies that *all* the peahens share the same aesthetic for peacock tails—this is what causes the tails to approach aesthetic perfection. The Darwinian narrative doesn't say, for example, that each peacock has her own personal aesthetic, some for blue spots, others for red spots, and so forth, leading to a diversity of peacock tails each of equal merit in the eyes of some beholder. Using instances of data on female mate choice for genetic compatibility as evidence for sexual selection means restating sexual-selection theory, a gut-and-amend tactic for saving a theory contradicted by the very data being claimed as support.

According to social selection, no hierarchy of genetic quality among males exists. If genes matter at all to female choice, females are choosing for genetic compatibility, not overall quality. All males are equivalent in genetic quality, excepting for a very rare fraction who obviously contain deleterious mutations and are present in a mutation-selection balance (say, 1 in 10^6).

Bateman's Principle

In 1948, the English geneticist Angus Bateman published laboratory experiments with *Drosophila* that were presented as confirming Darwin's theory of sexual selection. Bateman reported that, for a male, "fertility is seldom likely to be limited by sperm production but rather by the number of inseminations or the number of females available to him." Similarly, he claimed to have found in his flies an "undiscriminating eagerness in males and discriminating passivity in females" in accord with the sexual-selection narrative. More specifically, Bateman reported that male fitness (number of eggs bearing the male's paternity) increased with the number of mates, whereas for females, fitness was independent of number of mates beyond one—one male was sufficient to supply all the sperm needed to fertilize the eggs. Furthermore, the males were observed to have a higher variance in fitness than the females—that is, some males had paternity of many eggs and some of very few, whereas the females all produced about the same number of eggs and so had about the same fitness as one another. This observation was taken as evidence that the strength of sexual selection is stronger in males than in females, and later authors have even gone on to say that males are therefore more highly evolved than females. The Bateman experiments are a cornerstone of sexual-selection theory, and they have been widely cited in papers and textbooks in recent decades.

Over the last five years, however, many critiques have revisited the Bateman 1948 paper and have found that he overstated his results. Furthermore, sexual-selection advocates, including textbook writers, have quoted selectively from what Bateman did report and have sometimes made up quotations out of thin air, which they attributed to Bateman (Tang-Martinez 2000; Tang-Martinez and Ryder 2005; Dewsbury 2005; Parker and Tang-Martinez 2005; Gowaty and Hubbell 2005; Snyder and Gowaty 2007).

According to social selection theory, Bateman's principle is nonexistent. Instead, both males and females share the same definition of fitness, namely number of offspring successfully reared, and not different definitions—paternity for males and egg number for females.

Theory of Social Systems

Number of Tiers

Sexual selection approaches the modeling of social behavior using competitive evolutionary game theory (Maynard Smith 1982). Particular behaviors are viewed as strategies. The prisoner's dilemma game is an oft-cited example in which the strategies of play are to cooperate or to defect. The "payoff matrix" tabulates the payoff to each player for all combinations of these strategies. The solution to the game is an evolutionary stable strategy (ESS), which is a combination of strategies for both players such that a mutant allele for some other combination cannot increase when rare. This approach is a single-tier approach in the sense that particular behaviors are viewed directly as evolutionary strategies.

The problem with the evolutionary single-tier approach is that it requires thinking of particular behaviors in terms of a genetic basis, such as the "gene for" cooperating, for defecting, for shyness, for aggressiveness, etc. Behaviors rarely have much direct genetic basis. The single-tier approach forces narratives of genetic determinism.

Social selection approaches the modeling of social behavior as a two-tier problem (Roughgarden, Oishi, and Akçay 2006). Explaining social behavior is viewed as a topic in developmental biology, not evolutionary biology. Particular behaviors develop as animals interact with one another, similar to how morphology develops through cell-cell contact during embryogenesis. A social system is behavioral morphology—a system of phenotypes produced through interactive development. This perspective is shared somewhat by the literature on the evolution of social dominance (Capitanio 1991; Capitanio 1993; Dewsbury 1991; Dewsbury 1993; Barrette 1993; Moore 1991; Moore 1993; Moore, Brodie, and Wolf 1997; Moore, Wolf, and Brodie 1998; Wolf, Brodie, and Moore 1999; Moore, Haynes, Preziosi, and Moore 2002).

In social selection, the dynamics by which a behavioral system develops employs both cooperative and competitive game theory. In cooperative game theory, cooperative solutions may be attained by the parties playing with coordinated tactics and with the perception of shared goals made possible through animal friendships. These friendships are facilitated through behaviors that involve physical intimacy like reciprocal grooming, preening, same-sex sexuality, and interlocking vocalizations. Cooperative solutions can also be attained, though inefficiently, through continuing war-of-attrition conflict, resulting in standoffs that implement a division of resources. Other possibilities include noncooperative dynamics as well, in which the animals don't communicate in any way, and which winds up at a purely competitive outcome. Within the behavioral tier, all these dynamics are possible.

Social selection also envisions an evolutionary tier in which the payoff matrices and rules of play evolve. Thus, particular social behaviors evolve indirectly. First, the payoff matrix and rules of play evolve, which then lead to the development of particular behaviors.

Characteristics of Sexual Reproduction

The next issues pertain to typical behaviors that emerge during sexual reproduction. Sexual and social selection differ on how these are explained.

Parental Investment

According to sexual selection, the female has a higher parental investment than the male because the egg is bigger than the sperm. The sperm are considered "cheap" and the egg expensive. This initial difference is then extrapolated to an entire suite of female and male behaviors. Females are supposed to be coy because of the need to protect their high parental investment, and males promiscuous because their investment is supposedly low, and they have nothing to lose by playing the field.

According to social selection, male and female parental investments are more or less the same initially. An ejaculate might typically contain 10^6 sperm and an egg is typically 10^6 times as large as a sperm. So the size of the ejaculate and egg are often about the same order of magnitude (Dewsbury 1982). Therefore, according to social selection, the male and female begin with more or less equal investments in the offspring. Hence, males and females are not automatically promiscuous and coy, respectively, and sex roles emerge in local context, not as a matter of logical necessity tracing to gamete size.

Sexual Conflict

According to sexual selection, a male and female are always fundamentally in conflict. This stance begins with sexual selection viewing gametic dimorphism as the evolutionary result of conflict between egg and sperm over how much material each should include in the zygote. Sexual selection also postulates a basic conflict between the male and female over how much each should invest in the young. Behaviors in which males are supposed to confine or hurt females to prevent their being fertilized by rival males are highlighted as well. Sexual selection takes conflict as the baseline condition between male and female mates and sees male-female cooperation as a possible (and unlikely) secondary development.

The problem is that data do not confirm any ubiquitous male-female conflict in nature, although such conflict does occasionally happen. In avian biparental care of nestlings, for example, sexual selection predicts that each bird is trying to get the other to do most of the work. If one bird happens to "generously" feed the nestlings with an extra worm, the other is supposed to reduce the number of worms it supplies and use the saved energy to seek extra matings somewhere else. Experiments show, however, that the birds tending a common nest often try to help each other, and do not take advantage of each other (Wright and Cuthill 1990; Markman, YomTov, and Wright 1995; Schwagmeyer, Mock, and Parker 2002; Hinde 2006).

According to social selection, male and female mates fundamentally begin with a cooperative relationship because they have committed themselves to a shared account of evolutionary success. Their offspring represent a common investment whose evolutionary earnings cannot be subdivided. Hurting the other hurts oneself, and helping the other helps oneself, in terms of number of offspring successfully reared. As such, conflict only develops secondarily if a division of labor cannot be successfully negotiated.

Male Promiscuity

According to sexual selection, males are naturally and universally promiscuous, reflecting the low parental investment of a sperm compared to an egg. This rationale for male promiscuity seems patently obvious to sexual-selection theorists, a necessary consequence of defining a sperm as the smaller of the two gamete sizes. But upon reflection, it makes no sense for a male to abandon control of his reproductive destiny, leaving the rearing of his young to chance. A male's evolutionary success is not measured by the number of matings he enjoys but by the number of offspring he sires who are successfully reared.

According to social selection, male promiscuity is a strategy of last resort that occurs when males are excluded from control of offspring rearing. When excluded from a role in offspring rearing, the only strategy remaining is to fertilize as many females as possible. In social selection, male promiscuity is an anomaly, a strategy of last resort.

According to social selection, the widespread phenomenon of extra-pair paternity in birds is not promiscuity, but reflects a system of genetic transactions.

Monogamy

In sexual-selection theory, monogamy is an enigma, a violation of the basic dictate that males should be promiscuous. Therefore, sexual selection explains away the instances of species with monogamous pair bonds, including most birds and some mammals, as entrapment of males by females or as a default when no other mates are available or are otherwise impossible to obtain (Orians 1969; Emlen and Oring 1977; Clutton-Brock 1989; Gowaty 2006). An extensive nomenclature has been developed to classify the ecological conditions in which monogamy and various forms of polygyny are hypothesized to evolve in birds and mammals. In all these, the propensity of males is assumed to be to obtain as many matings as possible, subject to ecological constraints and social opportunity.

In social selection, two distinct forms of monogamy are distinguished: economic monogamy, which represents an agreement to carry out the work of offspring rearing in teams of one male and one female, and genetic monogamy, which represents an agreement not to mate outside the pair bond (Roughgarden 2004). Most monogamy is economic monogamy and nothing requires economic monogamy and genetic monogamy to coincide. In social selection, monogamy emerges in ecological situations where the work of offspring rearing is most efficiently done in male-female teams, rather than as solitary individuals or in teams of more than two individuals.

Extra-Pair Parentage

Extra-pair paternity (EPP) occurs when a male sires young in a nest other than the one he is working at with a female, and extra-pair maternity (EPM) occurs when a female deposits eggs at a nest other than the one she is working at with a male. Both EPPs and EPMs result in extra-pair parentage.

According to sexual selection, extra-pair parentage is "cheating" on the pair bond, the male is said to be "cuckolded," offspring of extra-pair parentage are said to be "illegitimate," and females who do not participate in extra-pair copulations are said to be "faithful" (Wagner 1992; Lifield and Robertson 1992; and Emlen, Wrege, and Webster 1998). This judgmental terminology reflects the failure to distinguish economic from genetic monogamy, and amounts to applying a contemporary definition of human marriage to animals. EPPs are assumed to reflect the inevitable outcome of basic male promiscuity, whereas EPMs are described as "sexual parasitism." The females who deposit eggs in a neighbor's nest are called "brood parasites." According to social selection, extra-pair parentage reflects a system of genetic side payments that stabilizes the social arrangement of economic monogamy when individuals differ in their capacities to contribute to offspring rearing (Akçay and Roughgarden 2007). Distributed parentage also spreads the risk of nest mortality across a network of nests, acting as a social insurance policy (Roughgarden 2007).

Secondary Sexual Characters

According to sexual selection, secondary sexual characters evolve to promote mating success. These traits, like the peacock's tail and the stag's antlers, are favored by females in mate choice so that their own sons will be similarly attractive and victorious when their time for mating comes in the next generation. The "beauty" a female perceives in a male's ornaments is how she apprehends a male's good genes. Male ornaments are "condition indicators" of genetic quality.

According to social selection, ornaments, both male and female, serve as "admission tickets" to power-holding cliques that control the opportunity for successful rearing of offspring, including the opportunity for mating, safety of the young from predation risk, and access of the young to food (Roughgarden, Oishi, and Akçay 2006). Accordingly, a peacock's tail, a rooster's comb, wattle and cockle-doodle-do facilitate male-male interactions, and females are indifferent to them. The reason admission tickets are expensive is that the advantage to membership in a clique resides in the power of monopoly, which is diluted when membership is expanded. By requiring a high price of admission, the monopolistic coalition is kept exclusive, maximizing benefit to those within.

Social selection envisions a class of traits called "social-inclusionary traits" that are needed to participate in the social infrastructure within which offspring are reared (Roughgarden 2004). Not possessing such traits, or not participating in social-inclusionary behaviors, is reproductively lethal. The strong natural selection imposed by the requirement of membership in power-holding cliques can produce the very fast evolution, including possibly runaway evolution, that has long been the signature of sexual selection. Admission tickets are examples of social-inclusionary (SI) traits. Other SI traits include those needed for the communication and cognition necessary to participate in the social infrastructure.

Admission tickets are not the only way to enter power-holding cliques. Conceivably, individuals might be recruited to join, and the admission ticket waived, if they supply capabilities or assets valued by the other members. But if the sole benefit from membership is monopolistic, then membership should require an expensive ticket.

Contradictions to Sexual Selection

Sexual Monomorphism

Species in which males and females are identical in appearance pose a direct contraction to the Darwinian templates that males should be showy and females drab. In many species from penguins to guinea pigs, it's impossible to tell the sexes apart without carefully inspecting their genital area. This phenomenon remains completely unexplained in sexual-selection theory—why in these species don't females bother to choose males on the basis of ornaments? Darwin dismissed the females in these species as lacking a sense of aesthetic.

In social selection theory, sexual monomorphism reflects the absence of same-sex power-holding cliques whose membership requires admission tickets. This should occur in ecological situations where the economically efficient coalition is the coalition of the whole.

Sex-Role Reversal

Species in which the male is drab and the female showy, the reverse of the peacock/peahen comparison, directly contradict the Darwinian "norm." Many such species are known, from sea horse and pipefish to jacanas. Sexual selection claims to "explain" sex role reversal by referring to the relation between parental investment and the operational sex ratio (OSR) (Trivers 1972; Clutton-Brock and Vincent 1991; Clutton-Brock and Parker 1992; Parker and Simmons 1996). The species with the higher parental investment is generally less available for mating than the sex with the lower investment, causing a net surplus of those from the low-investment sex relative to the high-investment sex. The ratio of those from one sex willing to mate to those of the other is called the operational sex ratio. In sex-role reversed species, the male happens to provide more parental investment than the female, say by carrying and/or tending the eggs, so the males are in short supply for mating relative to females. In this situation, the females may compete with one another for access to males, and become the showy sex, whereas the male remains drab. This account, however, is not an *explanation* of sex-role reversal, it is a *redescription* of the phenomenon. Sexual selection does not say why the male should happen to be the sex providing the higher parental investment in such species. Furthermore, the mere existence of sex-role reversed species contradicts the basic tenet of sexual selection that sex roles can be traced to gamete size, because pipefish and jacana males, like all other males, produce tiny sperm. Thus, gamete size does not dictate sex role, and this point represents an internal contradiction in sexual-selection theory.

According to social selection, reversed sex roles are not especially problematic because sex roles are always negotiated in local ecological situations anyway. According to social selection, it is in a male's interest to secure some control of the eggs, thereby retaining some control of his evolutionary destiny. In securing control of the eggs, in some ecological circumstances he might incidentally wind up with a higher parental investment than the female, and if so, sex role reversal might ensue.

Peripheral Narratives

Sexual-selection theory posits "norms" for gender expression and sexuality. Once "norms" are defined, departures from them can be labeled as "alternatives" or "exceptions." To accommodate natural variation in gender expression and sexuality, sexual-selection theory has developed many peripheral narratives for these "alternatives" and "exceptions" that extend sexual selection's central narrative and buttress sexual selection's authority to define normalcy. In contrast, social selection does not single out particular behavioral templates as normal, and theorizes all variation in gender expression and sexuality with the same form of explanation.

Multiple Male and Female Morphs

Many species have more than one type of male and of female, so that matching the males to one template and the females to another is impossible. Hence, updating the Darwinian profiles of "standard" male and female templates is impossible because there is more than one template per sex.

In many species of fish, lizards, and birds, for example, one genotype of male has a large body-size at reproductive age but must survive several years to attain that size, during which time it suffers a high cumulative risk of mortality. But once large, a male can command a territory and defend eggs laid in it. Other males with different genotypes reach reproductive age sooner, do not defend territories, and fertilize eggs that are in the territories defended by large males. This two-male pattern is common. The two males are competitive with each other—the large male chases the small one away when it attempts to fertilize the eggs in the territory he is defending (Howard 1978; Howard 1981; Howard 1984; Lance and Wells 1993; Bass 1992; Goodson and Bass 2000; Gross 1985; Lincoln, Youngson, and Short 1970).

A three-male pattern is also observed in some fish and birds, where the large male solicits the help of a medium-sized male. The pair of them together maintain the territory and participate in courtship. The large male allows the medium male to fertilize some of the eggs in the territory. The small male meanwhile remains a competitor and fertilizes some of the eggs in spite of the large and medium males' attempts to chase it away (Gross 1982; Gross 1991; Dominey 1980; Dominey 1981; Taborsky, Hudde, and Wirtz 1987; Alonzo, Taborsky, and Wirtz 2000; Oliveira and Almada 1998; Moore, Hews, and Knapp 1998; DeNardo and Sinervo 1994; Hogen-Warburg 1966; van Rhijn 1973; van Rhijn, 1983; van Rhijn 1985; van Rhijn 1991; Högland and Alatalo 1995; Lank and Smith 1987; Lank, Smith, Hanotte, Burke, and Cooke 1995; Hugie and Lank 1997; Widemo 1998; Bachman and Widemo 1999; Jukema and Piersma 2006). Furthermore, two or more types of females may also occur, laying different sizes of eggs and/or differing in aggressiveness (Kopachena and Falls 1993; Lowther 1961; Thorneycroft 1975; Houtman and Falls 1994; DeVoogd, Houtman, and Falls 1995).

These species with multiple male and female morphs all defy any attempt to directly apply sexual-selection theory because that theory posits only one template for male and female appearance and behavior. Therefore, sexual selection has been augmented with additional narratives to account for more than one morph per sex.

According to sexual selection, the large territory-holding male is taken as the reference male, and the other types of males are considered as "alternative mating strategies" and are defined as "sexual parasites." A pejorative language masquerades as description throughout the peripheral narratives of sexual selection. The small nonterritory-holding male is termed a "sneaker" who "steals" copulations that rightfully belong to the territory-holding male. The "sneaker" is depicted as stealthily entering the large male's territory through a back door. In fact, the small male is often more numerous than the large male, so he typifies "maleness" in the species more than the large male does, and the small males often band together in the open to chase away the large male and fertilize eggs in the territory rather than enter singly and in stealth.

Calling the nonreference males pejorative names fosters a research culture not open-minded or theoretically neutral. Time and again, "sneaker," as uttered by a teacher or researcher, is accompanied by snickering and guffaws from students and audience. I have never heard the term "sneaker" successfully used as pure description—it is always delivered with a snide joke. This research culture chills innovative inquiry into the social roles of the various types of males, camouflages locker-room banter as science, which then discourages a diverse research community, and enshrines sexual selection as doctrine, not through evidence and logic, but through coercive peer pressure.

I've often wondered what the gratuitous pejorative language of sexual-selection theorists says about themselves. Why are these fish, lizards, and birds so threatening to sexual-selection researchers that these animals cannot be described without descending to gutter humor?

Social selection views reproductive social groups in a completely different light. The various types of males are considered to be "genders," so that some species have two genders of males, or three, perhaps with two or three genders of females too—there may be multiple genders per sex (Roughgarden 2004). The dynamics of a "reproductive social group" extend economic theory for a simple male-female economic team to a larger team with more "social niches." A reproductive social group subsumes the concept of a "family"—which is a reproductive social group whose members happen to be genetically related. In a reproductive social group, some members are "prezygotic helpers"—animals that assist in bringing about courtship and mating, together with "postzygotic helpers"—members who remain at the nest to help rear the offspring that have already been born. Those not included in the reproductive social group's coalition form other arrangements to oppose it, either singly or in coalitions of their own. In this conceptualization, coalitions may form containing medium-sized males who assist in recruiting females to the nests of large males who are guarding the eggs and territory. A large-male-medium-male-female coalition may then be opposed by a small-male coalition that competes to fertilize the eggs. The complex social dynamics for this scenario can be approached with cooperative game theory (Roughgarden, Oishi, and Akçay 2006).

Feminine Males

In species with multiple male morphs (or genders), one of the morphs often has colors or markings shared with females (Gross 1982; Gross 1991; Dominey 1980; Dominey 1981; Taborsky, Hudde, and Wirtz 1987; Alonzo, Taborsky, and Wirtz 2000; Oliveira and Almada 1998; Moore, Hews, and Knapp 1998; DeNardo and Sinervo 1994; Hogen-Warburg 1966; van Rhijn 1973; van Rhijn, 1983; van Rhijn 1985; van Rhijn 1991; Högland and Alatalo 1995; Lank and Smith 1987; Lank, Smith, Hanotte, Burke, and Cooke 1995; Hugie and Lank 1997; Widemo 1998; Bachman and Widemo 1999; Jukema and Piersma 2006) . These "feminine males," as I term them for popular writing (Roughgarden 2004), are described as "female mimics" in sexual-selection theory. Female mimics are sexual parasites who steal the reproductive investment of territory-holding males through deceit. A female mimic is disguised as a female to fool the territory-holding male. The female mimic enters the territory-holding male's harem and mates with his females. This story has not been demonstrated and is implausible.

The concept of mimicry is borrowed from mimicry in predator-prey interactions. A Batesian mimic, for example, is a fly that resembles a bumble bee to avoid predation. The resemblance in Batesian mimicry is exact, and one must look closely through a magnifying glass to discern that a mimic is really a fly and not a bee. A bird on the wing has only a split second to decide whether to grab an insect, and doesn't have the luxury of time to discern what the bug's true identity is.

In contrast, the resemblance of a feminine male to a female is always far less than perfect, and the male invariably has a long time for inspection. The female mimic in the blue-gill sunfish is solicited to join the masculine territory-holding male through a long courtship ritual. Moreover, the behavior of masculine males to feminine males is never identical to the behavior of masculine males to females, indicating that the masculine male is aware of the difference between a female and a feminine male. Finally, deceit narratives require great asymmetry in gullibility. The masculine male is dumb enough not even to know the difference between a male and a female, whereas the feminine male is smart enough to deceive the masculine male on an elementary distinction. Yet fish, lizards, and birds are visual, diurnal animals who make a living with their eyesight and draw fine distinctions in their choice of prey. The entire concept of a female mimic is another locker-room fantasy of sexual-selection theorists—a guy fooled by a queen at a drag bar. In fact, some sexual-selection workers even borrow the phrase "she-male" from pornography to describe male-male courtship in garter snakes (Mason and Crews 1985; Shine, O'Connor, and Mason 2000; Shine, Harlow, Lemaster, Moore, and Mason 2000) instead of the professionally recommended term in entomology, "gynomorphic male" (Hilton 1987).

According to social selection, markings and colors on animals represent "body English"—how animals tell one another what their social role is, what their intentions are, and what activities they promise to perform. Feminine males are participating in a conversation on topics and with words used more frequently by females than by masculine males. The femininely-marked cooperator male hired by the masculine territory-holding male to assist in recruiting females to lay eggs in the territory is a "marriage broker" who uses his feminine markings to help attract and communicate with females. These markings, familiar to females, may encourage a sense of safety that it's OK to entrust her eggs to the territory. The female may already have met the feminine male if he was schooling with her while the masculine male was acquiring and defending the territory. In contrast, the masculine male is probably a complete stranger to her, and she may appreciate the recommendation of a feminine male before depositing eggs there. She needs to be assured that the masculine male will not abandon her eggs.

Masculine Females

In sexual-selection theory, masculine females are discussed under the rubric of "female ornaments"—hanging skin flaps (wattles), colored patches of feathers, antlers, and so forth usually considered as male ornaments (Amundsen 2000; Amundsen, Forsgren, and Hansen 1997; Craig 1996; Dean 1978; Donaldson and Doutt 1965; Wishart 1985; Wislocki 1954; Wislocki 1956; Wong and Parker 1988; Reimers 1993). Among insects, damselflies have been particularly well studied for masculine females (Fincke, Jdicke, Paulson, and Schultz 2005). And the frequency of both feminine males and masculine females has been jointly studied with hummingbirds (Bleiweiss 1992; Bleiweiss 2001).

Darwin dismissed out-of-place ornaments as male traits accidently expressed in females. The genetic system producing the ornaments in males is supposed not to shut off sufficiently in females, so that some females wind up with male traits by mistake, a developmental error.

According to social selection, masculine females are simply the reverse of feminine males, namely, a female using body English to converse on topics and with words used more frequently by males than by the feminine females. Such conversations might involve establishing and defending territories in species where this task is sometimes carried out by females. Masculine females appear underreported because feminine males draw more sensational attention.

Homosexuality

The extent of homosexuality as a natural part of the social systems of animals in their native habitats is just now being appreciated by biologists. However, biologists remain at fault for not publicizing this information—it still is not mentioned in any introductory textbooks, or on nature shows regularly shown on television, or in displays and exhibits at museums. Most people seem to think that homosexuality is "unnatural" and rare. In fact, surveys show over 300 reports now published in peer-reviewed literature documenting natural homosexuality just in birds and mammals (Bagemihl 1999), not to mention the other vertebrates, primates in particular (Fairbanks, McGuire, and Kerber 1977; de Waal 1995; Parish 1996; Vasey 1995; Vasey 1996; Vasey 1998a; Vasey 1998b), and all the invertebrates. I've reviewed and extended these surveys (Roughgarden 2004). In some species, homosexuality is mostly between males, in others, mostly between females, and, in still others, both. In some, homosexuality is relatively uncommon, occurring in about 10% of the matings, and in others as common as heterosexual matings, accounting for 50% of all matings.

According to sexual selection, homosexuality is explained away as an inadvertent mistake, or as deceit. If deceptive, the idea is that a small male sneaks into the territory of a large male, allows the large male to tire by acquiescing to a homosexual copulation, and then proceeds to mate with the females in the large male's harem. This and other lame excuses biologists have used to deny the reality of homosexuality in the animals they study are ludicrous. The sexual-selection narratives to explain homosexuality do, at least, credit the behavior as natural and adaptive to one of the participants, but view it in the pejorative light of exploitation—the "gay" animal has exploited the "straight" animal. Medical narratives of homosexuality do not even go this far, and even today consistently portray homosexuality as some genetic defect or maladaptive disease.

According to social selection, not only is homosexuality natural and adaptive, but the narrative to explain it focuses on positive contributions it brings to both parties (Roughgarden 2004). Homosexuality is grouped with many other social behaviors that involve physical intimacy such as mutual grooming, mutual preening, sleeping together, rubbing tongues together, and even making interlocking calls and other vocalizations. These behaviors allow two animals to work together as a team, to closely coordinate their actions so they make moves simultaneously. Furthermore, these behaviors allow animals a tactile sense of one another's welfare. Teamwork depends not only on coordinated activity, but on pursuit of a common goal, a goal of joint welfare that is the product of the fitness of both individuals. In social selection, the outcomes of cooperative game theory are realized through team play and the perception of team welfare, and homosexuality is merely one of the physically intimate behaviors between animals that enables teamwork (Roughgarden, Oishi, and Akçay 2006).

Application to Humans

Sexual-selection theory is taken as axiomatic to evolutionary psychology, and evolutionary psychologists have uncritically extrapolated the sexual-selection narratives to humans.

Human Attractiveness

Sexual selection is supposed to provide a biological explanation for beauty. Women are supposed to find men handsome who display traits indicating their genetic quality. Conversely, men are supposed to be promiscuous (Buss 1994; Thornhill and Palmer 2000).

According to social selection, males and females choose each other equally, with the criterion for both being compatibility of circumstance, temperament, and inclination that underlies effectiveness at raising offspring in the context of a human social infrastructure.

Human Brain

Sexual-selection theorists posit the human brain as a counterpart of the peacock's tail, an ornament used by men to attract women (Miller 2000). One imagines, I suppose, a man using his big brain to compose lovely sonnets to woo his mate. The problem with this theory is to explain why women have brains. According to sexual selection, a woman's brain is an imperfectly expressed male ornament, as out of place in a woman as a gaudy tail would be on a peahen. So, to explain the brain of women, sexual selection postulates that females use their brains to appreciate the brains of men—only a big-brained woman could appreciate the sonnet of a big-brained man.

Social selection views the human brain as a social-inclusionary trait, a trait needed to participate in the social infrastructure within which offspring are reared. Such a trait is equally necessary in both men and women.

Conclusion

As the table has summarized, social selection offers an evolutionary system of sex, gender, and sexuality as extensive and potentially as detailed as sexual selection does. Social selection offers this alternative even though sexual selection is regarded by many as established truth. Social selection differs from sexual selection not only in how specific behaviors are explained, but also in the worldview of nature that is implied. Social selection rejects the sexual selection's emphasis on conflict, promiscuity, deceit, and bad genes as self-serving locker-room bravado projected onto nature, and then retrieved from nature as though fact.

One might inquire whether the social selection and sexual selection are necessarily opposed. Could sexual selection be right sometimes and social selection right other times? Then we would need some metatheory to say which was correct and when. Or can one take pieces of each to construct a synthetic theory? I don't think so, but time will tell. Sexual selection appears incorrect at every single point, its supposed successes not demonstrated or misinterpreted, its logic riddled with flaws, and its worldview inaccurate. Thus the possibility of synthesizing both theories seems moot. Instead, an alternative evolutionary system of sex, gender, and sexuality is now available.

In her review of this article, Patty Gowaty urges me "not to overlook the work—often undercited or not cited—of a long list of scientists, some of whom are also feminists who have been making many of the same or similar criticisms as yours since at least 1980. Many of the names I thought of are women's . . . me . . . Marlene Zuk, Sarah Hrdy, Mary Jane West Eberhard . . . Zuleyma Tang-Martinez." I have indeed cited and acknowledged the important contributions to animal behavior by these researchers. Furthermore, I would love to present the social-selection alternative to sexual selection detailed in the table above as an extension of their work. But with one exception, all of the scientists Gowaty names have publicly either endorsed sexual selection, opposed my account of social selection, or remained pointedly silent. In contrast, Zuleyma Tang-Martinez recently took the lead in organizing a symposium critical of Bateman's principle, and has written privately to express support. Therefore, I can't present my account as an extension of the collective contributions of feminist evolutionary biologists.

Two reasons seem to explain why critiques of sexual selection from feminist evolutionary biologists have been largely ignored since the 1980s. First, the wording of the critiques is often so guarded and qualified that one can easily overlook the findings, or misinterpret their significance. As a typical example, Snyder and Gowaty's devastating reappraisal (Snyder and Gowaty 2007) of the 1948 Bateman experiments with *Drosophila melanogaster* that are basic to today's sexual selection, concludes with the sentence, "Based on our results, we believe that Bateman's principles are better expressed as hypotheses or questions" and, in the paper's abstract, they end with a "call for repetitions using modern statistical and molecular methods." Well, sure. These modest sentences hardly express any urgency for rejecting sexual selection theory and sound like yet another call for more research. Yet, demolishing the Bateman experiments undercuts the foundation of sexual-selection theory today, and this impact should have been clearly stated. Researchers since the 1980s can be forgiven for overlooking papers whose criticisms of sexual selection are expressed so obliquely.

The other reason why the critiques from feminist evolutionary biologists may have been ignored is that they read as scattered attempts to fix up sexual selection with a missing female perspective, leaving the skeleton intact. For example, Sarah Hrdy, in a review of *Evolution's Rainbow* for *Nature*, endorsed sexual selection by saying, "competition between those of one sex for reproductive access to the other remains a robust explanatory framework" (Hrdy 2004). Yet, this is precisely the issue being disputed. Similarly, the concept of social selection advanced by West-Eberhard extends sexual-selection reasoning to "resources" other than mates (1983). Zuk has signed letters opposing the need to replace sexual selection (Kavanagh 2006). I am grateful to Patty Gowaty for her support during the publication of *Evolution's Rainbow*, but she too has pointedly declined, on many occasions, to agree that sexual selection needs replacing. So, feminist evolutionary biologists almost unanimously support sexual selection as a whole, and no coherent stand has emerged in opposition to it. Their critiques then come to be perceived as a collage of small complaints each of which can be dealt with individually as needed.

Adding a more accurate female perspective to sexual selection is arguably an improvement over the status quo, but seems representative of feminist scholarship fifty years ago. Scholarship of that era accepted the male-female binary as a given and wished to define it better, and to insure women's rights. Later feminist scholarship, such as that of Gayle Ruben (1975), questions why there is a binary at all. I see this paper as an extension into science of this "third wave" of feminist theory.

Acknowledgments

I thank Patty Gowaty for providing detailed comments on this article and for sharing a prepublication manuscript reanalyzing the Bateman experiments. I also thank Erol Akçay for allowing me to quote an unpublished survey of extra-pair reproductive activity in birds together with unpublished results on a cooperative-game-theoretic transaction theory for extra-pair reproduction, and I thank Priya Iyer for allowing me to quote unpublished results on the evolution of anisogamy.

References

Akçay, E., and J. Roughgarden. 2007. Extra-pair reproductive activity: Review of data and a new theory based on transactions in a cooperative game. (In preparation).

Alonzo, S., M. Taborsky, and P. Wirtz. 2000. Male alternative reproductive behaviours in a Mediterranean wrasse, *Symphodus ocellatus*: Evidence from otoliths for multiple life-history pathways. *Evolutionary Ecology Research* 2:997–1007.

Amundsen, T. 2000. Why are female birds ornamented? *Trends in Ecology and Evolution* 15:149–155.

Amundsen, T., E. Forsgren, and L. T. T. Hansen. 1997. On the function of female ornaments: Male bluethroats prefer colourful females. *Proceedings of the Royal Soc.* B 264 (1388):1579–1586.

Bachman, G., and F. Widemo. 1999. Relationships between body composition, body size and alternative reproductive tactics in a lekking sandpiper, the ruff (*Philomachus pugnax*). *Functional Ecology* 13:411–416.

Bagemihl, B. 1999. *Biological Exuberance: Animal Homosexuality and Natural Diversity*. New York: St. Martin's Press.

Barrette, C. 1993. The "inheritance of dominance," or an aptitude to dominate? *Animal Behaviour* 46:591–593.

Bass, A. 1992. Dimorphic male brains and alternative reproductive tactics in a vocalizing fish. *Trends in Neurosciences* 15:139–145.

Bateman, A. J. 1948. Intrasexual selection in *Drosophila*. *Journal of Heredity* 2:349–368.

Bleiweiss, R. 1992. Widespread polychromatism in female sunangel hummingbirds (*Heliangelus: Trochilidae*). *Biological Journal of the Linnean Society* 45:291–314.

Bleiweiss, R. 2001. Asymmetrical expression of transsexual phenotypes in hummingbirds. *Proceedings of the Royal Society of London* B 268:639–646.

Bonduriansky, R. 2001. The evolution of male mate choice in insects: A synthesis of ideas and evidence. *Biology Review* 78:305–339.

Bullough, W. S. 1947. Hermaphroditism in the lower vertebrates. *Nature* 266:828–830.

Bulmer, M.G., and G.A. Parker. 2002. The evolution of anisogamy: A game theoretic approach. *Proceedings of the Royal Society of London* B 269:2381–2388.

Burley, N.T., and R. Symanski. 1998. "A taste for the beautiful": latent aesthetic mate preferences for white crests in two species of Australian grassfinches. *The American Naturalist* 152 (6):792–802.

Buss, D. 1994. *The Evolution of Desire*. New York: Basic Books.

Capitanio, J. P. 1991. Levels of integration and the "inheritance of dominance." *Animal Behaviour* 42:495–496.

Capitanio, J. P. 1993. More on the relation of inheritance to dominance. *Animal Behaviour* 46:600–602.

Chao, Lin. 1990. Fitness of RNA virus decreased by Muller's ratchet. *Nature* 348:454–455.

Charlesworth, B. 1978. The population genetics of anisogamy. *Journal of Theoretical Biology* 73:347–357.

Charnov, E. L. 1982. *The Theory of Sex Allocation*. Princeton, N.J.: Princeton University Press.

Clutton-Brock, T. H. 1989. Mammalian mating systems. *Proceedings of the Royal Society of London* B 236:339–372.

Clutton-Brock T. H., and G. A. Parker. 1992. Potential reproductive rates and the operation of sexual selection. *The Quarterly Review of Biology* 67:437–456.

Clutton-Brock T. H., and A. C. J. Vincent. 1991. Sexual selection and the potential reproductive rates of males and females. *Nature* 351:58–60.

Cockburn, A. 2006. Prevalence of different modes of parental care in birds. *Proceedings of the Royal Society* B 273:1375–1383.

Cowan, I. McT. 1946. Antlered doe mule deer. *Canadian Field-Naturalist* 60:11–12.

Cox, P.A., J.A. Sethian. 1984. Search, encounter rates and the evolution of anisogamy. *Proceedings of the National Academy of Science* USA 81:6078–6079.

Coyne, Jerry. 2004. "Charm schools" (Review of *Evolution's Rainbow* by J. Roughgarden). *Times Literary Supplement*. 30 July 2004.

Craig, A. 1996. The annual cycle of wing moult and breeding in the wattled starling *Creatophora cinera*. *Ibis* 138:448–454.

Darwin, Charles. 1871. *The Descent of Man, and Selection in Relation to Sex*. Princeton, NJ: Princeton University Press. (facsimile edition)

DeNardo, D., and B. Sinervo. 1994. Effects of corticosterone on activity and home-range size of free-ranging male lizards. *Hormones and Behavior* 28:53–65 and 273–287.

Dean, W. 1978. Plumage, reproductive condition, and moult in nonbreeding wattled starlings. *Ostrich* 49:97–101.

DeVoogd, T., A. Houtman, and J. Falls. 1995. White-throated sparrow morphs that differ in song production rate also differ in the anatomy of some song-related brain areas. *Neurobiology* 28:202–213.

de Waal, F. 1995. Bonobo sex and society. *Scientific American* March, 82–88.

Dewsbury, D. A. 1982. Ejaculate cost and male choice. *The American Naturalist* 119:601–610.

Dewsbury, D. A. 1991. Genes influence behaviour. *Animal Behaviour* 42:499–500.

Dewsbury, D. A. 1993. More on the inheritance of dominance relationships: extending the concept of the phenotype. *Animal Behaviour* 46:597–599.

Dewsbury, D. A. 2005. The Darwin-Bateman paradigm in historical context. *Integrative and Comparative Biology* 45:831–837.

Dominey, W. J. 1980. Female mimicry in bluegill sunfish—A genetic polymorphism? *Nature* 284:546–548.

Dominey, W. J. 1981. Maintenance of female mimicry as a reproductive strategy in bluegill sunfish (*Lepomis macrochirus*). *Environmental Biology of Fishes* 6:59–64.

Donaldson, J. C., and J. Doutt. 1965. Antlers in female white-tailed deer: A 4–year study. *Journal of Wildlife Management* 29:699–705.

Doust, J., and L. Doust. 1983. Parental strategy: Gender and maternity in higher plants. *BioScience* 33:180–186.

Emlen, S. T., and L. W. Oring. 1977. Ecology, sexual selection, and the evolution of mating systems. *Science* 197:215–223.

Emlen, S., P. H. Wrege, and M. S. Webster. 1998. Cuckholdry as a cost of polyandry in the sex-role reversed wattled jacana, *Jacana jacana*. *Proceedings of the Royal Society of London* B 265:2359–2364.

Fairbanks, L., M. McGuire, and W. Kerber. 1977. Sex and aggression during rhesus monkey group formation. *Aggressive Behavior* 3: 241–249.

Fincke, Ola M., Reinhard Jdicke, Dennis R. Paulson, and Thomas D. Schultz. 2005. The evolution and frequency of female color morphs in *Holarctic Odonata*: Why are male-like females typically the minority? *International Journal of Odonatology* 8:183–212.

Forsgren, E. 1997. Female sand gobies prefer good fathers over dominant males. *Proceedings of the Royal Society of London* B 264:1283–1286.

Geber, M., T. E. Dawson, and L. F. Delph, eds. 1999. *Gender and Sexual Dimorphism in Flowering Plants*. Berlin: Springer-Verlag.

Ghiselin, M. 1969. The evolution of hermaphroditism among animals. *Quarterly Review of Biology* 44:189–209.

Goodson, J. L., and Bass, A. H. 2000. Forebrain peptides modulate sexually polymorphic vocal circuitry. *Nature* 403:769.

Gowaty, P. 2006. Beyond extra-pair paternity: individual constraints, fitness components, and social mating systems. In *Essays on Animal Behavior: Celebrating 50 Years of Animal Behavior*, ed. Jeff Lucas and Lee Simmons, 221–256. Cambridge: Cambridge University Press.

Gowaty, P. A. 1996. Field studies of parental care in birds: New data focus questions on variation in females. In *Advances in the Study of Behav-*

iour, ed. C. T. Snowdon and J. S. Rosenblatt, 476–531. New York: Academic Press.

Gowaty, P. A., and S. P. Hubbell. 2005. Chance, time allocation, and the evolution of adaptively flexible sex role behavior. *Integrative and Comparative Biology* 45:931–944.

Gross, M. R. 1985. Disruptive selection for alternative life histories in salmon. *Nature* 313:47–48.

Gross, M. R. 1991. Evolution of alternative reproductive strategies: frequency-dependent sexual selection in male bluegill sunfish. *Philosophical Transactions of the Royal Society of London* B 332:59–66.

Gross, M. R. 1982. Sneakers, satellites and parentals: Polymorphic mating strategies in North American sunfishes. *Zeitschrift für Tierpsychologie*. 60:1–26.

Haigh, John, 1978. The accumulation of deleterious genes in a population—Muller's Ratchet. *Theoretical Population Biology* 14:251–267.

Hamilton, W.D., R. Axelrod, and R. Tanese. 1990. Sexual reproduction as an adaptation to resist parasites (A Review). *Proceedings of the National Academy of Sciences* (USA) 87:3566–3573.

Hilton, D. 1987. A terminology for females with color patterns that mimic males. *Entomological News* 98:221–223.

Hinde, C. 2006. Negotiation over offspring care?–A positive response to partner-provisioning rate in great tits. *Behavioral Ecology* 17: 6–12.

Hogen-Warburg, A. 1966. Social behavior of the ruff, *Philomachus pugnax* (L.) *Ardea* 54:109–229.

Högland, J., and R. Alatalo. 1995. *Leks*. Princeton, NJ: Princeton University Press.

Houtman, A., and J. Falls. 1994. Negative assortative mating in the white-throated sparrow *Zonotrichia albicollis*: The role of mate choice and intrasexual competition. *Animal Behavior* 48:377–383.

Howard, R. D. 1978. The evolution of mating strategies in bullfrogs, *Rana catesbeiana*. *Evolution* 32:859–871.

Howard, R. D. 1981. Sexual dimorphism in bullfrogs. *Ecology* 62:303–310.

Howard, R. D. 1984. Alternative mating behaviors in young male bullfrogs. *American Zoologist* 24:397–406.

Hrdy, S. 2004. Sexual diversity and the gender agenda. *Nature* 429:19–20.

Hugie, D., and D. Lank. 1997. The resident's dilemma: a female choice model for the evolution of alternative mating strategies in lekking male ruffs (*Philomachus pugnax*). *Behavioral Ecology* 8:218–225.

Iyer, P., and J. Roughgarden. 2007. Origin of male and female: Evolution of anisogamy reanalyzed. (In preparation)

Jukema, J., and T. Piersma. 2006. Permanent female mimics in a lekking shorebird. *Biology Letters* 2:161–164.

Kalmus, H. 1932. Ueber den Erhaltungswert den phaenotypischen Anisogamie und die Entstehung der ersten Geschlectsunterschiede. *Biol. Zentral*. 52:716.

Kavanagh, Etta, ed. 2006. Debating sexual selection and mating strategies. *Science* 312:689–697.

Kirkpatrick, Mark, and Ryan, Michael J. 1991. The evolution of mating preferences and the paradox of the lek. *Nature* 350:33–38.

Kopachena, J., and J. Falls. 1993. Aggressive performance as a behavioral correlate of plumage polymorphism in the white-throated sparrow (*Zonotrichia albicollis*). *Behavior* 124:249–266.

Kuamura, T., Y. Nakashima, and Y. Yogo. 1994. Sex change in either direction by growth-rate advantage in the monogamous coral goby, *Paragobiodon echinocephalus*. *Behavioral Ecology* 5:434–438.

Lack, D. 1968. *Ecological Adaptations for Breeding in Birds*. London, UK: Chapman & Hall.

Lance, S. L., and K. D. Wells. 1993. Are spring peeper satellite males physiologically inferior to calling males? *Copeia* 1993:1162–1166.

Lank, D., and C. Smith, 1987. Conditional lekking in ruff (*Philomachus pugnax*). *Behavioral Ecology and Sociobiology* 20:137–145.

Lank, D. B., C.M. Smith, O. Hanotte, T. Burke, and F. Cooke. 1995. Genetic polymorphism for alternative mating behaviour in lekking male ruff (*Philomachus pugnax*). *Nature* 378:59–62.

Lifjeld, J., and R. Robertson. 1992. Female control of extra-pair fertilization in tree swallows. *Behavioral Ecology and Sociobiology* 31:89–96.

Lincoln, G., R. Youngson, and R. Short. 1970. The social and sexual behavior of the red deer stag. *Journal of Reproduction and Fertility* suppl. 11:71–103.

Lowther, J., 1961. Polymorphism in the white-throated sparrow. *Zonotrichia albicollis* (Gmelin) *Canadian Journal of Zoology* 39:281–292.

Markman, S., Y. YomTov, and J. Wright. 1995. Male parental care in the orange-tufted sunbird–behavioural adjustments in provisioning and nest guarding effort. *Animal Behaviour* 50:655–669.

Mason, R., and D. Crews. 1985. Female mimicry in garter snakes. *Nature* 316:59–60.

Matsuda, H., and P.A. Abrams. 1999. Why are equally sized gametes so rare? The instability of isogamy and the cost of anisogamy. *Evolutionary Ecology* Res. 1:769–784.

Maynard Smith, John. 1982. *Evolution and the Theory of Games*. Cambridge University Press.

Miller, G. 2000. *The Mating Mind, How Sexual Choice Shaped the Evolution of Human Nature*. New York: Anchor Books (Random House).

Moore, A. J. 1991. Genetics, inheritance and social behaviour. *Animal Behaviour* 42:497–498.

Moore, A. J. 1993. Towards an evolutionary view of social dominance. *Animal Behaviour* 46:594–596.

Moore, A. J., E. D. Brodie III, and J. B. Wolf. 1997. Interacting phenotypes and the evolutionary process. I. Direct and indirect genetic effects of social interactions. *Evolution* 51:1352–1362.

Moore, A. J., J. B. Wolf, and E. D. Brodie III. 1998. The influence of direct and indirect genetic effects on the evolution of behavior: social and sexual selection meet maternal effects. In *Maternal Effects as Adaptations*, ed. T.A. Mousseau and C.W. Fox, 22–41. Oxford: Oxford University Press.

Moore, Allen J., Kenneth F. Haynes, Richard F. Preziosi, and Patricia J. Moore. 2002. The Evolution of Interacting Phenotypes: Genetics and Evolution of Social Dominance. *The American Naturalist* 160: S186–S197.

Moore, M., D. Hews, and R. Knapp. 1998. Hormonal control and evolution of alternative male phenotypes: generalizations of models for sexual differentiation. *American Zoologist* 38:133–151.

Mousseau, T.A., and C. W. Fox, eds. *Maternal effects as adaptations*. Oxford: Oxford University Press.

Munday, P., M. Caley, and G. Jones. 1998. Bi-directional sex change in a coral-dwelling goby. *Behavioral Ecology and Sociobiology* 43:371–377.

Oliveira, R., and V. Almada. 1998. Mating tactics and male-male courtship in the lek-breeding cichlid *Oreochromis mossambicus*. *Journal of Fish Biology* 52:1115–1129.

Orians, G. H. 1969. On the evolution of mating systems in birds and mammals. *The American Naturalist* 103:589–603.

Parish, A. 1996. Female relationships in bonobos (*Pan paniscus*), evidence for bonding, cooperation, and female dominance in a male-philopatric species. *Human Nature* 7:61–96.

Parker, G.A., R.R. Baker, and V.G.F. Smith. 1972. The origin and evolution of gamete dimorphism and the male-female phenomenon. *Journal of Theoretical Biology* 36:529–553.

Parker, G. A., and L. W. Simmons. 1996. Parental investment and the control of sexual selection: Predicting the direction of sexual competition. *Proceedings: Biological Sciences* 263:315–321.

Parker, P. G., and Z. Tang-Martinez. 2005. Bateman gradients in field and laboratory studies: A cautionary tale. *Integrative and Comparative Biology* 45:895–902.

Policansky, D. 1981. Sex choice and the size-advantage model in Jack-in-the-pulpit (*Arisaema triphyllum*). *Proceedings of the National Academy of Sciences* USA 78:1306–1308.

Radwan, Jacek. 2004. Effectiveness of sexual selection in removing mutations induced with ionizing radiation. *Ecology Letters* 7:1149–1154.

Randerson, J. P., and L.D. Hurst. 2001. The uncertain evolution of the sexes. *Trends in Ecology & Evolution* 16:571–579.

Reimers, E. 1993. Antlerless females among reindeer and caribou. *Canadian Journal of Zoology* 71:1319–1325.

Reynolds, J. D., N. B. Goodwin, and R. P. Freckleton. 2002. Evolutionary transitions in parental care and live bearing in vertebrates. *Philosophical Transactions of the Royal Society of London* B 357:269–281.

Ross, R. M. 1990. The evolution of sex change mechanisms in fishes. *Environmental Biology of Fishes* 29:81–93.

Roughgarden, Joan. 2004. *Evolution's Rainbow: Diversity, Gender and Sexuality in Nature and People*. University of California Press.

Roughgarden, Joan. 1991. The evolution of sex. *The American Naturalist* 138:934–953.

Roughgarden, Joan, Meeko Oishi, and Erol Akçay. 2006. Reproductive social behavior: cooperative games to replace sexual selection. *Science* 311:965–969.

Rowe, Locke, and David Houle. 1996. The lek paradox and the capture of genetic variance by condition-dependent traits. *Proceedings of the Royal Society of London* B 263:1415–1421.

Ruben, Gayle. 1975. The traffic in women: Notes on the 'political economy' of sex. In *Toward an Anthropology of Women*, ed. Rayna R. Reiter, 175–185. New York: Monthly Review Press.

Sadovy, Y., and D.Y. Shapiro. 1987. Criteria for the diagnosis of hermaphroditism in fishes. *Copeia* 136–156.

Schwagmeyer, P., D. Mock, and G. Parker. 2002. Biparental care in house sparrows: Negotiation or sealed bid. *Behavioral Ecology* 13:713–721.

Scudo, F. M. 1967. The adaptive value of sexual dimorphism: I, anisogamy. *Evolution* 21:285–291.

Shine, R., D. O'Connor, and R. Mason. 2000. Female mimicry in garter snakes: Behavioural tactics of "she-males" and the males that court them. *Canadian Journal of Zoolology* 78:1391–1396.

Shine, R., P. Harlow, M. Lemaster, I. Moore, and R. Mason. 2000. The transvestite serpent: Why do garter snakes court (some) other males? *Animal Behaviour* 59:349–359.

Snyder, B., and P. A. Gowaty. 2007. A reappraisal of Bateman's classic study of intrasexual selection in *Drosophila*. (In preparation).

Taborsky, M., B. Hudde, P. Wirtz. 1987. Reproductive behavior and ecology of *Symphodus (Crenilabrus) ocellatus*, a European wrasse with four types of male behavior. *Behavior* 102:82–118.

Tang-Martinez, Z. 2000. Paradigms and primates: Bateman's principle, passive females, and perspectives from other taxa. In *Primate Encounters: Models of Science, Gender, and Society*, ed. S. Strum and L. Fedigan. Chicago: University of Chicago Press.

Tang-Martinez, Z., and T. B. Ryder. 2005. The problem with paradigms. Bateman's worldview as a case study. *Integrative and Comparative Biology* 45:821–830.

Thorneycroft, H., 1975. A cytogenetic study of the white-throated sparrow, *Zonotrichia albicollis*. *Evolution* 29:611–621.

Thornhill, R., and C. Palmer. 2000. *A Natural History of Rape: Biological Bases of Sexual Coercion*. Cambridge, MA: MIT Press.

Tomkins, Joseph L., Jacek Radwan, Janne S. Kotiaho, and Tom Tregenza. 2004. Genic capture and resolving the lek paradox. *Trends in Ecology and Evolution* 19:323–328.

Trivers, R. L. 1972. Parental investment and sexual selection. In *Sexual Selection and the Descent of Man*, ed. B. Campbell, 136–179. Chicago: Aldine.

van Rhijn, J. G. 1973. Behavioural dimorphism in male ruffs *Philomachus pugnax* (L.). *Behaviour* 47:153–229.

van Rhijn, J. G. 1983. On the maintenance and origin of alternative strategies in the ruff *Philomachus pugnax*. *Ibis* 125:482–498.

van Rhijn, J. G. 1985. A scenario for the evolution of social organization in ruffs *Philomachus pugnax* and other Charadriiform species. *Ardea* 73:2537.

van Rhijn, J. G. 1991. *The Ruff*. Poyser.

Vasey, P. 1995. Homosexual behavior in primates: A review of evidence and theory. *International Journal of Primatology* 16:173–204.

Vasey, P. 1996. Interventions and alliance formation between female Japanese macaques, *Macaca fuscata*, during homosexual consortships. *Animal Behavior* 52:539–551.

Vasey, P. 1998a. Female choice and intersexual competition for female sexual partners in Japanese macaques. *Behaviour* 135:579–597.

Vasey, P. 1998b. Intimate sexual relations in prehistory: Lessons from the Japanese macaques. *World Archaeology* 29:407–425.

Wagner, R. 1992. The pursuit of extra-pair copulations by monogamous female razorbills: How do females benefit? *Behavioral Ecology and Sociobiology* 29:455–464.

Warner, R. R. 1982. Mating systems, sex change and sexual demography in the rainbow wrasse, *Thalassoma lucasanum*. *Copeia* 653–661.

West-Eberhard, Mary Jane. 1983. Sexual selection, social competition and speciation. *Quarterly Review of Biology* 58:155–183.

Widemo, F. 1998. Alternative reproductive strategies in the ruff (*Philomachus pugnax*): A mixed ESS? *Animal Behavior* 56:329–336.

Wishart, W. 1985. Frequency of antlered white-tailed does in Camp Wainright, Alberta. *Journal of Mammalogy* 35:486–488.

Wislocki, G. 1954. Antlers in female deer, with a report on three cases in *Odocoileus*. *Journal of Mammalogy* 35:486–495.

Wislocki, G. 1956. Further notes on antlers in female deer of the genus *Odocoeleus*. *Journal of Mammalogy* 37:231–235.

Wolf, J. B., E. D. Brodie III, and A. J. Moore. 1999. Interacting phenotypes and the evolutionary process. II. Selection resulting from social interactions. *The American Naturalist* 153:254–266.

Wong, B., and K. Parker. 1988. Estrus in black-tailed deer. *Journal of Mammalogy* 69:168–171.

Wright, J., and I. Cuthill. 1990. Biparental care: Short-term manipulation of partner contribution and brood size in the starling, *Sturmus vulgaris*. *Behavioral Ecology* 1:116–124.

22

It's a Boy!

Gender Expectations Intrude on the Study of Sex Determination

Caitilyn Allen

Femaleness is a deformity, albeit one that occurs in the ordinary course of nature.

—Aristotle

Introduction

The practice of science can be distorted by bias at multiple levels. Bias can intrude at any point along the process, spanning the formulation of a hypothesis, design of experiments testing that hypothesis, interpretation of the resulting data, and development of broad explanatory models. Less obvious, however, are the biases that limit which questions are asked in the first place. The absence of any research focus on a problem can go unnoticed. But who determines which issues are interesting and which are not worthy of scientific investigation? The study of human sex determination offers an example of this type of invisible bias. While extensive research has focused on the developmental pathway leading to a male baby, surprisingly little effort has gone into understanding how a human embryo becomes female.

Early in human development, male and female embryos cannot be distinguished from each other. In their abdomens are structures, charmingly named the indifferent gonads, which will eventually become the reproductive organs. Then, around forty days after conception, something happens. The indifferent gonads gradually become either ovaries or testes. In the female, a nearby structure called the Mullerian duct differentiates into the fallopian tubes, uterus, cervix, and upper vagina, while another nearby structure called the Wolffian duct degenerates. In the male, the Wolffian duct becomes the epididymis, the vas deferens, and the seminal vesicles, and the Mullerian duct degenerates. As the fetus continues to grow, an indeterminant nub of external tissue grows into either a clitoris, vulva, and lower vagina or a penis and scrotum (Fausto-Sterling 1992). What initiates these changes and determines the sex of the embryo? The answers depend on when and where the question is asked.

In this essay, I will argue that social gender expectations have confounded the study of biological sex determination. This confusion led to a distorted research focus wherein biologists essentially ignored half the process of human sex determination. Further, a combination of two forms of social bias generated an erroneous biological paradigm concerning this process. The first bias arose from the misclassification of intersex individuals as males or females, usually depending on whether or not a plausible penis was present. The second source of bias was the unexamined working assumption that, just as men are expected to be socially dominant and women are stereotypically socially passive, the biological process by which the developing embryo becomes male must be active, while the biological process of becoming female must be passive.

Humans have long wished to understand and control the sex of offspring. This desire arose from agricultural preferences for more productive cows and hens over bulls and roosters, as well as parents' desire for sons to work the land and inherit titles or property. Despite an abiding interest in the problem of how sex was determined, until the early 20th century, the process remained so obscure that the question gave rise only to what we would now regard as crackpot theories.

An early attempt at a scientific explanation was put forward by Aristotle, who believed that a woman was a defective male, an inherently flawed creature defined by what she lacked. "We must look upon the female character as being a sort of natural deficiency," he explained (1984); for a more thorough consideration of Aristotle's views on the biology of the sexes, see Tuana, 1989. The bilateral symmetry of the human form, combined with the existence of two sexes, suggested an obvious mechanism of sex determination. Perhaps one testicle gave rise to girl children and the other produced boys? Appalachian folk wisdom asserts that a man who loses one testicle will subsequently sire children of only one sex (Katherine Knotts, Leadmine, West Virginia, personal communication).

At the beginning of this century, Dr. William Kraft, editor of *The American Physician*, believed that he had "at last pierced the veil which surrounded the mystery of sex determination" (1908). Interestingly, Kraft attributed the power of determining sex to the ovum, which he believed went through different "seasons of sex" every six hours as the tide rose and fell. Dr. Kraft's argument, articulated in a book published in 1908, went as follows: 1. Males are better adapted to survive hardship than females (he felt this axiom was self-evident and needed no proof), therefore: 2. more males are conceived during times of hardship, while females predominate during conditions of plenty. He then postulated that: 3. in the evolutionary past, hardship corresponded to floods, while plenty resulted from low water, and he further postulated that: 4. the modern equivalent of flood and low water is the tidal cycle. Based on this line of reasoning, he asserted (without benefit of any empirical supporting data) that if "conjuction" between the sexes occurs during the rising tide in a given part of the world, male offspring will result. Fertilization while the tide is going out will produce female offspring (Kraft 1908).

The Feminizing X Chromosome

A more data-driven understanding of sex determination emerged with the discovery of chromosomes and the blossoming in the 1920s of cytogenetics (the study of chromosomes under the microscope). Cytogeneticists observed that cells from a male contain one X and one Y chromosome, while cells from females have two X chromosomes. Until 1959, scientists assumed that sex was determined in humans as it is in the fruit fly, the organism on which many pathbreaking genetic studies were done in the first half of the 20th century. In fruit flies, sex is determined by the ratio of X chromosomes to the other chromosomes. Thus, fruit flies are female if they carry a double dose of X chromosomes; conversely, if there is only one copy of the X chromosome, an individual is male.

This understanding was based on the discovery of fruit flies that carried abnormal numbers of sex chromosomes, rather than the usual pair of either XX or XY. Biologists observed that an XXY fruit fly is female, while a fruit fly is male if it has only one X chromosome and no Y chromosome (technically known as an XO individual). In other words, the number of X chromosomes controlled sex; the presence or absence of the Y chromosome had no apparent effect. This model was known as the balance theory of sex determination. In the absence of any direct evidence concerning sex determination in humans, biologists concluded: "We may assume that this balance theory of sex holds for man [*sic*]" (Stern 1949). Notwithstanding its evenhanded name, scientific discussions of the balance theory portrayed human sex determination in terms of a struggle between undefined male and female forces. A medical genetics text explained: "a preponderance of femaleness over maleness leads to a female, while a preponderance of maleness over femaleness leads to a male. . . . One X is insufficient to outweigh the influence of male determiners, but two X's are able to do so" (Stern 1949). A general biology text described the "masculinizing effect" which must be "sufficiently strong to override the feminizing influence of the X-chromosome" (Weisz 1959).

The Dominance of the Y Chromosome

The balance paradigm of sex determination in humans was overturned in 1959. In that year, scientists discovered several people with abnormal numbers of sex chromosomes who did not appear as the fruit fly–based balance model predicted. One paper identified an XXY person who had a penis and scrotum (Jacobs and Strong 1959), while a second described an XO individual with vulva and vagina (Ford, et al. 1959). These examples strongly suggested that, instead of being determined by the number of X chromosomes as it is in the fruit fly, human sex was determined by the presence or absence of the Y chromosome. Put another way, Aristotle was right: females could be defined by what they lacked.

This evidence spawned a powerful new developmental paradigm that has gone largely unchallenged to date. In this paradigm, maleness was *dominant* over femaleness, in the genetic sense of the word. What is genetic dominance? Humans carry two versions (alleles) of most genes; one copy is inherited from the mother and the other comes from the father. In some cases, one version is expressed while the other is silent; the silent version is said to be recessive, while the one that is expressed is called dominant. A familiar example is that of eye color, where brown eye color is usually dominant over blue. It is important to note that the relatively simple dominant/recessive model does not accurately describe the interactions between most pairs of human genes. Many genes are codominant or have complex or additive effects.

The idea that the presence of a Y chromosome was sufficient to produce a male made intuitive sense, possibly because maleness is implicitly defined in terms of the presence of a penis. Physicians treating children born with ambiguous genitals often assign a sex to the affected child based on the size and appearance of the penis/clitoris. If the structure is large enough, the child is labeled male; if the structure is considered too small to make a plausible penis, however, it is usually surgically removed and the child is surgically and hormonally constructed as a female (albeit often without the capacity for orgasm) (Wijngaard 1997). Even at the level of single-celled organisms, the operational definition of maleness is revealingly phallocentric. When microbiologists discovered that bacteria can take up DNA in a process that is far from analogous to sex (as no "baby bacterium" results—rather, a fragment of DNA passes from donor to recipient), the donor bacterium, which transfers DNA through a pilus tube, was immediately labeled male, while the recipient bacterium was labeled female (Spainer, 1995).

The dominant Y model of sex determination was also congruent with the concepts emerging from the study of hormones in male and female development. Marianne van den Wijngaard suggests that "scientists regarded development of the male as a modification of female development: male genitalia are created by the addition of 'something masculine' (i.e., androgens)" (1997). One researcher framed the issue more antagonistically:

> The "battle of the sexes" begins early in intrauterine life. Gestation takes place in a maternal estrogenic environment. Thus it becomes necessary, in dealing with male embryos, to have some factors to counteract this influence, else all individuals, irrespective of genotype, would be born as female. (Meyer 1966)

Textbook language suggests that biologists saw the Y chromosome as a surrogate for the male gender in a social setting. Keeton's widely-used introductory biology text explained: "The human Y chromosome bears genes with *strong* male-determining properties, and it is the presence of the Y that determines maleness and its absence that determines femaleness" (1972; emphasis added). Perhaps because it was consistent with the social ideal of active male leadership at the time, biologists embraced evidence that the active masculine Y chromosome was dominant over the passive feminine X chromosome. The virile Y chromosome set the agenda and controlled development. So appealing was this model that biologists ignored conflicting evidence suggesting the existence of a corresponding active female developmental pathway directed by a complex, interactive regulatory web of its own. The metaphor took over.

The developmental mechanism resulting in a female was not given much attention either in the literature or in the laboratory. Female development was described as the passive default outcome if no Y chromosome was present to activate the male developmental pathway. A 1973 biology text discussed sex determination without even mentioning the possibility of femaleness: "In humans, it is the presence or absence of the Y chromosome that determines maleness" (Wasserman 1973).

Possibly because it was seen as a passive process, research did not focus on the genetics of female development. Rather, it seemed logical and easier to try to identify the "gene for maleness" from the Y chromosome, since it conferred an attribute that was easily measured. It is also possible that unconsciously, biologists assumed that understanding male development was the more significant and interesting problem. The hypothetical gene for maleness on the Y chromosome was known as testis-determining factor, or TDF. As sophisticated molecular biology techniques made it possible to isolate and characterize individual human genes, several research groups entered a race to clone TDF.

The Discovery of SRY: The Dominant Y Paradigm Reinforced

After a false start (Page, et al. 1987), in 1990 a 10-member British team isolated a candidate gene from the Y chromosome and named it the sex determining region of Y or SRY (Sinclair, et al. 1990). This gene fit the expectations for the testis-determining factor in several important ways. First, the gene was both male-specific and evolutionarily conserved. Genes resembling SRY were found on the Y chromosomes of all mammals tested. Second, the presence or absence of SRY correlated with sex in humans. Some XY individuals who lacked SRY looked female, while a group of XX humans who appeared male or ambiguous carried a piece of the Y chromosome that included SRY, suggesting that SRY alone was enough to confer maleness or at least disrupt normal female development (Sinclair, et al. 1990). Third, the expression pattern of the gene was consistent with a role in sex determination. The mouse gene equivalent to SRY was turned on in the indifferent gonad just at the critical moment when the testes begin to develop (Koopman, et al. 1990). Finally, addition of the SRY gene to XX mice made them develop male reproductive structures (Koopman, et al. 1991). Collectively, these findings established that SRY plays a key role in the differentiation of males.

Further, the structure of the SRY protein hinted at its function. Part of SRY looks very much like a group of proteins known to control gene expression; in particular, it bears a striking resemblance to the proteins that regulate mating type in baker's yeast (which is somewhat analogous to sex as one organism of each mating type is required to form sexual spores) (Sinclair, et al. 1990). Consistent with such a regulatory role, some evidence suggests that SRY controls expression of a key male developmental pathway hormone called Mullerian Inhibiting Substance, or MIS (Haqq, et al. 1994). Recall that early in embryonic development, males and females have both Mullerian ducts and Wolffian ducts. In females, the Mullerian duct develops into the fallopian tubes and uterus, while the Wolffian duct degenerates. In males, the Wolffian duct eventually becomes part of the male reproductive system, while the Mullerian duct degenerates. Thus, MIS, the signal that causes Mullerian ducts to regress, is considered a crucial switch in male development.

All in all, the discovery of SRY gratified the expectations created by the dominant Y model and reinforced it substantially. If SRY was a regulatory protein that initiated testicular development and triggered MIS production, this explained perfectly how the Y chromosome could switch on the male developmental pathway (and perhaps switch off the female pathway). Several review articles were published, listing the convincing accumulated data in a satisfied tone that suggested a tidy conclusion to the study of sex determination (Goodfellow and Lovell-Badge 1993; Hawkins 1994). But despite all this detailed information about male sex determination and differentiation, very little was known about female sex determination or development. Was this because, as the textbooks suggested, there was nothing to learn? Or did the female developmental pathway remain in the shadows because of a lack of research into the area?

Intersexuality: The Data That Were Ignored

It would be hard to exaggerate the importance of assigning gender in human social interactions. We feel a pressing need to define each individual as either male or female, and any person whose gender is ambiguous makes us uncomfortable and even anxious (Fausto-Sterling 1992; Fausto-Sterling 1993). Just as there are people whose social gender is unclear, there are individuals whose biological sex is neither male nor female. Our social discomfort with ambiguous gender may be reflected in biological science's unwillingness to accept intersex individuals and incorporate them into working models about sex determination.

Such intersex people are relatively frequent; depending on the precise definition, about five children in every thousand are intersex (Blackless, et al. 2000; Fausto-Sterling 1993). When a child is identified as intersex, physicians typically plan a course of surgeries and hormone treatments to make the child appear either male or female to whatever degree is possible. (For further discussion of treatment of intersex individuals, see A.D. Dreger, this volume.) However, no matter what clinical approach parents and doctors choose for intersex babies, the fact of their existence has biological significance. By forcing intersex individuals into the bipolar conceptual categories of male or female, biologists miss the important clues their condition offers about normal human sex determination. Labeling XXY intersex individuals as male and XO intersex individuals as female generated a straightforward and appealing model in which sex was determined by the presence or absence of the Y chromosome. But, as we will see, this simplistic model glosses over a more complicated and interesting reality.

A rigidly bipolar view of the sexes underlies the "dominant Y" model of sex determination. This model is based on the assumption that an XO individual is a normal female. Indeed, the original paper describing the XO condition made this assumption explicit in its last sentence: "In conclusion, it should be emphasized that the XO patient should not be referred to as an instance of 'sex reversal,' as a 'chromosomal male,' or as a 'genetic male': she is a female, with an abnormal genotype" (Ford, et al. 1959). Were that actually the case, one could safely conclude that the absence of a Y chromosome results in a female even if there is only one copy of the X chromosome. However, this is at best an oversimplification and arguably wholly inaccurate. Genetic anomalies involving loss or gain of whole chromosomes or parts of chromosomes have serious global effects on human development. (A familiar example is Down Syndrome, in which duplication of a chromosome results in subnormal intelligence and a suite of associated physical disorders.) Thus, it is not surprising that XO individuals are *not* normal females.

In fact, long before the genetic nature of the affected individuals was known, the XO condition was recognized as a congenital condition called Turner Syndrome. In addition to lacking a second sex chromosome, people with Turner Syndrome are usually short in stature, have abnormal fingers and toes and do not go through puberty, menstruate, or develop secondary sex characteristics such as breasts and body hair. They are always sterile. Arguably, such individuals are not female, but rather intersex. Some characteristics associated with Turner Syndrome, like not menstruating, are more "male" than "female." Individuals with Turner Syndrome, however, have no penis, which sufficed for researchers to classify them as female.

Similarly, XXY individuals were described as male, and their supposed maleness was used to argue that even in the presence of two X chromosomes, a Y chromosome could confer maleness. But, like those with XO, XXY individuals are also intersex. Their condition, known as Klinefelter's Syndrome, results in little or no facial hair, a high-pitched voice, small testes, breast development, and abnormal testicular tissue that cannot produce sperm, making them sterile. In addition, affected individuals often also have abnormally tall stature and subnormal intelligence[1] (Jacobs and Strong 1959). Nonetheless, because they have a penis, XXY individuals were classified as biologically male.

---

[1]It is worth noting that the discovery of humans with sex chromosome anomalies also generated hypotheses about the role of these chromosomes in behavior. If the Y chromosome was the agent of biological maleness, might it also be responsible for causing stereotypic male social behavior? During the late 1960s, it was suggested that the extra Y chromosome carried by XYY individuals made them violent and aggressive. This theory arose from the observation that XYY individuals were more likely to be institutionalized than comparable XY men. A careful analysis revealed, however, that low intelligence, not unusual aggressiveness, explained the disproportionate representation of XYY individuals in institutions (Witkin, et al. 1976).

Thus, the dominant Y chromosome model of sex determination derived from a misclassification of XO individuals as female and XXY individuals as male, when they are in fact neither male nor female, but intersex.

The Invisible Genes for Femaleness?

The SRY/dominant Y model ignores a substantial body of contradictory data. Recently, it has become clear that female development is also an active process, directed by a regulatory circuit analogous to that controlling the male developmental pathway. Indeed, dating back to 1978, the scientific literature documents several dozen XY individuals who carried an intact SRY gene, but were nevertheless apparently female (Bardoni, et al. 1994; German 1978). How could such "sex reversed" people exist if SRY was the sole determinant of maleness and lacking SRY made you female? Until 1989, very little notice was taken of this incongruent phenomenon.

It emerged that these XY "females" resulted when a region of the X chromosome was duplicated. Males who inherit an abnormal X chromosome carrying two copies of a gene known as DAX-1 appear to be female, even though they still carry SRY on their Y chromosome (Bardoni, et al. 1994). Since one copy of DAX-1 is normally present on a male's X chromosome, DAX-1 appeared to cause male-to-female sex reversal only when there were two copies; in genetic terminology, it is a dosage-dependent gene.[2] Although duplication of DAX-1 causes sex reversal of XY individuals, XX individuals carrying an extra copy of DAX-1 are normal, fully fertile females (Muscatelli, et al. 1994).

Interestingly, just as development proceeds normally in females lacking SRY, deletion of DAX-1 doesn't appear to affect human male sexual development (Swain and Lovell-Badge 1999). If this gene had been discovered first, could it have been described as an active switch turning on the female developmental pathway and turning off the male developmental path? Could its existence have been used to ar-

[2]Readers may wonder why the two copies of DAX-1 carried by XXY males don't result in a female appearance. Understanding the dosage-dependent effect of DAX-1 is complicated by a phenomenon called X-inactivation. In females, one of the two X chromosomes is normally inactivated early in development, so that effectively each cell has just one working copy of the X chromosome. X-inactivation presumably explains why the second copy of DAX-1 carried by XXY "males" does not cause them to develop a female reproductive system.

gue that male development in the absence of two copies of DAX-1 must be passive and automatic?

As SRY is located on the Y chromosome, DAX-1 is found on the X chromosome in all mammals (though there is no obvious biological reason why this must be so; indeed, in marsupials, the DAX-1 equivalent lies on one of the nonsex chromosomes). DAX-1 is expressed early in the developing gonad at the same time and for the same period as SRY (Swain, et al. 1998). Finally, like SRY, DAX-1 encodes a regulatory protein that can control the expression of other genes (Iyer and McCabe 2004; Muscatelli, et al. 1994). It is clear that DAX-1 affects expression of a broad range of important developmental processes. One of the genes affected by DAX-1 encodes a recently-identified signaling protein called WNT4, which plays a role in female development somewhat analogous to that of MIS in male development.

WNT4 is not required for male development: XY mice lacking WNT4 have a normal male reproductive system. However, XX mice without WNT4 are abnormal and intersex. Their gonads produce testosterone as well as the usual estrogens, their Mullerian ducts do not form, and their Wolffian ducts do not regress as they normally should (Vainio, et al. 1999). These characteristics indicate that WNT4 is required for such crucial events in female development as formation of the Mullerian duct, regression of the unneeded Wolffian duct, and suppression of gonadal testosterone production. In short, WNT4 participates in an active female developmental pathway wholly inconsistent with the passive default system postulated in the dominant Y model (Fig. 22.1).

Biologists studying sex determination have not yet determined how, or even if, the genes controlling male and female development interact. Several models attempting to incorporate DAX-1 and WNT4 into the SRY-based model of sex determination suggest that the process of sex determination is a struggle between male and female forces, distinctly reminiscent of the contest between maleness and femaleness implied in descriptions of the balance theory of sex determination. Competitive or contentious mechanisms for these proteins' functions are proposed; biologists repeatedly use language suggesting hostility in what is, after all, an emotionless set of biochemical processes. One model proposed that selfish genes engage in intragenomic conflict, and that DAX-1 may suppress the putative selfish effects of SRY (Hurst 1994). Another paper, entitled

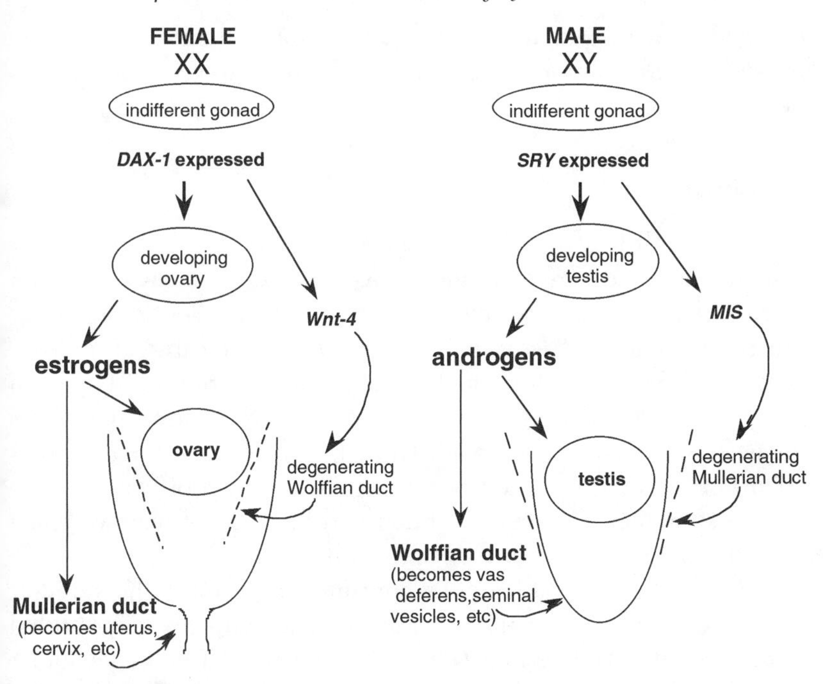

Figure 22.1 Parallel models of female and male embryonic sex development.

Dax1 antagonizes Sry in mammalian sex determination, remarks that "Dax1 functions more as an 'antitestis' gene rather than an ovary determinant" (Swain, et al. 1998). Yet another paper says: "In XX gonads . . . WNT4 gains control of the gonadal field . . . data strongly support the antagonistic relationship of these two signaling pathways" (Kim, et al. 2006). Interestingly, although there is no evidence that DAX-1 and SRY physically interact, these authors propose that the DAX1 and SRY proteins bind to each other. Protein-protein interactions are common, although it is not customary to suggest explicitly, as these authors do, that the two proteins "struggle for dominance in a molecular-level wrestling match" (Swain, et al. 1998). The emotional language continues in *Mammalian sex determination: A molecular drama* (Swain and Lovell-Badge 1999). Another favored metaphor is militaristic: a review on the genetics of gonad development was titled "*Battle of the Sexes*" (Jameson, et al. 2003), and a paper offers data consistent with a cellular "arms race" between DAX-1 and SRY (Dabovic, 1995). The language chosen by the biologists studying sex determination suggests an emotionally charged view of human sexual development as a struggle between masculine and feminine forces, complete with weapons, strategy, winners, and losers.

Conclusions

Biological data from human intersex individuals and laboratory research on genes like SRY and DAX-1 (as well as many others not discussed here) demonstrate that human female development requires a complex and active pathway, not merely the absence of SRY. It is, however, clear that much more scientific attention has been given to the details of male development, which is now understood in considerable molecular detail. The active male–passive female paradigm of sex determination persists, both in terms of where biologists bestow their research attention and in terms of how we teach students about human sex determination.

Standard biology textbooks continue to recount the familiar version of human sex determination unchanged except for added details about SRY function. The 1996 edition of Keeton's widely-used introductory text, *Biological Science*, still says: "The human Y chromosome bears a gene with strong male-determining properties. Its product is a DNA-binding protein that, early in fetal life, binds to numerous sites on the autosomes, initiating the developmental program that will produce a male. In the absence of this gene, a female develops" (Gould and Keeton 1996). Other texts form a unified chorus:

> If the embryo is a male, it will have a Y chromosome with a gene whose product converts the indifferent gonads into testes. In females, which lack a Y chromosome, this gene and the protein it encodes are absent, and the gonads become ovaries. (Raven and Johnson 1996)

> In general, a single Y chromosome is enough to produce maleness, even in combination with several X chromosomes. The absence of a Y chromosomes yields femaleness. (Campbell, et al. 2006)

> One of [the genes on the Y chromosome] is the master gene for male sex determination. Its expression leads to the for-

mation of testes, the primary male reproductive organs. In this gene's absence, ovaries form automatically. (Starr and Taggart 2001)

Most of the facts given in these texts are solid enough (although functional ovaries are *not* formed in XO or XY:SRY-deleted individuals as they imply). However, these facts do not tell the whole story. There is now good evidence documenting what should have seemed likely from the beginning: that female sex determination requires an active developmental pathway, just as male sex determination does. Arguably the details of this particular issue are less important than the larger truth they reveal: expectations regarding gender behavior can affect scientists' research focus, data interpretation, and working models, even at the molecular levels.

The study of sex determination is certainly not the only case where biological data were seen through a distorting lens of gender mythology. In a parallel example, until relatively recently biologists assumed that during human fertilization, the male and female gametes exhibit the same behaviors that we expect in men and women. Sperm cells were seen as active, competitive, even rapacious, while the egg was seen as a passive element in the process, a prize that went to the fittest sperm (Biology and Gender Study Group 1989). An evenhanded experimental approach that examined the roles of the egg and the sperm revealed that in fact the active participation of both cells is required for successful fertilization (Schatten and Schatten 1983).

The study of primate social behavior offers another example of the question that was not asked. The significant fact that most primate societies are fundamentally matriarchal went unnoticed for decades because field researchers focused exclusively on male-male and male-female interactions, ignoring the crucially revealing female-female interactions (Gowaty 1997; Hrdy 1981). Indeed, paternity analysis found that status in a male hierarchy is unrelated to reproductive success in many primates (Kolata 1976).

Feminist biologists have recognized that rigid cultural distinctions between male and female can interfere with scientific objectivity by layering cultural expectations about masculinity and femininity onto biological data concerning the sexes, or even onto biological systems seen as analogous to the sexes (Keller 1985). Ruth Bleier notes that:

> . . . the historical separation of human experience into mutually contradictory realms, female and male, engendered our culturally inherited dualistic modes of thought, and that male-female dichotomy was built into our ways of perceiving truth. (1984)

It would be hard to dispute that every person carries deep internal preconceptions of a personal nature about sex and gender. The challenge facing biologists who work either directly with sex or on systems with historical or current gender associations (whether of a metaphorical or practical nature), is consciously to ferret out and eliminate these sources of bias. A received truth about sex that is intuitively obvious must always be suspect. If biologists rigorously reflect on the questions that are going unasked, as well as those that form almost automatically in the mental foreground, the resulting science will be more complete. Such careful application of the scientific method, together with thoughtful critique and analysis by other biologists, offers our best opportunity to avoid or detect the failures of objectivity that too often taint biological studies of the sexes.

References

Aristotle. 1984. The generation of animals. In *The complete works of Aristotle,* ed. J. Barnes, 8. Princeton: Princeton University Press.

Bardoni, B. et al. 1994. A dosage sensitive locus at chromosome Xp21 is involved in male to female sex reversal. *Nature Genetics* 7:497–501.

Biology and Gender Study Group. 1989. The importance of feminist critique for contemporary cell biology. In *Feminism and science,* ed. N. Tuana, 172–187. Bloomington: Indiana University Press.

Blackless, M. et al. 2000. How sexually dimorphic are we? Review and synthesis. *American Journal of Human Biology* 12:151–166.

Bleier, R. 1984. *Science and gender: A critique of biology and its theories on women.* New York: Teachers College Press.

Campbell, N. A. et al. 2006. *Biology: Concepts and connections*, 5th ed. San Francisco: Benjamin Cummings.

Dabovic, B, et al. 1995. A family of rapidly evolving genes from the sex reversal critical region in Xp21. *Mammalian Genome* 6:571–580.

Fausto-Sterling, A. 1992. *Myths of gender: Biological theories about women and men*, 2nd ed. New York: Basic Books.

Fausto-Sterling, A. 1993. The five sexes: Why male and female are not enough. *The Sciences* 33:20–25.

Ford, C. E., et al. 1959. A sex chromosome anomaly in a case of gonadal dysgenesis (Turner's Syndrome). *The Lancet* 1:711–713.

German, J. 1978. Genetically determined sex-reversal in 46,XY humans. *Science* 202:53–56.

Goodfellow, P. N., and R. Lovell-Badge. 1993. SRY and sex determination in mammals. *Annual Review of Genetics* 27:71–92.

Gould, J. L., and W. T. Keeton. 1996. *Biological science*. New York: W. W. Norton and Co.

Gowaty, P. A., ed. 1997. *Feminism and evolutionary biology*. New York: Chapman and Hall.

Haqq, C. M. et al. 1994. Molecular basis of mammalian sexual determination: activation of Mullerian inhibiting substance gene expression by SRY. *Science* 266:1494–1500.

Hawkins, J. R. 1994. Sex determination. *Human Molecular Genetics* 3:1463–67.

Hrdy, S. B. 1981. *The woman that never evolved*. Cambridge, MA: Harvard University Press.

Hurst, L. D. 1994. Embryonic growth and the evolution of the mammalian Y chromosome. II. Suppression of selfish Y-linked growth factor may explain escape from X-inactivation and rapid evolution of *Sry*. *Heredity* 73:233–243.

Iyer, A. K., and E. R. McCabe. 2004. Molecular mechanisms of DAX1 action. *Molecular Genetics and Metabolism* 83:60–73.

Jacobs, P. A., and J. A. Strong. 1959. A case of human intersexuality having a possible XXY sex-determining mechanism. *Nature* 183:302–303.

Jameson, J. L. et al. 2003. Battle of the sexes: New insights into genetic pathways of gonadal development. *Transcripts of the American Clinical and Climatological Association* 114:51-63; discussion 64–65.

Keeton, W. T. 1972. *Biological science*. New York: W. W. Norton and Co.

Keller, E. F. 1985. *Reflection on science and gender*. New Haven: Yale University Press.

Kim, Y. et al. 2006. Fgf9 and Wnt4 act as antagonistic signals to regulate mammalian sex determination. *PLoS Biology* 4:e187.

Kolata, G. 1976. Primate behavior: Sex and the dominant male. *Science* 191:55–56.

Koopman, P. et al. 1990. Expression of a candidate sex-determining gene during mouse testis differentiation. *Nature* 348:450–452.

Koopman, P. et al. 1991. Male development of chromosomally female mice transgenic for *Sry*. *Nature* 351:117–121.

Kraft, F. 1908. *Sex of offspring: A modern discovery of a primeval law*, vol. B. Cleveland, OH: Barsuette, Inc.

Meyer, M. M. 1966. Tumors of dysgenetic gonads in intersexes: Case reports and discussion regarding their place in gonadal oncology. *Bulletin of the New York Academy of Medicine* 42:396–404.

Muscatelli, F. et al. 1994. Mutations in the DAX-1 gene give rise to both X-linked adrenal hypoplasia congenita and hypogonadotropic hypogonadism. *Nature* 372:672–676.

Page, D. C. et al. 1987. The sex-determining region of the human Y chromosome encodes a finger protein. *Cell* 51:1091–1104.

Raven, P. H., and G. B. Johnson. 1996. *Biology*, 4th ed. Dubuque, IA: William C. Brown.

Schatten, G., and H. Schatten. 1983. The energetic egg. *The Sciences* 23:28–34.

Sinclair, A. H. et al. 1990. A gene from the human sex-determining region encodes a protein with homology to a conserved DNA-binding motif. *Nature* 346:240–4.

Spanier, B. 1995. *Im/partial science: Gender ideology in molecular biology.* Bloomington Indiana University Press.

Starr, C., and R. Taggart. 2001. *Biology, the unity and diversity of life*, 9th ed. New York: Brooks/Cole, Inc.

Stern, C. 1949. *Principles of human genetics*. San Francisco: W. H. Freeman and Co.

Swain, A., and R. Lovell-Badge. 1999. Mammalian sex determination: A molecular drama. *Genes and Development* 13:755–767.

Swain, A. et al. 1998. *Dax1* antagonizes *Sry* action in mammalian sex determination. *Nature* 391:761–767.

Tuana, N. 1989. The weaker seed: The sexist bias of reproductive theory. In *Feminism and science,* ed. N. Tuana. pages 147–171. Bloomington: Indiana University Press.

Vainio, S. et al. 1999. Female development in mammals is regulated by Wnt-4 signalling. *Nature* 397:405–409.

Wasserman, A. O. 1973. *Biology*. New York: Appleton-Century-Crofts.

Weisz, P. B. 1959. *The science of biology*. New York: McGraw-Hill Book Company.

Wijngaard, M. v. d. 1997. *Reinventing the sexes: The biological construction of femininity and masculinity*. Bloomington: Indiana University Press.

Witkin, H. et al. 1976. Criminality in XYY and XXY men. *Science* 193:547–555.

23

Sex Beyond the Karyotype

Alice Dreger

When I teach a course on concepts of sex, I often begin by asking students to write down answers to two questions:

1. Do you think you're a man or a woman?
2. What makes you think that?

These two questions were inspired long ago by my own seventh grade "health" class. The teacher was the indomitable Mr. Beijer, a man who never failed simultaneously to embarrass and amuse us. When we came to the first day of the sex education unit, we found ourselves voluntarily divided into girls on one side of the room and boys on the other—as if we thought talking about sex could get you pregnant. Mr. Beijer stood before us for a few minutes, silently observing with raised eyebrows the gender territorialism, all the while making us even more nervous. Then he asked:

"How can you tell if someone is a boy or a girl?"

We giggled and blushed, and no one answered. So he repeated:

"How can you tell if someone is a boy or a girl?"

More giggles.

Finally one nerdy guy stuck his hand up and answered confidently: "By the genes." We all breathed a sigh of relief. Until . . .

"That's right!" answered Mr. Beijer. "You pull those jeans down!"

My own students tend to answer my questions along these same two lines. They either think that they are men or women because of their genes—or more generally their "sex chromosomes" (I'll explain shortly why that term is in scare-quotes)—or because of their genitals. Less often, they point to some aspect of physiology: "I menstruate, so I must be a woman." Or they point to a vague sense of gender identity: "I seem to feel like other men feel, so I must be a man." Occasionally someone makes a reference to sexual orientation. A few years before the "Age of Will and Grace," one student answered, "I know I'm a man because I'm sexually attracted to women." Another student put me in stitches by responding: "Well, I know I'm a man because my boyfriend is gay."

So what's the right answer? How do you know if you're a man or a woman? Or, put a bit differently, what makes someone a man or a woman? Does being male make you a man? Does being a woman make you female? For that matter, what makes a person male or female?

Who Cares?

First, a metaquestion: Why does all this matter? Well, for a lot of reasons. For one, as Mr. Beijer would want me to remind you, knowing whether you're male or female (or somewhere in between) can give you important clues about your body, like whether you can get pregnant, whether you're at risk for cervical or prostate cancer, whether frequently riding a bicycle with a racing seat might make you infertile (watch out, guys!). At a psychological level, whether you think of yourself as a man or a woman affects your sense of self, including your sense of how you fit in the world. This is partly because your social status as a man or woman matters very much. To use just a few examples, you can't be perceived as a mother if you're not perceived as female, and mothers and fathers tend to hold different kinds of places in cultural folklore and sentimentality. Men are frequently discouraged from becoming elementary school teachers or day-care providers because people more readily associate sexual abuse of young children with men, even though the vast majority of men are not sexual predators and some women are. If you're a woman, you're significantly less likely than a man to become a member of Congress, a top executive, or a physicist, and significantly more likely to be raped. People who think of themselves as straight aren't going to be sexually interested in you if you appear to be of the same gender as them; same goes for people who think of themselves as gay if you appear to be of the "opposite" gender.

Children in our culture (and indeed in every culture) are taught what it means to be members of their genders. In the United States, many boys are actively discouraged from crying, i.e., they're taught to suppress emotions that might make them appear weak, and many girls are actively taught to attend quickly to the needs of others. Boys are also frequently taught that they're supposed to be attracted to (and partner with) girls, and vice versa. For example, our local community center sometimes holds "mother-son" or "father-daughter" dances, but never mother-daughter or father-son dances because, as I was once told by a horrified neighbor to whom I suggested the idea, "that would be gay."

Similarly, legally your gender or sex identity matters a lot. In the United States, only men are allowed to marry women and vice versa. I couldn't have married my husband if the state of Indiana had believed he was female or I was male. A few states now allow civil unions, but even those are restricted by gender, because they are restricted to men partnering with men, and women with women. There's a federal law that prohibits genital cutting done on girls for cultural reasons—a practice sometimes called "female circumcision" or "female genital mutilation"—but a parent can easily and legally have a boy's foreskin cut off for cultural reasons. (It's known as "routine neonatal male circumcision.") Title IX was specifically designed to protect girls and women from discrimination in athletics programs, because historically they have been discriminated against. Only men can be drafted into the military, even though many women now serve in combat zones. In short, your sex and gender identities matter a lot.

Do Your "Sex Chromosomes" Make You a Man or a Woman?

So, given how much your gender identity matters, what makes you a man or a woman? Many people assume that what makes you a man is having XY chromosomes and what makes you a woman is having XX chromosomes. This is because they think XY will turn an embryo into a male, and XX will turn an embryo into a female. The truth is that most people don't know their karyotypes, but they assume they know what "sex chromosomes" they have in their cells. But guess what? You could be a man with an XX karyotype or a woman with XY without knowing it.

All embryos start off basically the same—with "protogonads" that might eventually become ovaries or testes (or ovotestes, i.e., organs with both testicular and ovarian tissue), and with the first rudiments of genital organs. In most cases, whether the fetus has a Y chromosome will determine whether it becomes male or female. The region of the Y chromosome that initiates male-typical development is called *SRY*. But here's the kicker—the first of several: *SRY* can be translocated (moved over) to an X chromosome. As a consequence, an embryo may have an XX combination with the *SRY* gene, and in this instance it will develop along a male-typical pathway. The man who ultimately results from this development will have male genitalia, testes, and undergo a male puberty. He may never find out he's XX because he's pretty much like men who are XY. Meanwhile, an XY combination with a non-functioning *SRY* results in an embryo developing along the female-typical path. The woman with this genetic background will develop as female. She may never find out she's XY because she is pretty much like women who are XX.

So, wait, you say; maybe then a functioning *SRY* makes you male, and the lack of a functioning *SRY* makes you female? Well, a functioning *SRY* gene is a necessary but not a sufficient condition to developing as a typical male. That's because *many* genes besides those located on the "sex chromosomes" contribute to sex development, including *WT-1* on chromosome 11, *SOX9* on 17, and *SF-1* on 9. That's why "sex chromosome" is a misnomer for the X and Y; there are many genes on the other chromosomes that make important contributions to sex development, and there are genes on the X chromosome that code for traits having nothing to do with sex.

SRY certainly is important. In the first trimester of prenatal life, *SRY* causes the undifferentiated protogonads to morph into testes, and the testes start making the "masculinizing" hormones that contribute to other parts of the body (genitals, brain, etc.) becoming male-typical. That's why a person with an XXY karyotype (a variation called Klinefelter syndrome) will develop mostly along a male-typical pathway—because he has an operational *SRY*. A person with one X and no Y in her karyotype (Turner syndrome) or XXX (Triple X syndrome) will develop mostly along a female-typical pathway—because she has no *SRY*. What happens to folks with mosaic karyotypes—where some cells may have XX and others XY, for example—depends on the particulars of their situation. There are even some people with three different kinds of karyotypes in their cells—in other words, each cell has one of three different "sex chromosome" combinations.

Now, there are also some women, like my friend Jane Goto, who have a functioning *SRY* gene (on a Y chromosome) but develop much more female-typical than male-typical. Jane and other women like her have a condition called complete androgen insensitivity syndrome (CAIS).[1] As the name implies, they lack the receptors that would ordinarily respond to androgens ("masculinizing" hormones), so although they have testes, prenatally their genitalia form as female-typical—so typical that CAIS is often not diagnosed until the teen years or sometimes even later. At puberty a CAIS girl's body naturally changes to look like a typical woman's, with rounded breasts, hips, female-typical fat and muscle patterns, etc. This happens because the excess testosterone produced by their testes gets converted to estrogen through a process called aromatization. (The same process makes some men who take too many steroids end up with female-typical breasts and shrunken testicles.) Because the girl has no ovaries or uterus, she does not menstruate, even though it is clear she is undergoing a feminizing puberty. A gynecologist puts a speculum in the girl's vagina and discovers it is shorter than average and ends in a pouch, without a cervix or uterus; a karyotype confirms that she has XY chromosomes. In terms of their bodily conformation and genitals, CAIS women look like other women, except that they never get hairy in the pubic region, armpits, or on their arms and legs like most girls and women, because that kind of body hair requires androgen-sensitivity. (Most other women make some testosterone in their ovaries and androgens in their adrenal glands, so they get relatively hairy after puberty. This is presumably why I'm hairier than Jane unless I shave.) CAIS women have female self-identities, are raised as girls, frequently marry men, and often adopt children because, like many other women, they wish to be mothers.

Then Maybe Your Internal Anatomy is What Makes You Male or Female?

But aren't CAIS women really men inside women's bodies? I don't think it makes any sense to think of them that way. Yes, they

---

[1]You can read about Jane and her advocacy work at the web site of the Intersex Society of North America at http://www.isna.org.

have testes, but remember that they're having none of the obvious responses to androgens ("masculinizing" hormones). And it doesn't really make sense to assign people a sex or gender identity based simply on what gonads they do or don't have. People have tried, though. I trace the history of this in *Hermaphrodites and the medical invention of sex* (1998). Would a man like Lance Armstrong who gets testicular cancer and has his testicles removed no longer be a man? Would your mother or adult daughter no longer be a woman if her ovaries were removed for some medical reason? Did my friend Cheryl Chase magically go from *not* being a girl to *being* a girl when doctors removed the testicular portions of her ovotestes when she was a child?[2]

Well, then beyond the gonads—maybe CAIS girls and women have "masculine" brains? Not by the usual scientific meanings of that. In prenatal life, several small regions of the brain differentiate according to sex. So far as we know, this happens because, although in prenatal life both typical females and typical males have their brains exposed to androgens, typical females are exposed to fewer androgens than typical males, because the body of an average female fetus makes fewer androgens than the body of an average male fetus. Now, as a result of having the *SRY* gene, fetuses with CAIS have the male-typical hormone mix coursing about them prenatally, *but* remember that they don't have androgen receptors, so those "masculinizing" hormones seem to have no effect on their brains. In other words, if you lined up in a spectrum the "masculinization" of the prenatal brains from androgen exposure, and you put the most "feminine" (least androgenized) on the left and most "masculine" (most androgenized) on the right, the order would go like this: CAIS women (like Jane); sex-typical women (like me, so far as I know); sex-typical men (like my husband, so far as he and I know). Because CAIS girls are identified and raised as girls, it is safe to assume that whatever sex differentiation happens in their brains because of gender-based nurture probably happens in the female-typical way.

Okay, so here's another sex development variation to consider: Some men have male-typical external anatomy, male-typical prenatal brain development, and ovaries. This can happen when an individual has XX chromosomes (with no *SRY* translocation) and an extreme form of virilizing congenital adrenal hyperplasia (CAH). The

[2]You can read much more about Cheryl and her work at http://www.isna.org. Use the Bibliography link to read Cheryl's publications.

lack of the *SRY* gene means the protogonads morph into ovaries. So you would expect female-typical development to cascade from there. But the adrenal glands in these fetuses are in overdrive, making lots of androgens. As a consequence, the genitals and the brain get masculinized, enough so that a few of these children are not identified as having ovaries until they are adults. Relatively late diagnosis is rarer nowadays because of widespread screening for CAH in newborns. They're raised as boys with no one being the wiser. I got a call a few years back from a nineteen-year-old man who had just discovered he had ovaries and a uterus, XX chromosomes, and virilizing CAH. Let's call him Matthew. Matthew had learned he needed medical treatment for the metabolic dangers of his CAH (it can be quite dangerous), but beyond that, had some good questions about his identity: "How do I tell my parents and my girlfriend? And should I become a woman, and get the equivalent of sex-change surgery—get my male genitals changed to female, and change my endocrine system to be more like that of a female?" I suggested he tell his loved-ones soon and carefully, with the help of a counselor. As for his last question—should he become a woman?—I asked him, "Did you ever think of yourself as a woman before?" He said no, he had always been a boy and a man. "So why," I asked him, "would you let your ovaries tell you who you are? I don't let my ovaries tell me who I am." He said his doctor had told him that he might be fertile as a female, if he underwent surgery and medical management, but I asked him, "Do you want to be a mother?" "No," he said, "I've always wanted to be a father." I gently suggested to him he would have to think long and hard about this, and maybe he would have to consider grieving the loss of his fertility as a man and having his ovaries removed. He would need absolutely no surgery to convince anyone he was a man; he was a man.

So Maybe Your Genitals Are What Make You a Man or a Woman?

When your gender gets identified by someone else during a prenatal sonogram, just after birth, or at the showers of your local gym, it is usually because of your genitals. But genitals don't actually come in just two flavors (so to speak). The clitoris and the penis are homologues; that means that they develop from the same protoorgan in prenatal life, and so, as it turns out, you can have pretty much every variation between a classic penis and a classic clitoris. The same is true for the labia majora and the scrotum, which are also homologues. And on and on with homologous sex organs. Now, you can say, "Well, those people are the exceptions," but as it turns out, it is really hard to figure out who exactly "those people" are. What I mean by that is this: How small does a penis have to be before it counts as something other than a penis? How big does a clitoris have to be?

Because sex development is complicated—as I'm sure you've gathered by now!—some people seem to be born with sex-atypical genitals (sex-atypical when compared to their karyotypes) but sex-typical internal anatomies and brain development So, you can have a man like Hale Hawbecker who is male-typical except for being born with a very small penis. Hale tells his story in *Intersex in the age of ethics* (1999). You can also have women with CAH who are born with large clitorises but seem to be female-typical in terms of internal anatomy and brain development. And again, beyond these congenital variations, I'd ask the after-birth-change question: if a boy lost his penis in an accident, would he no longer be a male or a boy? Genitals seem like a very odd way to define for sure a person's sex or gender identity.

To sum up where we've been so far: 1. most people don't know their karyotypes, yet they seem to think of themselves and function as males/men or females/women anyway; in other words, although we may act like karyotypes are the basis for our social gender system, in practice they're not; 2. because *SRY* can be translocated or nonfunctional, you can be an XX male/man or an XY female/woman; 3. because more than the "sex chromosomes" are needed for sex development, you can't just look at a karyotype and guess a phenotype; 4. you can't just look at a person's gonadal status and guess the rest of his or her body type or gender identity, and it doesn't seem to make any sense to say that a person is the sex his or her gonads are, regardless of everything else going on; 5. ditto with genitals; 6. prenatal brain development involves some "sex differentiation" but that differentiation is not simply a result of one's karyotype, and brain development doesn't always match gonad or genital type.

So Are Social Roles What Make You a Man or a Woman?

Keep in mind that, for thousands of years, people knew very little about gonads and nothing about genes or chromosomes, yet they managed to have gender identities. That's in part because, in the past, as now, in practice how you are treated in terms of your gender (and many other things, for that matter, like race and age and ability) is based on rough assumptions about your apparent biology. In the case of gender, these are rough assumptions that we consciously or more often unconsciously apply to people because they look like boys or men, girls or women.

Because researchers have figured out that a person's social role and self identity and biology are not the same things, they use the terms "gender identity" to talk about a person's self identity and/or social identity, and "sex" to talk about a person's biology. A person's gender usually starts with the announcement "It's a boy!" or "It's a girl!" (an announcement based on perceived sex, in a prenatal sonogram or at birth), and the establishment of your gender happens again and again throughout your life when people refer to you as a boy/man or a girl/woman, or treat you as a boy/man or a girl/woman, simply because of how they perceive the clothed you. For example, a person assigns you a gender identity by using the masculine (he) or feminine (she) pronoun to talk about you, or by inviting you to play in a women's soccer league, or by walking or talking or flirting or working with you in gender-specific ways. A man is regendered when he is identified as a "male nurse," and a woman undergraduate is regendered when she is labeled a "coed." Many (if not all) social encounters involve implicit or explicit gender attribution.

In practice, gender identity also gets implied or reinforced in encounters understood to be "homosexual" or "heterosexual." That's because what "homosexual" means is that same-gendered people are connecting in some sexual way, and what "heterosexual" means is that the same is happening for "opposite"-gendered people. In this sense, my student was right when he said he knew he was a man because his boyfriend is gay. Think also about the way genders get established when a woman driver tries to flirt her way out of a traffic ticket from a male cop—she's appealing as a woman to his presumed straight masculinity.

Sometimes someone ascribes the wrong gender to you. This can happen now and then by accident, when you are misidentified in a quick encounter, or it can happen all the time if you're in the situation where the gender you were assigned when you were a child isn't the one you feel. Max Beck was born with "ambiguous genitalia." As was the practice then, doctors recommended that his parents raise him as a girl, which they did, naming their child Judy. The doctors removed Judy's testes and surgically altered her genitals to make them look more like a girl's. To make a complicated life story too short, Judy ultimately became a lesbian, falling in love and partnering with another lesbian woman named Tamara, before figuring out that she really felt like a man. Max changed his name and his legal gender and so was able to marry Tamara. (In this sense, I think my student was wrong when he said "I know I'm a man because my boyfriend is gay." I think Tamara is still a woman even though her lover is now a man.)

Only a small percentage of people who are transgender—i.e., who do not feel themselves to be the gender they were assigned as children—seem to have some intersex condition like Max does. Most appear to be born with male-typical or female-typical bodies but feel so wrong in their social gender identities they are willing to undergo much personal struggle to achieve a social gender identity that matches their sense of self.

In this way, many people who are transgender express the same thing as my student who said this: "I think I'm a man because I seem to feel like other men feel." What does that mean? Well, Lisa Lees expressed it to my sex class this way: When I was socially identified as a man, and people related to me as a man, I always felt like there was something wrong in that interaction. When I became socially identifiable as a woman, the interactions felt right; I felt like who they were seeing was who I was seeing in myself.[3]

So Does Biology Not Matter At All?

Hmm, not so fast. If biology didn't matter at all to most people's gender identities, then it wouldn't make sense that every culture in the world maintains some gender role expectations for boys/men versus girls/women, would it? By the way, some cultures have more than two sets of roles, but all have at least two based on male-typical and female-typical. It also wouldn't make sense that, all over the world, *on average* young girls tend to be more interested in role-playing of supportive social relationships (like pretending to be a mother or the cook for the family) and *on average* young boys tend to be more in-

[3]Read Lisa's own writings about gender at http://www.lisalees.com.

terested in playing with objects that they can manipulate (like blocks, balls, and trucks).

When I started doing work on sex development variations, like a lot of other feminists I believed that boys and girls were different because they were *taught* to be different. I thought gender identity was all a result of nurture. But that no longer makes sense to me, in part because of cross-cultural studies, but also because I've met and learned of enough people (like Max Beck and, for that matter, Lisa Lees) who, in spite of intense gender training, ended up with a very different gender identity than the ones they were taught.

Sometimes gender warriors talk about wanting to get rid of gender altogether, but I really doubt that is ever going to happen. I think most people—for complex reasons of nature and nurture—feel some kind of gender. It's important not to forget that, even while naive beliefs about gender are sometimes used to oppress people, gender is also frequently pleasurable for people. Three women chatting in a bathroom are often enjoying their genders. A dad playing ball with his son is often enjoying his gender. A woman having sex with a man as a woman having sex with a man is enjoying her gender. The idea of getting rid of gender strikes me as not only naive, but pretty oppressive, too.

What Now?

I think, given all this, we can come to at least five conclusions about sex and gender identity in humans: 1. We tend to act like biology is the basis for our social gender distinctions, but in fact in day-to-day interactions, it is your social presentation that matters. (The state of Indiana did not require me and my husband to submit evidence of our karyotypes, gonadal types, or even genitals to get hitched. They just made me get a rubella titer.) 2. It is safe to assume that nature (biology) and nurture (environment) contribute to every individual's experience of sex and gender identity. 3. It is equally safe to assume that, for any given individual, you're not going to be able to predict with certainty by looking at her or him what a karyotype will show, what organs she or he has got inside, or what gender identity she or he will end up with. 4. You can, in practice, try to divide people into two sexes by picking some biological marker and deciding that's what really, really, *really* matters. So you can decide to use possession or lack of a Y chromosome, or possession or lack of an operative *SRY* gene, or gonadal tissue type, or phallus size, or whatever you want. But you're always going to find someone who, by this system, ends up in a category that doesn't seem to make sense because it doesn't line up with the rest of what you have in mind when you say "male" or "female" or "man" or "woman." People tend to like their social categories neat. Nature is a slob. Any attempt to divide people neatly into two sex types is ultimately a social action, not a scientific (or natural) one, and it should be recognized as such.

5. Just because something is natural doesn't mean it is good. Just because boys around the world tend to play with weapons more than girls doesn't mean only men should get drafted; just because girls on average may be more attentive to the needs of others than boys doesn't mean men shouldn't be allowed and expected to cook and clean for their families; just because there are lots more types of sex than two doesn't mean we have to have hundreds of gender types in our culture. We should not try to line up our social systems with what (we assume) is happening in the natural world, but should, rather, base our social systems on a developed sense of justice and fairness. Knowing more and more about the nature of sex doesn't tell us what to do about men, women, and other people. It just helps us begin to think about what to do.

Acknowledgments

I would like to gratefully acknowledge the help of Jane Goto, Lisa Lees, Aron Sousa, Eric Vilain, and Bruce Wilson in thinking about and writing this article.

References

Dreger, Alice. 1998. *Hermaphrodites and the medical invention of sex*. Cambridge: Harvard University Press.

Dreger, Alice, ed. 1999. *Intersex in the age of ethics*. Hagerstown, MD: University Publishing Group.

24

Ambiguous Bodies and Deviant Sexualities: Hermaphrodites, Homosexuality, and Surgery in the United States, 1850–1904

Christina Matta

In 1903, a 20-year-old woman approached Dr. J. Riddle Goffe, a New York gynecologist, for an examination. The woman, whom he identified only as E. C., had an interesting blend of sexual characteristics: she had facial hair and a vagina—though Goffe could find no internal generative organs—and a clitoris that resembled a penis in size and shape. E. C. wanted this structure removed; she claimed that she found it "annoying" and that "it made her different from other girls." Goffe obliged by removing her clitoris and using its skin to expand her vagina. Eight months later, he examined her again and marveled at her physical recovery and her buoyant mood (Goffe 1903).

From these clinical details, Goffe's report appears to be straightforward and not at all unusual. There is little to distinguish this from approximately 110 other reports of hermaphroditism (to use a 19th-century term) that American physicians published in medical journals between 1808 and 1904.[1] Indeed, Goffe's report was significant not so much for its content as for its context: he published in the midst of profound changes in how physicians explained and classified different forms of sexual behavior and gender expression. E. C. and the issues her anatomy raised for Goffe therefore provide an entry into an exploration of the complex relationship between the medicalization of homosexuality and the perceived need for normalizing surgeries on patients with ambiguous genitalia.

In some ways, the medical history of hermaphroditism and the medical history of homosexuality are parallel stories. Hermaphroditism and homosexuality raised similar questions about gender categories and behaviors; hermaphrodites were an abnormality in a society that thought nearly exclusively in terms of absolute sexual dichotomies, and legal and medical authorities alike struggled to identify what hermaphrodites really were and what rights they might possess. As Jennifer Terry (1999) has noted, homosexual intercourse was a direct challenge to a binary system of human sex in which "men and women, as opposites, were thought to be naturally attracted to one another" (Dreger 1998). It was this double challenge—first to dimorphic sex, then to normal sexual behavior—that led late-19th-century physicians to believe that surgery—whether reconstruction or outright removal of the sexual organs—was a necessary treatment for hermaphroditism.

Because hermaphrodites violated traditional sex categories and gender expectations by the very shape of their genitals, it is not sur-

[1]This tally, from a survey of United States-published articles listed under "hermaphroditism" or "hermaphrodites" in the *Surgeon General's Index*, First through Fourth Series, represents individual publications, but does not necessarily reflect the number of individual patients examined. Different physicians may have written about the same patient, for example, and some physicians or publications included descriptions of more than one patient. Both these sorts of publications were infrequent, however, and effectively balanced each other out. The tally does not include translations of foreign articles, reprints from foreign journals, or papers duplicated verbatim in multiple American journals (though it does include multiple publications about the same patient). Including papers published in foreign journals pushes these numbers considerably higher—the Second Series, for example, lists well over 200 papers under hermaphroditism alone. I have also excluded articles on animal or experimental hermaphroditism and cases of ambiguous sex listed under cross-referential topics such as gynandria or hypospadias from this count. I have left out intersexuality in particular (though it seems counterintuitive), because the term does not appear in the *Index* until the Fourth Series, which reflects its appearance in medical literature beginning in the 1930s.

prising that their sexual behaviors drew physicians' attention in the same way their manner of dress, their occupation, or their name captivated curious doctors. Physicians had taken note of hermaphrodites' sexual lives even in the early part of the 19th century, well before the introduction of homosexuality to medical literature. In 1808, Dr. William Handy examined a hermaphrodite with a penis, testicles, labia, and an overwhelmingly feminine personal presentation, and he was surprised to discover that "there has never existed [in this person] an inclination for commerce with the female" (Handy 1808). Other reports throughout the century took similar note of hermaphrodites' physical attractions. Physicians often alluded to sexual attractions by mentioning marital status, and both were among the criteria physicians used to determine their patients' true sex. Pointing out hermaphrodites' sexual behavior, then, was not new, and such references were in fact quite common.

At the same time, physicians had only limited means for treating hermaphrodites—that is to say, for establishing them within a dimorphic anatomical and social norm—and they responded to these medical anomalies with social prescriptions. For example, physicians would inform their patients that they were not men as they had previously believed, but women, and that they should behave accordingly by wearing dresses and quitting their jobs, for example. These instructions were not easily enforced, however, and there is little reason to believe that hermaphrodites changed their customs or habits simply because a doctor told them to. Social solutions were temporary at best, and depended entirely upon patients' own willingness to conform to one set of gender roles or the other. Many hermaphrodites likely continued about their normal ways unimpeded.

Only in 1852, when surgeon Samuel D. Gross published an account of what he claimed was the first corrective surgery on a patient of ambiguous sex, did a second possibility for managing hermaphroditism appear in medical journals. Gross had performed the surgery in 1849 on a three-year-old girl whose parents were concerned when she developed "boyish" interests—she had abandoned "dolls and similar articles of amusement" for "boyish sports." Upon examination, Gross discovered that she did not have a penis or a vagina, but possessed a small clitoris, an indentation covered by an imperforate membrane, and testes. Gross removed her testes, and when he examined her again he was pleased to note that her interests had returned to more feminine pursuits—he claimed she found "great delight in sewing and housework" and that her "habits [had] materially changed and [were] now those of a girl" (Gross 1852).

The need to uphold standard gender roles figured prominently in Gross's justifications for performing the surgery. More important, he also speculated about the girl's future sexual activity. He claimed that surgery would remove "that portion of the genital apparatus which, if permitted to remain until the age of puberty, would be sure to be followed by sexual desire." An operation would spare the child the otherwise inevitable humiliation of attempting sexual intercourse with a male only to have penetration be unsuccessful and embarrassing. (He did not, however, describe any attempts to open her "indentation" or to create a vagina.) But his concern for this particular patient was indicative of a larger concern with the sociosexual consequences of hermaphroditism: he claimed that it could lead to moral and social degradation and, if left unattended, could lead "to the ruin of character and peace of mind." He argued:

> A defective organization of the external genitals is one of the most dreadful misfortunes that can possibly befall any human being. There is nothing that exerts so baneful an influence over his moral and social feelings, which carries with it such a sense of self-abasement and mental degradation, or which so thoroughly "maketh the heart sick" as the conviction of such an individual that he is forever debarred from the joys and pleasures of married life, outcast from society, hated and despised, and reviled and persecuted by the world. (Gross, 1852)

The editors of *Beck's Medical Jurisprudence*, published 11 years later, likewise claimed that the operation Gross performed was a preemptive strike against a pubescent development of sexual urges "which could only be productive of evil" (Beck and Beck 1863). From these comments, it is clear that Gross and the editors all regarded corrective surgery as a means of curtailing immoral sexual behavior as well as a means of correcting hermaphroditism. The first surgery on a hermaphrodite, then, was also the first successful attempt to control hermaphrodites' sexual behavior in the name of maintaining social propriety.

Although Gross expressed hope that his case would provide an example for future physicians faced with similar patients, surgery did not fully replace existing practices of prescribing social behaviors. Indeed, of the 41 reports of hermaphroditism published from 1852 to 1879, only three mentioned surgery as a treatment option: Gross's report; one published in 1868 in which a woman requested the removal of what turned out to be a testicle; and a third from 1869 detailing an operation to correct hypospadias (a condition in which the urethra opens on the underside or at the base of the penis instead of at the tip) in a child with a penis and labia (Gross 1852; Avery 1868; Logan 1869). For the most part, physicians during this period still recommended exile, buying a new, gender-appropriate wardrobe, or other social solutions to establish hermaphrodites within normal dimorphic sex and gender paradigms; the efficacy of these recommendations remained dependent on patients' willingness to follow them.

Physicians who proposed surgery also often found that their patients had different ideas about their bodies. Physicians could no more force patients to undergo surgery than they could force them to wear trousers instead of dresses, and, as a result, patients often shrugged off their uncommon bodies as harmless and went about their business as usual. One example of this stands out in particular: in 1880, 46-year-old Mary O'Neill approached Dr. Edward Swasey at the New York Hospital for the Ruptured and Crippled for a truss to contain a double hernia (Swasey 1881). Upon examining her, Swasey discovered two labial tumors that were connected to cords resembling spermatic cords and suspected that they were testicles, even though he also found "nothing in feature, form, or deportment to suggest that she is not a woman." He offered to remove the tumors, but O'Neill refused and left the hospital with her truss.

Swasey accepted O'Neill's decision without moral or medical criticism, in contrast to many of his surgical colleagues who believed that surgery was necessary to alleviate a pathetic condition or to prevent the possibility of hermaphrodites engaging in sexual activity. The potential futility of suggesting an operation to an unwilling patient did not stop surgically-minded physicians from trying, however; from 1880 to 1904, the suggestion of surgery appeared in 13 of 49 published case studies of hermaphroditism—almost five times as many references as between 1852 and 1879, and in three fewer years. This increase may in part reflect other factors, including advances in surgical techniques (such as more effective anaesthetics and the introduction of antiseptic and aseptic methods) and the rise of gynecology and urology as medical specialties. Doctors who encouraged surgery for patients with ambiguous genitalia did not cite these factors in their published work; instead, the justifications that surgeons offered after 1880 suggest that, like Gross, they found hermaphrodites' sexual behavior to be a serious threat. It is no coincidence that surgeries increased in frequency after homosexuality began to be discussed in medical journals. Physicians tacitly agreed that surgery was a potential means of correcting sexual ambiguity while also preventing sexual depravity.

It was this emphasis on sexuality as a legitimate reason to promote surgery that distinguished American medical responses to hermaphroditism from those of their European counterparts. European doctors also performed normalizing operations and were aware that their hermaphroditic patients may have harbored sexual attractions. As Alice Dreger (1998) has detailed, French and British physicians "seem to have made something of a habit" of removing errant gonads or sex organs from their patients. Likewise, French experts on hermaphroditism expressed a keen interest in theorizing their patients' sexuality, and although British experts were not as inclined towards psychoanalyzing hermaphrodites' sexual tastes, they agreed that "testicles naturally meant desire for women, and ovaries naturally meant desire for men" (Dreger 1998). American physicians shared this assumption but conflated hermaphroditism with homosexuality to a far greater degree. As a result, a medical distaste for homosexuality played a much more prominent role in physicians' prescriptions of surgery in the United States than in France or Britain.

From their first appearance in American medical literature, homosexuals had been included in the broader category of sexual inverts—a category that encompassed transvestites; women who smoked, whistled, or preferred sports and masculine dress; and men who were "fond of looking in the mirror" or who had effeminate voices (Chauncey 1989). Hermaphrodites' behavior fell under the rubric of sexual inversion because of their ambiguous genitalia, but this was not the only connection between homosexuality and hermaphroditism. In their effort to understand homosexuality and to identify what could be done to correct it, physicians and sexologists employed the term "psychical hermaphroditism." How else to accurately describe patients who looked like one sex but behaved like the other in their sexual lives than to claim their bodies were of one sex and their minds of the other? Furthermore, the same causal explanation applied to both phenomena. Physicians struggling to identify the causes of hermaphroditism turned to embryology and developmental biology for answers. Embryonic bisexuality, cellular differentiation, and Ernst Haeckel's insistence that ontogeny recapitulated phylogeny all appeared in medical reports of hermaphroditism with varying degrees of relevance and accuracy (e.g., Cochran 1878; Cummings 1883). Harry Oosterhuis (2000) noted that physicians and sexologists invoked these same explanations in their analyses of homosexuality. Whereas early 19th-century physicians had regarded hermaphrodites as a curious and disturbing anatomical puzzle to be solved (and often explained by causes such as conception on a state line or startling a pregnant woman), introducing the language of development redefined hermaphroditism and homosexuality as the results of a biological process gone horribly wrong. Both hermaphrodites and homosexuals, therefore, were biological deviants from a dimorphic, heterosexual norm; treating their condition consisted of normalizing the outward behavioral and anatomical effects of that process.

The medicalization of hermaphroditism, already concerned with gender boundaries, became entangled in the medical profession's redefinition of homosexuality as a matter of sexual object choices, and therefore fell doubly victim to the perceived need to define appropriate sexual and social behaviors. In 1896, Dr. Samuel E. Woody refused to operate on a 20-year-old patient who presented herself as a woman. She claimed to have had sexual attractions to both males and females, and that she had "copulated with males with success and satisfaction," though at the time she was attracted mostly to women. Woody described her anatomy, which included a clitoris that "became" a penis "on titillation," testicles, a small vagina-like pouch, and possibly a uterus, and concluded that his patient was not a woman, but a young man. Yet to Woody, the patient's dual sexual attractions posed an additional problem: regardless of her true sex, she could not possibly be heterosexual. He saw this as a telling commentary on hermaphrodites' sexualities in general and took his interventionist duty one step further than simply refusing to operate. He asserted that, because they were "so ill-fitted for the generative function and so prone to psychical perversions and moral degradation, such cases should be castrated in early life." Under this reasoning, physicians had a moral obligation to correct their patients' bodies to forestall inevitable tendencies towards sexual deviance, even if it meant complete removal of their sexual organs.

Woody's arguments appeared only five years after urologist G. Frank Lydston proposed surgical treatments as a means of curtailing homosexual behavior (Terry 1999). Lydston believed that homosexuality was a form of perversity caused either by a defective brain or defective genitals, but he also believed these perversities could be corrected permanently. Correcting the defective gonads, whether by complete removal or by replacing them with "better" ones, should therefore eliminate homosexuality, just as Woody argued that castrating hermaphrodites would prevent them from ever having erotic attractions. Indeed, Lydston reported success in eliminating homosexuality in men by transplanting a testicle from a heterosexual man into a homosexual man (who then, according to Lydston, began to feel sexual attractions toward women), and he likewise claimed that ovariotomy and clitorectomy were an effective means of preventing the expression of lesbian sexuality. In addition, he explicitly linked homosexual behavior with physical abnormalities. Lydston included "sexual perversion with defect of genital structure, e.g., hermaphroditism" as one possible form of "congenital, and perhaps hereditary sexual perversion," and opined that

> it is often difficult to draw the line of demarcation between physical and moral perversion. Indeed, the one is so often dependent upon the other that it is doubtful whether it were wise to attempt the distinction in many instances. But this does not effect (*sic*) the cogency of the argument that the sexual pervert is generally a physical aberration—a *lusus naturae*. (Lydston 1889a; see also Lydston1889b)

By Lydston's description, then, homosexuality and hermaphroditism—representing the moral and physical sides of sexuality respectively—were both deviant conditions. The sexual pervert, regardless of the exact nature of his or her perversity, was a trick of nature, and stood in clear contrast to normal sexuality. Lydston's language indicates that physicians intent upon eliminating sexual inversion constructed a direct connection between homosexuality and hermaphroditism through both the categories they invoked to describe their patients and the surgeries they claimed could normalize them.

This equation of the two conditions also problematized surgery as well—Goffe's 1903 treatment of E. C., for example, drew criticism from Fred J. Taussig, a St. Louis gynecologist. Again, E. C. appeared to be a woman; she had a vagina and a clitoris that resembled a penis—a fairly mundane set of characteristics as far as hermaphrodites went. Even her sexual desires suggested femininity: she confided in Goffe that she had "never had any girl love affairs or been attracted passionately to any girl, but has been attracted by boys." Goffe, thus convinced that he had before him a woman with an enlarged clitoris, asked her if she would prefer to be a man or a woman. She answered that she wished to be a woman, and Goffe performed the surgery that "placed the young patient safely in the ranks of womankind, where she desired to be."

Taussig (1904) expressed his dissatisfaction with this course of treatment in the *American Journal of Obstetrics and Diseases of Women and Children*, where Goffe had initially published his report. While he was not as explicitly antihomosexual in his reasoning as Woody, Taussig likewise argued that physicians needed to intervene to control hermaphrodites' sexual desires. He explained that sexual urges were socialized behavior and argued "that the patient in this case had the sexual desires of a woman must, therefore, be looked upon, more as a result of her education, of suggestion and imitation than as in any way conclusive evidence of her true sex." Taussig also thought it likely that E. C. had undescended testicles, in which case her sexual desire was definitely misdirected. If she had testicles, she should rightly have been attracted to women, and it was Goffe's duty to ensure that E. C. did not act upon her attraction to other men. With this in mind, Taussig found Goffe's treatment inappropriate. Although he did not offer a moral condemnation of homosexual behavior as had Woody, Taussig's fears reflected the medicalization of homosexuality as a deviant condition. Homosexuality, like ambiguous genitalia, stood on the wrong side of the divide between normal and abnormal.

The disagreement did not end there, however. In a second set of articles, published in the *Interstate Medical Journal* in 1904, Goffe defended himself against Taussig's criticisms (Goffe, et al. 1904). He acknowledged that he asked E. C. if she would prefer to be a man or a woman, but he claimed he did so as the final stage of his examination after having already concluded she was female. By asking what E. C. would prefer, he argued, he was merely obtaining her consent for the operation, not allowing her to base her decision upon her sexual tastes. Such an act would clearly be detrimental to both the patient and society at large, since she was trapped between her sexual inclinations and her anatomy. At any rate, Goffe stated, this point was now irrelevant, as E. C. was menstruating at the time of her follow-up examination, thereby proving his judgment correct.

Taussig dismissed this as unimportant to the discussion at hand (Goffe, et al 1904). He claimed that, while it was all well and good that Goffe had turned out to be correct, there was no way he could have known this at his initial examination. Goffe had therefore taken quite a risk, as Franz Neugebauer (Europe's leading authority on hermaphroditism) had published many examples of patients with the same presentation who turned out to be men. Waiting until the patient had either ejaculated or menstruated, or examining the mature gonads through laparotomy, Taussig continued, were the only fail-safe means of determining sex. Aside from this, however, Taussig had nothing else to say, given that Goffe had assessed E. C. accurately, and he closed by claiming that he only wished to provoke discussion, not to criticize. There appears to have been no further public discussion between the two physicians.

By 1904, then, normalizing surgery was already justifiable for its supposed ability to reestablish hermaphrodites within the boundaries of dimorphic sex, both physically and behaviorally. The recognition that hermaphrodites' bodies could dictate—and, more to the point, misinform—sexual desire was already present in Gross's 1852 report, yet surgery did not immediately catch on as the preferred treatment for nondimorphic bodies. Physicians concerned about hermaphrodites' liminal state promoted surgery as an appropriate intervention with apparent urgency only after the introduction of homosexuality into American medical literature. Again, between 1852 and 1879, only three physicians even mentioned surgery in their reports of hermaphroditism, and only two performed it—compared to 13 discussions of surgery between 1880 and 1904.

This suggests that simply knowing surgery was possible could not convince physicians that it was the best means of treating or correcting hermaphrodites. After 1880, when abnormal genitals became strongly associated with abnormal sexual behavior (as was evident in medical terminology and explanation, as well as in moral belief), surgeons began to express interest in more invasive ways of enforcing standard sexual categories. Discomfort with homosexuality, therefore, was among the most pronounced influences that contributed to the establishment of surgery as a necessary medical treatment for hermaphroditism in the early 20th century. While the refinement of surgical practices and techniques may very well have contributed to physicians' overall willingness to perform increasingly complex and intrusive gynecological and abdominal surgeries, understanding the rising popularity of genital surgeries in the late 19th and early 20th centuries requires first understanding the assumptions about gender, sex, and sexual behavior that informed medical practice. Reconsidering the link between the medicalization of homosexuality and of hermaphroditism, then, offers a new perspective from which to explain how these assumptions contribute to physicians' eagerness to promote normalizing surgery as a therapeutic option in the 20th and 21st centuries.

Acknowledgments

This paper was initially published in *Perspectives in Biology and Medicine* 48:1 (Winter 2005):74–83, and is reprinted here with kind permission from the The Johns Hopkins University Press. I would like to thank Caitilyn Allen, Rima Apple, Alice Dreger, Judith Houck, Ron Numbers, Elizabeth Reis, and Micaela Sullivan-Fowler for their expert advice and guidance during the preparation of this paper. I am especially grateful to Lynn Nyhart, who patiently read and commented on many drafts of this paper.

References

Avery, H. N. 1868. A genuine hermaphrodite. *Medical Surgery Report* 19:144.

Beck, T. R., and J. B. Beck. 1863. *Elements of medical jurisprudence*, 12th ed., vol. 1. Review by C. R. Gilman. Philadelphia: J. B. Lippincott.

Chauncey, G., Jr. 1989. From sexual inversion to homosexuality: The changing medical conceptualization of female "deviance." In *Passion and power: Sexuality in history*, ed. K. Peiss and C. Simmons, 87–117. Philadelphia: Temple University Press.

Cochran, J. 1878. Hermaphroditism. *Transactions of the Medical Association of Alabama* 209–239.

Cummings, E. F. 1883. A case of congenital malformation of the genital organs. *Boston Medical Surgical Journal* 108(9):195–196.

Dreger, A. D. 1998. *Hermaphrodites and the medical invention of sex*. Cambridge: Harvard University Press.

Goffe, J. R. 1903. A pseudohermaphrodite, in which the female characteristics predominated: Operation for removal of the penis and the utilization of the skin covering it for the formation of a vaginal canal. *The American Journal of Obstetrics and Diseases of Women and Children* 48(6):755–763.

Goffe, J. R. et al. 1904. Hermaphroditism and the true determination of sex. *Interstate Medical Journal* 11:314–318.

Gross, S. D. 1852. Case of hermaphroditism, involving the operation of castration and illustrating a new principle in juridical medicine. *American Journal of Medical Science* 24:386–390.

Handy, W. 1808. Account of an hermaphrodite; communicated by Dr. Wm. Handy, of New York, to Dr. Miller. *Medical Repository* 6(1):36–37.

Logan, S. 1869. Case of supposed hermaphroditism: Operation. *New Orleans Medical Surgery Journal* 22:453–454.

Lydston, G. F. 1889a. Sexual perversion, satyriasis, and nymphomania. *Medical Surgery Report* 61(10):253–258.

Lydston, G. F. 1889b. Sexual perversion, satyriasis, and nymphomania. *Medical Surgery Report* 61(11):281–285.

Oosterhuis, H. 2000. *Stepchildren of nature: Krafft-Ebing, psychiatry, and the making of sexual identity*. Chicago: University of Chicago Press.

Swasey, E. 1881. An interesting case of malformation of the female sexual organs, representing either a rare variety of hermaphroditism, or of double congenital ovarian hernia with absence of uterus. *The American Journal of Obstetrics and Diseases of Women and Children* 14:94–105.

Taussig, F. J. 1904. Shall a pseudo-hermaphrodite be allowed to decide to which sex he or she shall belong? *The American Journal of Obstetrics and Diseases of Women and Children* 49(1):162–165.

Terry, J. 1999. *An American obsession: Science, medicine, and homosexuality in modern society.* Chicago: University of Chicago Press.

Vaughan, G. T. 1891. A case of hermaphroditism. *New York Medical Journal* 125–126.

Woody, S. E. 1896. An hermaphrodite. *Louisville Medical Month* 2:418–419.

25

The Larry Summers Hypothesis: Innate Inability or Bias?

Ben A. Barres

Few tragedies can be more extensive than the stunting of life, few injustices deeper than the denial of an opportunity to strive or even to hope, by a limit imposed from without, but falsely identified as lying within.

—Stephen Jay Gould

When I was about 14 years old, I had an unusually talented math teacher. One day after school, I excitedly pointed him out to my mother. To my amazement, she looked at him with shock and said with disgust: "You never told me that he was black." I looked over at my teacher and, for the very first time, realized that he was an African-American. I had somehow never noticed his skin color before, only his spectacular teaching ability. I would like to think that my parents' sincere efforts to teach me prejudice were unsuccessful. I don't know why this lesson takes for some and not for others. But, now that I am 51 years old, as a female-to-male transgendered person, I still wonder about that sometimes, particularly when I hear male gym teachers telling young boys "not to be like girls" in that same derogatory tone.

In 2005, Harvard University President Larry Summers suggested that differences in innate aptitude rather than discrimination were more likely to be to blame for the failure of women to advance in scientific careers (Summers 2005). Harvard professor Steven Pinker put forth a similar argument in an online debate (Pinker 2005), and an almost identical view was elaborated in a 2006 essay by Peter Lawrence entitled "Men, women, and ghosts in science" (2006). Whereas Summers prefaced his statements by saying he was trying to be provocative, Lawrence did not. Whereas Summers talked about "different availability of aptitude at the high end," Lawrence talked about average aptitudes differing. Lawrence argued that, even in a utopian world free of bias, women would still be underrepresented in science because they are innately different from men.

Lawrence draws from the work of Simon Baron-Cohen in arguing that males are "on average" biologically predisposed to systematize, to analyze, and to be more forgetful of others, whereas females are "on average" innately designed to empathize, to communicate, and to care for others (Baron-Cohen 2003). He further argues that men are innately better equipped to aggressively compete in the "vicious struggle to survive" in science. Similarly, Harvard professor Harvey Mansfield states in his new book, *Manliness*, that women don't like to compete, are risk adverse, less abstract, and too emotional (2006).

I will refer to this view—that women are not advancing because of innate inability rather than because of bias or other factors—as the Larry Summers Hypothesis. It is a view that seems to have resonated widely with male—but not female—scientists. Here, I will argue that available scientific data do not provide credible support for the hypothesis but instead support an alternative one: which is that women are not advancing because of discrimination. I have no desire to make men into villains (as Henry Kissinger once said, "Nobody will ever win the battle of the sexes; there's just too much fraternizing with the enemy"). As to who the practitioners of this bias are, I will be pointing my finger at women as much as men. I am certain that all the proponents of the Larry Summers Hypothesis are well-meaning and fair-minded people who agree that treatment of individuals should be based on merit rather than on race, gender, or religion stereotypes.

Like many women and minorities, however, I am suspicious when those who are at an advantage proclaim that a disadvantaged group of people is innately less able. Historically, claims that disadvantaged groups are innately inferior have been based on junk science and intolerance (Gould 1996). Despite powerful social factors that discourage women from studying math and science from a very young age (Steele 1997), there is little evidence that gender differences in math abilities exist, are innate, or are even relevant to the lack of advancement of women in science (Spelke 2005). A study of nearly 20,000 math scores of children aged 4 to 18, for instance, found no differences of any magnitude between the genders (Leahey and Guo 2001), one-third of the winners of the elite Putnam Math Competition in 2005 were women. Moreover, differences in math test results are not correlated with the gender gap in who chooses to leave science (Xie and Shauman 2003). I will explain why I believe that the Larry Summers Hypothesis amounts to nothing more than blaming the victim, why it is so harmful to women, and what can and should be done to help women advance in science.

If innate intellectual abilities are not to blame for women's slow advance in science careers, then what is to blame? The foremost factor, I believe, is the societal assumption that women are innately less able than men. Many studies, summarized in Virginia Valian's excellent book *Why so slow?* (1998), have demonstrated a substantial degree of bias against women—more than is sufficient to block advancement in many professions. Here are a few examples of bias from my own life as a young woman. As an undergrad at the Massachusetts Institute of Technology (MIT), I was the only person in a large class of nearly all men to solve a hard math problem, only to be told by the professor that my boyfriend must have solved it for me. I was not given any credit. I am still disappointed about the prestigious fellowship competition I later lost to a male contemporary when I was a PhD student, even though the Harvard dean who had read both applications assured me that my application was much stronger (I had published six high-impact papers whereas my male competitor had only published one). Shortly after I changed sex, a faculty member was heard to say, "Ben Barres gave a great seminar today, but then his work is much better than his sister's."

Anecdotes, however, are not data, which is why gender-blinding studies are so important (Valian 1998). These studies reveal that in many selection processes, the bar is unconsciously raised so high for women and minority candidates that few emerge as winners. For instance, one study found that women applying for a research grant needed to be five times more productive than men in order to be considered equally competent (Wennerås and Wold 1997). Even for women lucky enough to obtain an academic job, gender biases can influence the relative resources allocated to faculty, as Nancy Hopkins discovered when she and a senior faculty committee studied this problem at MIT (Lawler 1999). The data were so convincing that MIT president Charles Vest publicly admitted that discrimination was responsible. For talented women, academia is all too often not a meritocracy.

Despite these studies, very few men or women are willing to admit that discrimination is a serious problem in science. How is that possible? Valian suggests that we all have a strong desire to believe that the world is fair (1998). Remarkably, women are as likely as men to deny the existence of gender-based bias (Rhode 1997). Accomplished women who manage to make it to the top may "pull up the ladder behind them," perversely believing that if other women are less successful, then one's own success seems even greater. Another explanation is a phenomenon known as "denial of personal disadvantage," in which women compare their advancement with other women rather than with men (Valian 1998).

My own denial of the situation persisted until last year, when several events opened my eyes, at the age of 50, to the barriers that women and minorities still face in academia. In addition to the Summers speech, the National Institutes of Health (NIH) began the most prestigious competition they have ever run, the Pioneer Award, but with a nomination process that favored male applicants (Carnes, et al. 2005). To their credit, in response to concerns that 60 out of 64 judges and all nine winners were men, the NIH has revamped their Pioneer Award selection process to make it fairer. I hope that the Howard Hughes Medical Institute (HHMI) will address similar problems with their investigator competitions. When it comes to bias, it seems that the desire to believe in a meritocracy is so powerful that, until a person has experienced sufficient career-harming bias themselves, they simply do not believe it exists.

My main purpose in writing this commentary is that I would like female students to feel that they will have equal opportunity in their scientific careers. Until intolerance is addressed, women will continue to advance only slowly. Of course, this feeling is also deeply personal. I am tired of powerful people using their positions to demean me just because I am different from them. The comments of Summers,

Mansfield, Pinker, and Lawrence about women's lesser innate abilities are all wrongful and personal attacks on my character and capabilities, as well as on my colleagues' and students' abilities and self-esteem. I will certainly not sit around silently and endure them.

Lawrence, Mansfield, and others claim that women are innately less competitive, less rational, and more emotional than men. There is absolutely no science to support this contention. On the contrary, it is overwhelmingly men who commit violent crimes in anger—for example, 25 times more murders than women. If anyone ever sees a woman with road rage, they should write it up and send it to a medical journal. The only hysteria that exceeded MIT professor Nancy Hopkins's (more than well-founded) outrage after Larry Summers's comments was the shockingly vicious news coverage by male reporters and commentators. Hopkins also received hundreds of nasty, hateful, and even pornographic messages, nearly all from men, that were all highly emotional.

There is no scientific support, either, for the contention that women are innately less competitive (although I believe powerful curiosity and the drive to create sustain most scientists far more than the love of competition). Many girls are discouraged from sports for fear of being labeled tomboys, yet the data shows that male and female tennis players do not differ in their competitiveness (Sherman, et al. 2000). A 2002 study did find a gender gap in competitiveness in financial tournaments, but the authors suggested that this was due to differences in self-confidence rather than ability (Gneezy, et al. 2003). Indeed, again and again, self-confidence has been pointed to as a factor influencing why women "choose" to leave science and engineering programs. When women are repeatedly told they are less able, their self-confidence falls and their ambitions dim (Fels 2004). This is why Valian has concluded that simply raising expectations for women in science may be the single most important factor in helping them make it to the top (1998).

Steven Pinker has responded to the critics of the Larry Summers Hypothesis by suggesting that they are angry because they feel the idea that women are innately inferior is so dangerous that it is sinful even to think about it and it should be taboo (2005). Harvard Law School professor Alan Dershowitz sympathizes so strongly with this view that he plans to teach a course next year called "Taboo." At Harvard we must have *veritas*; all ideas are fair game. I completely agree. I welcome truth and any future studies that will provide a better understanding of why women and minorities are not advancing at the expected rate in science and so many other professions.

But it is not the idea alone that has sparked anger. Disadvantaged people are wondering why privileged people are brushing the truth under the carpet. If a famous scientist or a president of a prestigious university is going to pronounce in public that women are likely to be innately inferior, would it be too much to ask that they actually be aware of most of the relevant data? It would seem that just as the bar goes way up for women applicants in academic selection processes, it goes way down when men are evaluating the evidence for why women are not advancing in science. That is why women are angry. It is incumbent upon those proclaiming gender differences in abilities to address rigorously whether suspected differences are really innate, or even relevant, before suggesting that a whole group of people is innately wired to fail.

What happens at Harvard and other universities serves as a model for many other institutions, so I would like to see us get it right. To anyone who is upset at the thought that free speech is not fully protected on university campuses, I would like to ask, as did third-year Harvard Law student Tammy Pettinato (2005): what is the difference between a faculty member calling their African-American students lazy and one pronouncing that women are innately inferior? Some have suggested that those who are angry at Larry Summers's comments should simply fight words with more words (hence this essay). In my view, when faculty tell their students that they are innately inferior based on race, religion, gender, or sexual orientation, they are crossing a line that should not be crossed—the line that divides free speech from verbal violence—and it should not be tolerated at Harvard or anywhere else. In a culture where women's abilities are not respected, women cannot effectively learn, advance, lead, or participate in society in a fulfilling way.

Although I have argued that the Larry Summers Hypothesis is incorrect and harmful, I still feel that the academic community is one of the most tolerant around. But, as tolerant as academics are, we are still human beings influenced by our culture. Comments by Summers and others have made it clear that discrimination remains an under-recognized problem that is far from solved. The progress of science increasingly depends on the global community, yet only 10% of the world's population is male and Caucasian. To paraphrase Martin Luther King, a first-class scientific enterprise cannot be built upon a foundation of second-class citizens. If women and minorities are to achieve their full potential, all of us need to be far more proactive. So what can be done?

First, enhance leadership diversity in academic and scientific institutions. Diversity provides a substantially broader point of view, with more sensitivity and respect for different perspectives, which is invaluable to any organization. More female leadership is vital to lessening the hostile working environment that young women scientists often encounter. In addition to women and underrepresented minority groups, we must not forget Asians and lesbian, gay, bisexual, and transgendered folks. There are enough outstanding scientific leaders in these racial and gender groups that anyone with a will to achieve a diverse leadership in their organization could easily attain it.

Second, the importance of diverse faculty role models cannot be overstated. There is much talk about equal opportunity, but, in practice, serious attention still needs to be directed at how to run fair job searches. Open searches often seem to be bypassed entirely for top leadership positions, just when it matters most— search committees should not always be chaired by men and the committee itself should be highly diverse (Carnes, et al. 2005; Moody 2004). Implementation of special hiring strategies and strong deans willing to push department chairs to recruit top women scientists are especially effective (Lawler 2006). It is crucial in the promotion process that merit be decided by the quality and not by the quantity of papers published.

Women faculty, in particular, need help from their institutions (and their partners) in balancing career and family responsibilities. In an increasingly competitive environment, women with children must be able to compete for funding and thrive. Why can't young faculty have the option of using their tuition benefits, in which some universities pay part of the college tuition fees for the children of faculty, for day care instead? Tuition benefits will be of no help if female scientists don't make tenure. And institutions that have the financial capability, such as HHMI, could help by making more career transition fellowships available for talented women scientists.

Third, there should be less silence in the face of discrimination. Academic leadership has a particular responsibility to speak out, but we all share this responsibility. It takes minimal effort to send a brief message to the relevant authority when you note a lack of diversity in an organization or an act of discrimination. I don't know why more women don't speak out about sexism at their institutions, but I do know that they are often reluctant, even when they have the security of a tenured faculty position. Nancy Hopkins is an admirable role model, and it is time that others share the burden. It doesn't only have to be women who support women. I was deeply touched by the eloquent words of Greg Petsko following Summers's comments (2005). And it has been 30 years since I was a medical student, but I still recall with gratitude the young male student who immediately complained to a professor who had shown a slide of a nude pinup in his anatomy lecture.

Fourth, enhance fairness in competitive selection processes. Because of evaluation bias, women and minorities are at a profound disadvantage in such competitive selection unless the processes are properly designed (Valian 1998; Wennerås and Wold 1997; Carnes, et al. 2005; Moody 2004). As the revamped NIH Pioneer Award demonstrates, a few small changes can make a significant difference in outcome. By simply changing the procedure so that anyone can self-nominate and by ensuring a highly diverse selection committee, the number of women and minority winners went up to nearly 40% from zero. This lesson can and should now be applied to similar processes for scientific awards, grants, and faculty positions. Alas, too many selection committees still show a striking lack of diversity—typically greater than 90% of their members are white males. When selection processes are run fairly, reverse discrimination is not needed to attain a diverse and fair outcome.

Finally, we can teach young scientists how to survive in a prejudiced world. Self-confidence is crucial in advancing and enjoying a research career. From an early age, girls receive messages that they are not good enough to do science or will be less liked if they are good at it. The messages come from many sources, including parents, friends, fellow students, and, alas, teachers. When teachers have lower expectations of them, students do worse. But we are all at fault for sending these messages and for remaining silent when we encounter them. Teachers need to provide much more encouragement to young people, regardless of sex, at all stages of training from grade school to graduate school. Occasional words of encouragement can have enormous effects.

All students, male and female, would benefit from training in how to be more skillful presenters, to exert a presence at meetings by asking questions, to make connections with faculty members who may help them to obtain grants and a job, and to have the leadership skills necessary to survive and advance in academia. Because women and minorities tend to be less confident in these areas, their mentors in particular need to encourage them to be more proactive. I vividly recall my PhD mentor coming with me to the talks of famous scientists and forcing me to introduce myself and ask questions, while he sat in the audience watching me to make sure that I did so. There is a great deal of hallway mentoring that goes on for young men that I am not sure many women and minorities receive (I wish someone had mentioned to me when I was younger, that life, even in science, *is* a popularity contest—a message that Larry Summers might have found helpful as well). It is incumbent on all of us who are senior faculty to keep a lookout for highly talented young people, including women and minority students, and help them in whatever way possible with their careers.

Acknowledgments

This chapter was originally published as "Does gender matter?"—a commentary in *Nature* 442 (July 13, 2006):133–136 and is reprinted with kind permission from the *Nature* Publishing Group, London, United Kingdom.

References

Baron-Cohen, S. 2003. *The essential difference: Men, women, and the extreme male brain.* London: Allen Lane.

Carnes, M. et al. 2005. NIH Director's Pioneer Awards: Could the selection process be biased against women? *Journal of Women's Health* 14:684–691.

Fels, A. 2004. *Necessary dreams*. New York: Pantheon Press.

Gneezy, U. et al. 2003. Pay Enough or Don't Pay At All. *Quarterly Journal of Economics* 18:1049–1074.

Gould, S. J. 1996. *The mismeasure of a man*. New York: W. W. Norton & Co.

Lawler, A. 1999. Tenured women battle to make it less lonely at the top. *Science* 286:1272–1278.

Lawler, A. 2006. Women in science. Progress on hiring women faculty members stalls at MIT. *Science* 312:347–348.

Lawrence, P. A. 2006. Men, Women, and Ghosts in Science. *PLoS Biology* 4:13–15.

Leahey, E. and G. Guo. 2001. Gender differences in mathematical trajectories. *Social Forces* 80:713–732.

Mansfield, H. 2006. *Manliness*. New Haven: Yale University Press.

Moody, J. 2004. *Faculty diversity: Problems and solutions*. New York: Taylor and Francis.

Petsko, G. A. 2005. Feet in mouth disease. *Genome Biology* 6:105.

Pettinato, T. 2005. Sexism by any other name. . .*The Record*, Feb. 3.

Pinker, S. 2005. Sex ed: Views of L. Summers. *New Republic*, February 14.

Rhode, D. L. 1997. *Speaking of sex: The denial of gender inequality*. Cambridge: Harvard University Press.

Sherman, N. et al. 2000. Trait competitiveness among elite high school tennis palyers. *Medicine and Science in Sports and Exercise* 32(5) supplement abstract:408.

Spelke, E. S. 2005. Sex differences in intrinsic aptitude for mathematics and science: A critical review. *American Psychologist* 60:950–958.

Steele, C. M. 1997. A threat in the air: How stereotypes shape the intellectual identities and performance of women and African-Americans. *American Psychologist* 52:613–629.

Summers, L. 2005. Letter to the faculty regarding NBER remarks. http://www.president.harvard.edu/speeches/summers/2005/facletter.html.

The science of gender and science. 2005. Pinker vs. Spelke: A debate. http://www.edge.org/3rd_culture/debate05/debate05_index.html.

Valian, V. 1998. *Why so slow?* Cambridge: MIT Press.

Wennerås, C., and A. Wold. 1997. Nepotism and sexism in peer-review. *Nature* 387:341–343.

Xie, Y., and K. Shauman. 2003. *Women in science: Career processes and outcomes*. Cambridge: Harvard University Press.

Contributors

CAITILYN ALLEN is professor of plant pathology and women's studies at the University of Wisconsin–Madison. Dr. Allen teaches a course on biology and gender that examines popular and scientific claims about cognitive and behavioral differences between the sexes and has published a variety of articles regarding gender as a social construction and on what gender stereotypes mean for women—and men—practicing science. She is the winner of multiple teaching awards and was the founding faculty director of the Women in Science and Engineering Residential Program at the University of Wisconsin. Her laboratory research focuses on the plant pathogen *Ralstonia solanacearum*.

ANNIE I. ANTÓN is an associate professor of computer science in the College of Engineering and a Senior Research Ethics Fellow at North Carolina State University. She received her PhD in computer science at the Georgia Institute of Technology in Atlanta. She was awarded an NSF CAREER Award, named a CRA Digital Government Fellow, selected for the 2004–2005 IDA/DARPA Defense Science Study Group, named the "2005 Woman of Influence in the Public Sector" by *CSO Magazine* and the Executive Women's Forum. She is a member of the International Association of Privacy Professionals, a senior member of the IEEE as well as a member of the ACM US Public Policy Executive Committee. She currently serves on several boards, including the NSF Computer & Information Science & Engineering Directorate Advisory Council, The US Department of Homeland Security Data Privacy and Integrity Advisory Committee, the Computing Research Association (CRA) Board of Directors and the CRA-W board, and the US ACM Public Policy Committee's Executive Committee. She serves on the editorial board of *IEEE Transactions on Software Engineering*, the *Requirements Engineering Journal*, and *Computer & Security*. Dr. Antón is director of ThePrivacy

Place.Org (http://www.theprivacyplace.org). Her URL is: http://www.csc.ncsu.edu/faculty/anton.

BEN A. BARRES is a professor of neurobiology, neurology, neurological sciences, and developmental biology at Stanford University. Dr. Barres is active in promoting equality for a diversity of scientists, including convincing the National Institutes of Health to change the selection process for its Director's Pioneer Awards. His laboratory and clinical research interests include the development and function of glial cells in the mammalian central nervous system. Dr. Barres's work has been published in *Science, Nature,* and *Cell.*

MICHAEL BUGEJA directs the Greenlee School of Journalism and Communication at Iowa State University, where he also serves on the board of the Institute of Science and Society. Dr. Bugeja is the author of over 100 articles and 20 books, including *Interpersonal divide: The search for community in a technological age*, which won the 2005 Clifford G. Christians Award for research in media ethics.

KAREN CLOUD-HANSEN is an assistant researcher in the Handelsman Laboratory at the University of Wisconsin–Madison. Dr. Cloud-Hansen received her PhD in 2004 and currently studies antimicrobial resistance in commensal bacteria from various environments.

JAMES X. DEMPSEY is policy director at the Center for Democracy and Technology. His areas of expertise include privacy, electronic surveillance, and national security issues. Prior to joining CDT, Mr. Dempsey was Deputy Director of the Center for National Security Studies, special counsel to the National Security Archive, and assistant counsel to the House Judiciary Subcommittee on Civil and Constitutional Rights. He is coauthor of *Terrorism and the Constitution: Sacrificing civil liberties in the name of national security* (2006).

RABBI ELLIOT N. DORFF directs the Rabbinical and Master's programs at the University of Judaism, where he is currently Rector and Distinguished Professor of Philosophy. He also teaches Jewish law at the University of California–Los Angeles School of Law, and has advised multiple national organizations on bioethics. Rabbi Dorff has published over 150 articles on Jewish thought, law, and ethics. He is the author of ten books, including *Contemporary Jewish ethics and morality: A reader* and *Matters of life and death: A Jewish approach to modern medical ethics*.

KAREN M. DOWNS is a professor in the Department of Anatomy at the University of Wisconsin–Madison. Dr. Downs's research focuses on development of the umbilical cord, and its developmental relationship to the embryo. Her work has been published in a number of peer-reviewed journals including *Genes and Development*, *Development*, *Developmental Biology*, and *Placenta*.

ALICE DREGER is a medical humanist and bioethicist at the Feinberg School of Medicine at Northwestern University in Chicago. Her scholarship and patient advocacy have focused on the social and medical treatment of people born with norm-challenging body types, including intersex, conjoinment, dwarfism, and cleft lip. For seven years she served as Chair of the Board, Chair of Fundraising, and Director of Medical Education for the Intersex Society of North America, a policy and advocacy group for people born with atypical sex anatomies. Dr. Dreger is the author of numerous articles and has published three books, most recently *One of us: Conjoined twins and the future of normal* which has received positive reviews in the *New Yorker*, *Nature*, the *London Review of Books*, and the *New England Journal of Medicine*. Her essays on science, medicine, and life have been featured in the *New York Times*, the *Wall Street Journal*, the *Washington Post*, and the *Chicago Tribune*. She has appeared on over three dozen broadcast venues, including "Good Morning America," HBO, "Discovery Health," National Public Radio, and CNN International.

PATRICIA ADAIR GOWATY is Distinguished Research Professor in the Institute of Ecology at the University of Georgia. Her research focuses on natural selection and variation in mating systems, sex allocation, and sex-differentiated behavior. Dr. Gowaty is also involved in efforts to improve the teaching of ecological concepts at the precollege level. She is the author of many research articles and editor of *Feminism and Evolutionary Biology: Boundaries, Intersections, and Frontiers*.

CARL E. GULBRANDSEN is the managing director of the Wisconsin Alumni Research Foundation, a private, nonprofit technology transfer organization that supports the University of Wisconsin–Madison. Before assuming that position, Gulbrandsen led the organization's patent and licensing team and orchestrated the intellectual property strategy surrounding the first human embryonic stem cells that were isolated and cultured in James Thomson's laboratory. He holds a PhD in physiology and a law degree, both from the University of Wisconsin–Madison.

JO HANDELSMAN is Howard Hughes Medical Institute Professor and chair of the Department of Bacteriology and codirector of the Women in Science and Engineering Institute at the University of Wisconsin–Madison. Her work has appeared in a wide array of scientific journals, and she is coauthor of *Biology brought to life: A guide to teaching students to think like scientists* and *Entering mentoring: A seminar to train a new generation of scientists.*

DANIEL LEE KLEINMAN is a professor in the Department of Rural Sociology at the University of Wisconsin–Madison, where he is director of the Holtz Center for Science and Technology Studies and is affiliated with the Integrated Liberal Studies Program. He is the author of *Science and technology in society: From biotechnology to the Internet; Impure cultures: University biology and the world of commerce;* and *Politics on the endless frontier: Postwar research policy in the United States*. He is the editor of *Science, Technology and Democracy*.

TRAVIS KRIPLEAN received his BS in computer science and sociology in 2005 from the University of Wisconsin–Madison. He is currently a graduate student in computer science at the University of Washington, and his research interests include the relationship between computer technology and civic engagement.

JILL O. LADWIG is the senior writer for the Wisconsin Alumni Research Foundation, a private, nonprofit technology transfer organization that supports the University of Wisconsin–Madison. She has been a science communicator at the University of Hawaii-Manoa, the University of California–Irvine, and the University of Wisconsin. She holds a bachelor's degree from the University of Wisconsin–Madison and a master's degree in journalism from the Medill School at Northwestern University.

JAMES LATTIS is the director of the University of Wisconsin Space Place, the goal of which is to educate the public about astronomy and space exploration. Dr. Lattis has an MS degree in astrophysics and a PhD in the history of science with a focus on astronomy.

ROGER D. LAUNIUS is chair of the Division of Space History at the Smithsonian Institution's National Air and Space Museum in Washington, DC. Between 1990 and 2002 he served as chief historian of the National Aeronautics and Space Administration. A graduate of Graceland College, he received his PhD from Louisiana State University, Baton Rouge, in 1982. He has written or edited more than 20 books on aerospace history, including *Critical issues in the history of spaceflight*; *Space stations: Base camps to the stars*, which received the American Institute of Aeronautics and Astronautics' history manuscript prize; *Reconsidering a century of flight*; *To reach the high frontier: A history of U.S. launch vehicles*; *Imagining space: Achievements, possibilities, projections, 1950–2050*; *Reconsidering Sputnik: Forty years since the Soviet satellite*; *Innovation and the development of flight*; *Frontiers of space exploration*; *Spaceflight and the myth of presidential leadership*; and *NASA: A history of the U.S. civil space program*. He is frequently consulted by the electronic and print media for his views on space issues, and has been a guest commentator on National Public Radio, CNN, "The News Hour with Jim Lehrer," ABC, CBS, NBC, and others.

DAVID R. LEGATES is an associate professor in climatology and the director of the Center for Climate Research at the University of Delaware. He also serves as the Delaware Climatologist. Dr. Legates's analysis of climate change data has been sought by many including the United States Justice Department and several climate modeling groups for the evaluation of climatic models. Dr. Legates is the author of numerous research articles, book chapters, monographs, and opinion pieces about climate change and variability.

NANCY G. LEVESON is professor of aeronautics and astronautics and of engineering systems at the Massachusetts Institute of Technology. Dr. Leveson is a pioneer in the field of software safety, and studies how computer software affects the safety of major instruments, including the space shuttle. Dr. Leveson is a member of the National Academy of Engineering. She has published over 200 research papers and is the author of *Safeware: System safety and computers*.

PATRICIA R. MASTRANDREA is cofounder of Delphi International, with extensive experience partnering with federal, state, and local organizations. She founded Delphi's social responsibility and environment program, working with corporations, communities, and government representatives on environmental issues. The program has produced a series of public outreach products concerning ecologically, socially, and economically sustainable practices. Agencies including the American Lung Association (ALA), California Air Resources Board, California Air Pollution Control Officers Association, National Park Service Air Resources Board, and the California Department of Education have recognized her program, *The Clean Air Primer*, for excellence as an interdisciplinary clean air curriculum. The Primer received a 2005 Clean Air Award from the ALA. As an educator, Ms. Mastrandrea, an administrator, writer, and curricula expert, has extensive teaching experience and has conducted master classes for teachers. She has worked with Stephen Schneider as a researcher, writer, and editor since 2005.

CHRISTINA MATTA is a 2007 graduate in the history of science at the University of Wisconsin–Madison. Her Ph.D. dissertation examines the emergence of botanical bacterial bacteriology from changes in academic botany in 19th-century Germany. She is the author of "Ambiguous bodies and deviant sexualities" and coauthor of "It takes a village: Role of indigenous microbial communities in infectious disease."

CLARK A. MILLER is an associate professor in the Consortium for Science, Policy, and Outcomes and the Department of Political Science at Arizona State University. He has written extensively on the role of science and expertise in democratic governance and international affairs, with an emphasis on science and technology policy and the human dimensions of global environmental change. He has also worked with the United Nations Environment Programme and other international institutions to find ways to enhance the linkages between global environmental science and policy. He is coeditor of *Changing the atmosphere: Expert knowledge and environmental governance*.

REV. DR. TADEUSZ PACHOLCZYK serves as the Director of Education for The National Catholic Bioethics Center. He earned a PhD in neuroscience from Yale University where his research focused on cloning genes for neurotransmitter transporters, and worked for several years as a molecular biologist at Massachusetts General Hospital/Harvard Medical School. Fr. Tad studied for five years in Rome, where he did advanced work in dogmatic theology and in bioethics, examining the question of delayed ensoulment of the human embryo. He has testified before members of the Massachusetts, Wisconsin, Virginia, North Carolina, and Pennsylvania state legislatures during deliberations over stem cell research and cloning, given presentations and participated in roundtables and debates on contemporary bioethics throughout the United States, Canada, and in Europe, and done numerous media commentaries, including appearances on CNN International, "ABC World News Tonight," and National Public Radio.

ROGER PIELKE, JR. is a professor in the Environmental Studies Program at the University of Colorado at Boulder. Dr. Pielke is also a Fellow of the Cooperative Institute for Research in the Environmental Sciences, where he serves as Director of the Center for Science and Technology Policy Research. Dr. Pielke's current areas of interest include understanding disasters and climate change, the politicization of science, decision-making in the face of uncertainty, and policy education for scientists. He is the author of a forthcoming book titled: *The Honest broker: Making sense of science in policy and politics*, to be published by Cambridge University Press in 2007.

JOAN ROUGHGARDEN is professor of biological science and geophysics at Stanford University, where her research focuses on evolutionary biology. Dr. Roughgarden is the author of a number of articles and several books, including *Evolution's rainbow: Diversity, gender, and sexuality in nature and people*.

ABDULAZIZ SACHEDINA is Frances Myers Ball Professor of Religious Studies at the University of Virginia. Dr. Sachedina is currently working on a book entitled *Islamic Biomedical Ethics*. He has been a visiting professor and lecturer at campuses worldwide. Dr. Sachedina is a Senior Associate at the Center for Strategic and International Studies and a member of multiple religious and bioethical advisory boards.

DANIEL SAREWITZ is Director of the Consortium for Science, Policy, and Outcomes and professor of science and society in the School of Life Sciences at Arizona State University. His work focuses on the ways that social outcomes derive from political and policy decisions about science and technology. His research questions include: How does the distribution of the social benefits of science relate to the way that we organize scientific inquiry? What are the origins of human know-how? How does science influence political controversies? How do the normative commitments of scientists influence the trajectories of science and innovation? He writes widely on these issues for scholarly and general audiences. His most recent book is *Living with the genie: Essays on technology and the quest for human mastery* (coedited with Alan Lightman and Christina Desser).

STEPHEN H. SCHNEIDER is the Melvin and Joan Lane Professor of interdisciplinary environmental studies, professor of biological sciences, and professor (by courtesy) of civil and environmental engineering. He is codirector of the Center for Environmental Science and Policy in the Freeman Spogli Institute for International Studies and a Senior Fellow in the Woods Institute for the Environment at Stanford University. A past recipient of the prestigious MacArthur Fellowship, Dr. Schneider is a member of the U.S. National Academy of Sciences and has been a consultant to the federal government since the Nixon administration. He has authored over 600 publications, including the books *The Genesis strategy: Climate and global survival*; *The Coevolution of climate and life*; *Global warming: Are we entering the greenhouse century?*; and *Laboratory Earth: The planetary gamble we can't afford to lose*. Currently, Dr. Schneider is counseling policy makers about the importance of using risk management strategies in climate-policy decision-making, given the uncertainties in future projections of global climate change. In addition to continuing to serve as a noted advisor to decision makers, he consults with corporate executives and other stakeholders in industry and the nonprofit sectors regarding possible climate-related events and is actively engaged in improving public understanding of science and the environment through extensive media communication and public outreach.

EUGENE H. SPAFFORD is a professor of computer sciences at Purdue University, and is executive director of the Center for Education and Research in Information Assurance and Security (CERIAS). Additionally, he has courtesy appointments as a professor of philosophy, of communication, and of electrical computer engineering. He has been working in areas of software engineering, information security, cybercrime investigation, privacy, and social impact of computing for over 25 years. He has served in numerous professional roles, including on the President's Information Technology Advisory Committee (PITAC), Air Force Scientific Advisory Board, and the FBI Regional Computer Forensic Laboratory Advisory Board. As of spring 2007, he is a member of the Computing Research Association board of directors, and is the chair of the ACM's US Public Policy Committee. He is a Fellow of the ACM, the IEEE, and the AAAS and has won a number of awards and recognitions for his contributions to science and public policy. His personal web page is www.http://www.homes.cerias.purdue.edu/~spaf/.

DIANE VAUGHAN is a professor of sociology and international and public affairs at Columbia University. Dr. Vaughan's research interests include "the dark side of organizations": mistake, misconduct, and disaster. Her books include *Controlling unlawful organizational behavior*; *Uncoupling*; and *The Challenger launch decision*. The latter was awarded the Rachel Carson Prize by the Society for Social Studies of Science and the American Sociological Association's Robert K Merton Award. *The Challenger launch decision* was also nominated for the Pulitzer Prize. In 2003, Dr. Vaughan joined the staff of the *Columbia* Accident Investigation Board to work on the investigation and official report, for which she authored a chapter. Currently she is doing research on air-traffic control.

MARK WARSCHAUER is a professor in the Department of Education and the Department of Informatics at the University of California, Irvine. Warschauer also serves as Associate Director of the Ada Byron Research Center for Diversity in Computing & Information Technology and as Director of UCI's Ph.D. in Education program. Previously, Warschauer has taught or conducted research at the University of Hawaii, Moscow Linguistic University, and Charles University in Prague. He also spent three years in Egypt as director of educational technology within a United States–funded development project. He has conducted field research in China, Brazil, India, and Singapore. Warschauer's research focuses on the relationship of information and technology use to language and literacy development, educational reform, and social inclusion. His recent books include *Technology and social inclusion: Rethinking the digital divide* and *Laptops and literacy: Learning in the wireless classroom*. His work has also appeared in *Scientific American* magazine and a wide range of academic journals, such as *Educational Policy*, *The Information Society*; *Journal of Computer-Mediated Communication*, and *Anthropology & Education Quarterly*. Further information on Dr. Warschauer and copies of his recent papers are available at http://www.gse.uci.edu/faculty/markw.

LEWIS H. ZISKA currently is an ecologist with the USDA's Agricultural Research Service who has been studying the impacts of rising carbon dioxide and global climate change on weed biology in managed and natural ecosystems for the past 15 years. He began his career as a Smithsonian fellow, then took up residence as the Project Leader for global climate change at the International Rice Research Institute in the Philippines before joining USDA. At present he is investigating the role of rising carbon dioxide and changing climate on weed-crop competition, invasive/noxious weeds and weeds and public health. Dr. Ziska's research has appeared in *National Geographic*, the *New York Times*, *USA Today*, the *Washington Post*, *Newsweek*, *US News and World Report* and *CNN Headline News*.

Index